Springer Series in Electronics and Photonics 22

Edited by D. H. Auston

Springer Series in Electronics and Photonics
Editors: D.H. Auston W. Engl T. Sugano

Volume 21 **Picosecond Electronics and Optoelectronics**
Editors: G.A. Mourou, D.M. Bloom, C.-H. Lee

Volume 22 **High-Speed Electronics**
Basic Physical Phenomena and Device Principles
Editors: B. Källbäck and H. Beneking

This series was originally published under the title
Springer Series in Electrophysics
and has been renamed starting with Volume 22.

Volumes 1–20 are listed on the back inside cover

High-Speed Electronics

Basic Physical Phenomena and Device Principles

Proceedings of the International Conference
Stockholm, Sweden, August 7–9, 1986

Editors: B. Källbäck and H. Beneking

With 195 Figures

Springer-Verlag Berlin Heidelberg New York
London Paris Tokyo

Dr. Bengt Källbäck
RIFA AB, S-163 81 Stockholm, Sweden

Professor Dr. Heinz Beneking
Institut für Halbleitertechnik, Rhein.-Westf. Technische Hochschule,
Sommerfeldstraße, D-5100 Aachen, Fed. Rep. of Germany

Series Editors:

Dr. David H. Auston
AT & T Bell Laboratories, 600 Mountain Avenue, Murray Hill, NJ 07974, USA

Professor Dr. Walter Engl
Institut für Theoretische Elektrotechnik, Rhein.-Westf. Technische Hochschule,
Templergraben 55, D-5100 Aachen, Fed. Rep. of Germany

Professor Takuo Sugano
Department of Electronic Engineering, The Faculty of Engineering,
The University of Tokyo, 7-3-1, Hongo, Bunkyo-ku, Tokyo, 113, Japan

ISBN-13:978-3-642-82981-9 e-ISBN-13:978-3-642-82979-6
DOI: 10.1007/978-3-642-82979-6

Softcover reprint of the hardcover 1st edition 1986

2153/3150-543210

Preface

In the past, a number of Satellite Conferences have been held in connection with the International Conference on Physics of Semiconductors, covering selected fields of interest. In 1986, when the main conference was held in Stockholm, Sweden, new phenomena had to be discussed: superlattices, hot electron phenomena and new device structures for high-speed applications. The aim was to select topics which would be of interest to physicists as well as to electronics engineers. Therefore a Satellite Conference on High-Speed Electronics, Basic Physical Phenomena and Device Principles, was arranged at Saltjöbaden, a coastal resort near Stockholm. An organizing committee was established after the first suggestion made by Professor Grimmeiss from the University of Lund, Sweden, and some preliminary discussions on the Conference format. A Program Committee was established to be responsible for the further selection of the invited talks, the regular papers and poster presentation. The aim was to have a broad spectrum of contributions to attract physicists as well as device-oriented engineers and to stimulate discussions among the participants.

These Proceedings contain all oral and poster presentations, with emphasis on the invited talks, which give a competent overview of the field. The fast publication by Springer-Verlag has permitted the presentation of an up-to-date survey of the principles of high-speed electronics. Incorporation in the Springer Series in Electronics and Photonics will enable the book to be distributed worldwide and to reach all interested scientists.

We have to thank all those who made the conference a successful meeting, those who attended as well as the members of the local organizing committee.

Finally, our sincere thanks go to RIFA and the Ericsson group, Sweden, especially to Dr. Ramqvist, for generous support of the conference.

Stockholm
Aachen, September 1986

B. Källbäck
H. Beneking

Contents

Part I **Hot-Electron Dynamics**

Modelling of High Electron Velocity Effects for Devices
By R. Castagné (With 7 Figures) 2

Ballistic Transport and Electron Spectroscopy in Tunnelling Hot Electron Transfer Amplifiers (THETA)
By M. Heiblum (With 6 Figures) 11

Hot Electron Transistors, By J.R. Hayes, A.F.J. Levi, A.C. Gossard, and J.H. English (With 4 Figures) 19

Tunneling Through III–V Low-Barrier Heterostructures
By P. Guéret, A. Baratoff, S. Bending, H. Meier, E. Marclay, and M. Py (With 3 Figures) 24

Velocity Overshoot and Suppression of Diffusivity and Microwave Noise in Short n^+-n-n^+ Structures of GaAs
By V. Bareikis, K. Kibickas, J. Liberis, A. Matulionis, R. Miliušytė, J. Paršeliūnas, J. Požela, and P. Sakalas (With 4 Figures) 28

Mobility Overshoot of Hot Electrons
By V.N. Freire, A.R. Vasconcellos, and R. Luzzi 32

Monte-Carlo Simulation of the Effects Induced by Real-Space Transfer in a HEMT . By M. Mouis, P. Dolfus, B. Mougel, J.-F. Pône, and R. Castagné (With 7 Figures) 35

Application of the Integral Boltzmann Equation to the Hot Electron Problem in an Inhomogeneous Submicron Structure
By P. Kuivalainen and K. Lindberg (With 5 Figures) 40

Three Picosecond Oscillations in Avalanche Electron- Hole Plasma Induced by Energy Relaxation Phenomena
By D. Lippens and J.L. Nieruchalski (With 4 Figures) 44

Part II Heterostructures, Superlattices and Quantum Wells

Resonant Tunneling Transistors, Tunneling Superlattice Devices and New Quantum Well Avalanche Photodiodes
By F. Capasso (With 12 Figures) 50

Novel Real-Space Transfer Devices
By A. Kastalsky (With 8 Figures) 62

Transport Characteristics in Heterostructure Devices
By S. Muto, S. Hiyamizu, and N. Yokoyama (With 8 Figures) ... 72

Technical Issues of High-Speed Heterostructure Devices
By T. Ikegami, K. Yamasaki, T. Mizutani, T. Ishibashi, and M. Fujimoto (With 10 Figures) 79

Hot-Carrier-Excited Two-Dimensional Plasmon in Selectively Doped AlGaAs/GaAs Heterointerface Under High Electric Field Application
By T. Kobayashi, N. Okisu, and Y. Sambe (With 5 Figures) 88

Optical High-Field-Transport Experiments in GaAs Quantum Wells. By J. Shah and R.A. Höpfel (With 3 Figures) 93

Optical Time-of-Flight Investigation in Ambipolar Carrier Transport in Specially Designed GaAs/GaAlAs Quantum Well Structures. By H. Hillmer, G. Mayer, A. Forchel, K.S. Löchner, and E. Bauser (With 3 Figures) 97

An Ultra-Fast Optical Modulator: The Double-Well GaAs/GaAlAs Superlattice (DWSL). By B. Deveaud, A. Chomette, A. Regreny, J.L. Oudar, D. Hulin, and A. Antonetti (With 2 Figures) 101

High-Velocity Vertical Transport in Graded Gap GaAs/GaAlAs Superlattices. By B. Lambert, B. Deveaud, A. Chomette, A. Regreny, and R. Romestain (With 3 Figures) 104

Modelling of Mobility Degradation in Submicron MOSFETs After Electrical Stressing
By B. Cabon-Till and G. Ghibaudo (With 2 Figures) 108

Negative Differential Mobility and Drift Velocity Overshoot in a Single Quantum Well of AlGaAs/GaAs/AlGaAs Heterostructure
By C. Hamaguchi, K. Miyatsuji, and H. Tanimoto (With 5 Figures) 112

Monte Carlo Study of Hot Electron Transport in GaAs-AlGaAs Quantum Wells
By S.M. Goodnick and P. Lugli (With 3 Figures) 116

Part III High-Speed Electronic Devices

High-Speed Bulk Unipolar Structures in Silicon
By J.M. Shannon 122

Silicon Bulk Barrier Diodes Fabricated by LPVPE
By H. Beneking, L. Vescan, A. Gruhle, J.M. Cloos, and M. Marso
(With 4 Figures) 123

Impact Ionization Breakdown of GaAs Current Limiters
By K. Lehovec (With 2 Figures) 127

The New High Speed Devices: The Barrier Transistor and the TEG-Base Transistor
By C.Y. Chang, W.C. Liu, and Y.H. Wang (With 6 Figures) 132

Monte Carlo Investigation of the High Electron Mobility Transistor. By U. Ravaioli and D.K. Ferry (With 4 Figures) 136

Excess Gate Current Due to Hot Electrons in GaAs-Gate FETs
By D.J. Frank, P.M. Solomon, D.C. La Tulipe, Jr., H. Baratte, C.M. Knoedler, and S.L. Wright (With 3 Figures) 140

Potential Barriers in Doped GaAs by OM-VPE
By P. Roentgen, G. Fernholz, and H. Beneking (With 6 Figures) . 144

Low Noise High Electron Mobility Transistors Grown By MOVPE
By M. Wolny, P. Chambery, A. Briere, and J.-P. Andre
(With 3 Figures) 148

Emitter-Coupled Logic Ring Oscillators Implemented with GaAs/GaAlAs Single and Double Heterojunction Bipolar Transistors: A Comparison. By C. Dubon-Chevallier, P. Desrousseaux, A.M. Duchenois, C. Besombes, J. Dangla, C. Bacot, and D. Ankri (With 4 Figures) 151

Self-Aligned Technology Using Refractory Ohmic Contacts for GaAs/GaAlAs Heterojunction Bipolar Transistors
By K. Daoud-Ketata, C. Dubon-Chevallier, J. Etrillard, and J.F. Bresse (With 3 Figures) 156

Novel Cryoelectronic Device Concept Based on Magnetically Controlled Current Flow in Bulk Semiconductors
By J. Mannhart, J. Parisi, and R.P. Huebener (With 7 Figures) . 160

InGaAlAs/InGaAs/InGaAlAs NnpnN Double Heterojunction Bipolar Transistors: Experimental Characteristics and Monte-Carlo Interpretation. By J.L. Pelouard, P. Hesto, J.P. Praseuth, and L. Goldstein (With 5 Figures) 164

Lattice-Strained Double Heterojunction InGaAs/GaAs Bipolar Transistors. By L.P. Ramberg, P.M. Enquist, Y.K. Chen, F.E. Najjar, L.F. Eastman, E.A. Fitzgerald, and K.L. Kavanagh . 168

Cryogenic GaAs Integrated Circuits Using a Lightly Doped GaAs FET Structure. By J. Kamdem, M. Henry, J.P. Klein, F. Alexandre, and M. Gloanec (With 8 Figures) 172

GaAs-on-Insulator Structure Prepared by Heteroepitaxy of Fluorides and GaAs. By T. Asano, K. Tsutsui, H. Ishiwara, and S. Furukawa (With 5 Figures) 177

Part IV **High-Speed Opto-Electronics**

Optoelectronic Generation of Very High Speed Electromagnetic Transients. By D.H. Auston, M.C. Nuss, and P.R. Smith 184

Picosecond Electro-Optic Sampling
By G.A. Mourou (With 9 Figures) 191

High-Speed Integrated Circuit Testing by Time-Resolved Photoemission. By R. Clauberg, H.K. Seitz, A. Blacha, J.A. Kash, and H. Beha (With 2 Figures) 200

Overview of Optical Switching and Bistability
By N. Peyghambarian, H.M. Gibbs, D. Hulin, A. Antonetti, A. Migus, and A. Mysyrowicz (With 2 Figures) 204

Monte Carlo Investigation of High-Speed GaAs Schottky Barrier Photodiode
By M.A. Osman, U. Ravaioli, and D.K. Ferry (With 5 Figures) . . 210

A High-Speed Au/$In_{0.53}Ga_{0.47}As$/InP Schottky Barrier Photodiode for 1.3–1.65 μm Photodetection
By J.H. Kim and S.S. Li (With 3 Figures) 214

High Sensitivity Picosecond Optical Pulse Detection by Semiconductor Laser Amplifiers Via Cross-Correlation
By K. Ketterer, E.H. Böttcher, and D. Bimberg (With 5 Figures) 218

GaAs Photoconductors to Characterize Picosecond Response in GaAs Integrated Devices and Circuits. By R.B. Hammond, N.G. Paulter, and A.J. Gibbs (With 2 Figures) 223

Characterization of On-Chip Polycrystalline Silicon Photoconductors. By D.K. Fitzpatrick, D.R. Bowman, and R.B. Hammond (With 3 Figures) 226

Index of Contributors 231

Part I

Hot-Electron Dynamics

Modelling of High Electron Velocity Effects for Devices

R. Castagné

I.E.F. Université de Paris Sud, Orsay, France

1. INTRODUCTION

High-speed devices are needed to drive high-speed integrated circuits or to amplify at frequencies up to 100 GHz. They are characterized by a very short transit time of the electrons across the active region. The control currents have also to be high enough to cope with stray capacitances. In addition to these performances the devices must satisfy the absolute necessity of a reduced internal feedback and of low-impedance access zones.

The quest for short transit times, in addition to a large technological effort in the reduction of device dimensions, has given rise to many investigations for the proper use of materials with low electron effective mass and low scattering rate. This has been done in two possible ways.

The first way is to deal with the delay in the heating of the electron population in a high electric field, in order to obtain a drift velocity in excess of its equilibrium value (a "velocity overshoot"). Strictly speaking, the excess velocity concept is attached to the existence of the collision-limited regime and assumes that the mean momentum of the electron population is relaxed. So it will be observed in materials where the mean energy relaxation time τ_ε is, by a significant amount, larger than the mean momentum one τ_m.

The second way is the so-called ballistic regime.In this case the transit time t_t across the active region must be less than the mean time of free flight between collisions . The ballistic idea is easy to understand for one electron ; to be effective, the electron initial velocity must be directed along the electric field and as high as possible. For a whole electron population, the concept is valuable only if each one enters the active zone with the same initial velocity. So a selection mechanism must be set up at the input, generally devolved to a potential barrier.

In all circumstances, a correct prediction of the device performance and device function requires careful modelling which takes injection phenomenon and non-stationary transport phenomenon correctly into account, together with field dimensional effects.

2. THE MODELLING TOOLS

High-speed device modelling, especially for III-V devices, needs to take into account the non-stationarity of the transport phenomenon, and then to introduce the electron energy as an independent variable. Many powerful tools are now accessible to most laboratories. Different theoretical approaches have been followed : direct solution of Boltzmann Transport

Equations (BTEs) /1-4/,the multiparticle Monte Carlo procedure /5-7/, use of momentum and energy relaxation equations /8-13/. We will concentrate here on the last two techniques.

The multiparticle Monte Carlo procedure (MMC) is certainly the most powerful and exciting technique for advanced device simulation, keeping close to the underlying quantum physical principles. Advanced algorithms allow the time-dependent simulation of 2D real-space devices. A set of N independent particles is introduced, each representing a given number of electrons. N is generally limited by the computer resources. The motion of each particle is modelled as a result of its acceleration by the 2D electric field and the various collision processes. The particles have also to be confined in the closed space of the device according to boundary conditions accounting for exterior surfaces, contacts or interfaces. Finally, the particle displacements create a space charge which must be computed in order to solve the Poisson equation. The various details of a MMC can be found in the literature (see for instance /5/ or /6/). Table 1 gives a typical partition of the CPU time among the various steps of a MMC program.

Table 1. Typical partition of the CPU time in a MMC program

	Scattering parameters	Electron concentrat.	Particle motion	Poisson	Other
%CPU Time	40%	18%	30%	10%	2%

The implementation of a MMC procedure generally encounters 4 major difficulties. 1. The electrical charge balance must be respected. A false balance leads to artefacts such as false electron temperature and false current. 2. In low field regions, for instance in N^+ regions surrounding contacts, the free adjustment of the electron temperature to the lattice temperature makes necessary the introduction of a non-elastic formulation of the interaction with acoustic phonons. 3. An artificial pseudo-random motion of the particles generally arises due to time discretisation and sequential numerical treatment of the Poisson equation and of the particle displacement algorithm. /6, 14/. This is responsible for an artificial heating. 4. Noise is always present in the MMC simulation. Either "artificial noise" linked to the artificial motion, as above, or "natural noise" amplified by the restriction in the total number of particles. The necessity to obtain mean values against time for the various physical quantities drastically increases the computing time. This is the main limitation to MMC.

Simulation algorithms based on the momentum and the energy relaxation equations have been developed /14-16/ for both 1D and 2D geometry devices. The relaxation equations were first proposed by Blotekjaer /11/. Their equivalence to BTEs has been proved to hold under certain, not truly restrictive, conditions /10/. The main parameters : electron mean effective mass $m^*(\varepsilon)$, mean temperature $T(\varepsilon)$, mean momentum and mean energy relaxation times, respectively $\tau_m(\varepsilon)$ and $\tau_\varepsilon(\varepsilon)$, can be determined as functions of the mean energy ε by use of a MMC simulation of the material under stationary transport conditions for various values of the electric field /10/. Simulation of the ballistic phenomenon is not truly accessible to these techniques, especially because the parameter $\tau_m(\varepsilon)$ is a collective property, representative of the effect of a great number of collisions. On the other hand, in polar semiconductors in which $\tau_m \ll \tau_\varepsilon$ the excess velocity phenomena can be correctly modelled with only three equations by assuming that the mean momentum is relaxed. The current density is then expressed by a classical drift-diffusion law, using a mobility $\mu(\varepsilon) = q\tau_m(\varepsilon)/m^*(\varepsilon)$ depending on the electron population mean energy ε. The set of equations to be solved is then : the Poisson equation, a charge balance equation and an energy balance equation. The solution of this system gives access to the distributions of the energy, the electron concentration and the electric field, and to the various currents at electrodes. The main advantage of this technique is the absence of noise, which allows an easy determination of the various small-signals parameters. On the other hand, like any hydrodynamic-scheme, the system is sensitive to high gradients in the extensive variables. It is often preferable to solve it by a total matrix formulation, which again needs large computer resources and much CPU time.

3. MODELLING THE CONDITIONS LEADING TO HIGH VELOCITIES

High transport velocities can be obtained for an electron population in a device provided steps are taken (i) to limit the effect of collisions; (ii) to set all the electron velocities in the right direction; (iii) to avoid transfer into the upper valleys of the conduction band (for III-V compounds); and (iv) to bring the electron velocities to their maximum accessible value as quickly as possible.

The velocity for an individual electron with energy ε is $v = \hbar^{-1}(\partial\varepsilon/\partial k)$, which is a well-defined characteristic of the material. The mean velocity of an electron population is always less than this individual value, due to collision process. A compromise must then be found between the maximum accessible velocity v_M and the collision rate, which both depend on the electron energy. Figure 1 gives the total scattering probability corrected by the scattering angle factor $\langle 1-\cos\theta\rangle$ vs ε (the electron energy) for GaAs at 300 K and 77 K. It is clear that the point of practical interest is near the transfer threshold $\Delta\varepsilon_{\Gamma L} \approx 0.3$ eV. The maximum accessible velocity is $v_M \approx 10^8$ cm.s^{-1}, the mean time between collisions is $t_c \approx 3\times10^{-13}$s so that electrons can travel a maximum distance $d \approx 0.3$ µm. Table 2 gives the same parameters for some common semiconductors.

In practice,however,electrons must get their kinetic energy from a potential energy gradient (i.e. a gradient in the bottom ε_c of the conduction band). Device modelling can help to solve two problems : (i) to find an optimum field profile; (ii) to find a structure giving it

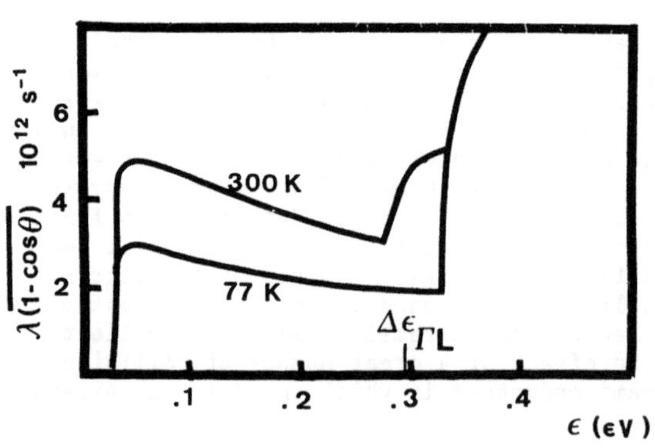

Fig.1. Total scattering probabilities in GaAs as a function of the electron energy.

Table 2. Parameters of electrons in some common semiconductors

	GaAs(300K)	GaAs(77K)	InP(300K)	GaInAs(300K)
VM/cm-1/	108	108	1.1×10^{8}	1.2x108
tc/s/	2.5×10^{-13}	5×10^{-13}	1.8×10^{-13}	2×10^{-13}
d /µm/	0.25	0.5	0.2	0.25

practically. The first problem can be solved "in abstracto", i.e. considering pure electric field variation without diffusion effects, by modelling the electron population behaviour under variation of the field with time. Figure 2 gives simulation results for the GaAs case. It gives the maximum mean electron velocity for an electron population having to travel a distance d for 3 electric field profiles /17/. In the first two cases maximum field E_0 is chosen so that electrons reach the transfer energy at the end of the distance d. In the last case E_0 is an impulse whose amplitude and duration are chosen so that the electron energy reaches $\Delta\varepsilon_{\Gamma L}$. One can see that the step function (1) leads to an excess velocity regime (velocity overshoot) which remains for about 0.6 µm. The ramp profile approaches the field "seen" by an electron travelling in the channel of a FET. It is clearly less favourable than the step function. The last case corresponds to ballistic launching. Over short distances this seems to be the best way to obtain very high velocities.

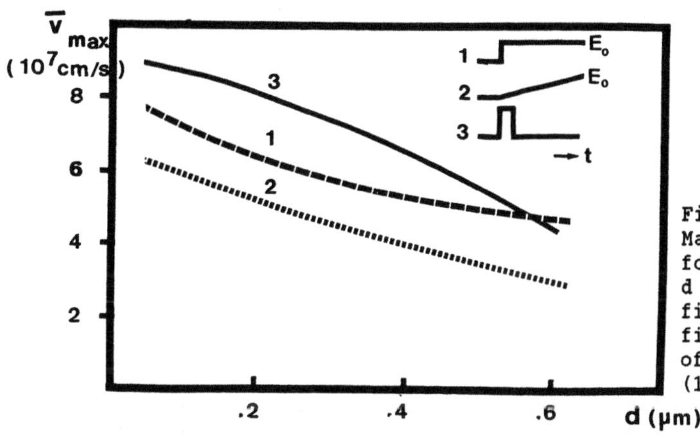

Fig.2.
Maximum mean velocity for a given distance d and different profiles of the electric field as a function of time.
(17) GaAs 300 K.

For practical use in devices, on the other hand, one must consider that the electric field varies in space, and not in time. Spatial variation of the field means generally a gradient in the electron concentration. Transport velocity is then the result of both the electrostatic (long-range) force and of the diffusion (short-range) process. To take these two tendencies correctly into account, even in a very simple geometry, quite elaborate tools are needed.y . The situation depicted by Figs.3 and 4 is a typical one. A GaAs N^+NN^+ structure operating at 77 K has been modelled using MMC. Note that due to the presence of the N^+N potential barriers the electric field profile is not rectangular. Electrons are thermionically emitted at the cathode with zero initial velocities. They first have a ballistic flight until the polar optical phonon emission appears, giving rise to collision-limited transport in an excess velocity regime, up to the anode. The n.v product

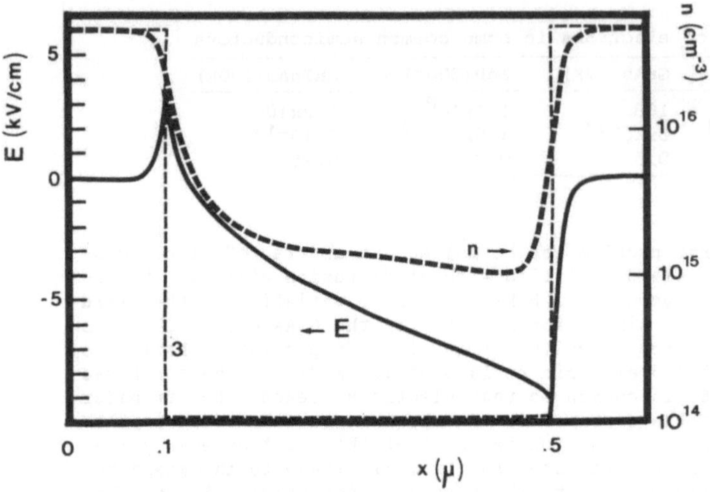

Fig.3.
Electron density and electric field in a N÷NN+ structure (Monte Carlo simulation) Gas GaAs 77 K, N= 5.10^{15} cm

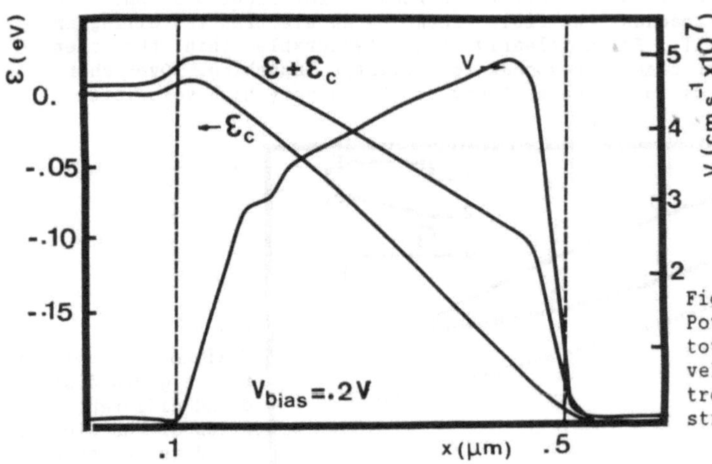

Fig.4.
Potential energy ε_c, total energy $\varepsilon_c+\varepsilon$ and velocity of the electrons in the same structure as Fig.3.

at any point in the device is essentially limited to low values by the thermionic emission process, or, which is equivalent, by the fact that the selection operated by the barrier leads to very low initial velocities along the axis. An ideal condition for high speed would be then to obtain a high gradient in the potential energy without occurrence of the diffusion process.

It was soon recognized that the conduction band discontinuity at an heterojunction can be used to launch electrons with a high initial velocity along the axis of the device. Two essential conditions must be respected : (i) the electrons must be emitted inside the central Γ valley of semiconductor 2. They must then be extracted from the Γ valley of semiconductor 1; (ii) the energy discontinuity $\Delta\varepsilon_c$ must be less than the transfer energy $\Delta\varepsilon_{\Gamma L}$ in semiconductor 2. It is not evident that these conditions can be easily realized in the case of a GaAlAs/GaAs heterojunction. On the other hand they are fulfilled by the InP/GaInAs system.

4. MODELLING OF HIGH-SPEED DEVICES

Device modelling must be able to predict or verify the functionality of the various parts working together in a device. For instance, a precise modelling of the access zones can be essential when electron injection controls the active zone. A careful modelling of the internal feedback suppression is also, at any time, a necessity. We shall illustrate these considerations by some typical examples.

Field effect devices : A significant ballistic regime is not possible in uniformly doped FETs at room temperature. On the other hand, the excess velocity regime is present, increasing the drain current and modifying the current saturation mechanism. The 2D energy relaxation models are convenient. Figure 5 gives the results obtained for a GaAs MESFET using this technique at Lille University. One can notice in these figures that the $\varepsilon = 0.30$ eV equienergy line superposes on an equipotential line roughly near 0.4V. This result has been verified in numerous simulations of submicron gate MESFETs. It corresponds to the fact that electrons travelling with excess velocity lose only a small amount of their energy in emitting optical phonons. This allows a very simplified ("thumb rule") analytical approach as follows : (i) Determine, in the channel, the position of the 0.4 V equipotential line. (ii) Treat the first part of the channel with a constant mobility μ_o, the second part being treated with a saturated velocity.

A ballistic conduction can be observed on the other hand in a self-aligned implanted N^+NN^+ MESFET when the doping level in the central zone is low /15/. Figure 6a shows the mobile carrier repartition in a structure realized on a GaAlAs buffer layer. Figure 6b shows that the drain current does not saturate even for drain potentials higher than the transfer energy. This drawback is due to the fact that the space charge of transferred electrons in the drain N^+ region is negligible with respect to the thermal carrier density. In this case triode-like characteristics are obtained. Internal feedback makes the device not suitable.

Heterojunction (HJ) bipolar devices :Figure 7 gives simulation results of ballistic launching in the case of an InP/GaInAs HJ. In the presence of ballistic transport MMC modelling must be used /19 /. Figure 7a gives the velocity distributions in the planes at 0.01 µm (1) and 0.1 µm (2) from the HJ. The decay along the base of the currents associated with ballistic (1) and quasi-ballistic (2) populations are given in Fig. 7b. Again a simplified analytical model of the heterojunction bipolar transistor (HBT) can been derived from these results. Two electron populations are considered in the base: A ballistic population with an exponentially decreasing associated current density $J_1(x)$ which feeds a thermalized population. The latter is treated by a classical balance equation with an additional term representing the feeding from the ballistic flux. The equation leads to a Ebers-Moll-like model of the HBT which can be used in circuit simulation.

CONCLUSION.

The quest for optimum conditions giving the highest velocity or lowest transit time in the active zone of a device, calls for models able to take into account non-stationary transport phenomenon. At present those most often used are the MMC models and the energy relaxation models. These last give a good description of the excess velocity regime and

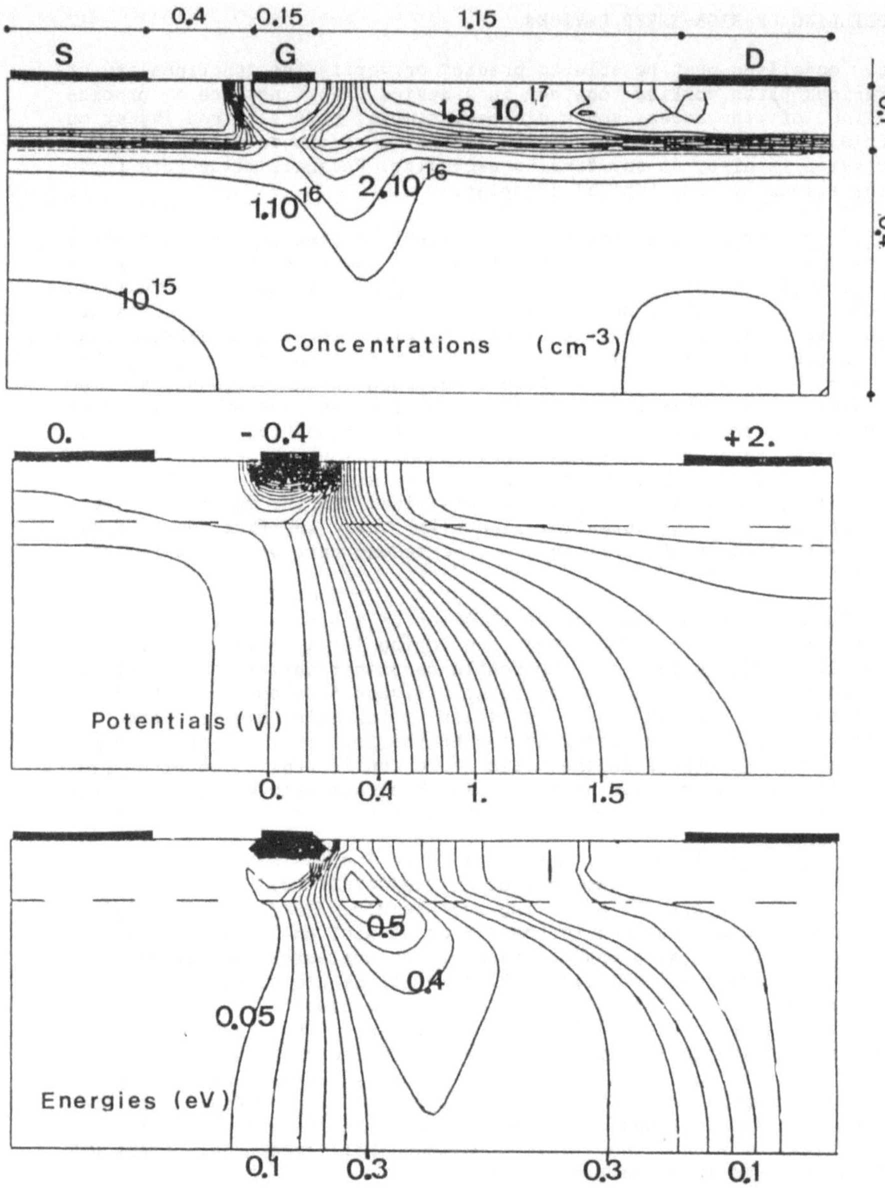

Fig.5a-c. Simulation résults by the energy relaxation method (CHS. Lille Univ.) GaAs MESFET. a) equiconcentration, b) equipotential, c) equienergy

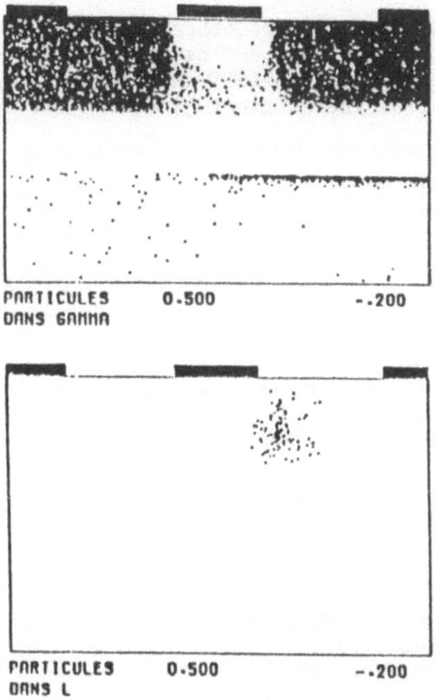

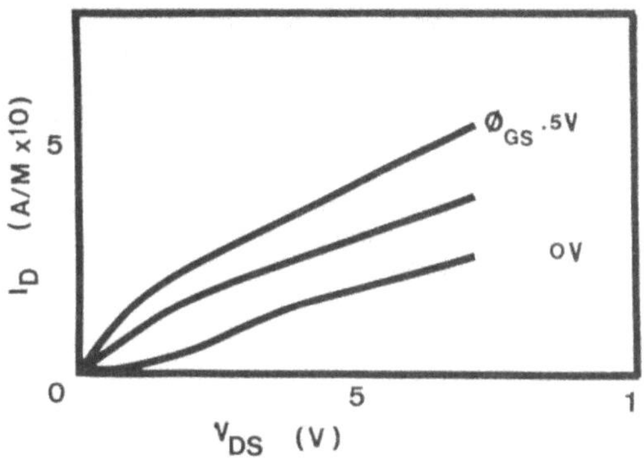

Fig.6. MMC simulation for an N+NN+ GaAs MESFET
77K   N = $5x10^{15}$ cm^{-3}

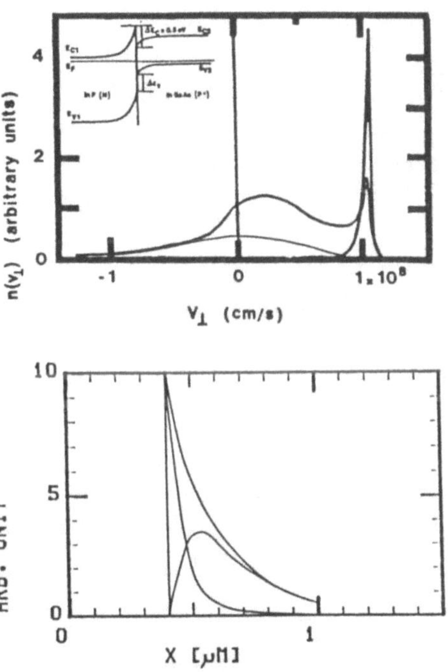

Fig. 7: MMC simulation results of electron injection at an InP/GaInAs heterojunction /19/.

give easy access to any small-signal parameter. On the other hand, ballistic regimes need more physical insight, which only MMC can afford. In every case the access zones of the device must be carefully modelled in insofar as their behaviour determines that of the active region.

Acknowledgment. The author gratefully acknowledges fruitful discussions with H. Baranger, now at ATT Lab, and with his partners from French Greco-microondes.

References

1. J-P. Nougier, M. Rolland, Phys. Rev. B8 5728 (1973)
2. H.L. Grubin, D.K. Ferry, G.J. Iafrate, J.R. Barker,in VSLSI Electronics, Vol.3, Chap. 6, p.197 (1982)
3. J-C. Vaissière, Thèse de doctorat d'Etat, USTL Montpellier, France)
4. H. Baranger, J.H. Wilkins, Phys. Rev. B30, n° 12 (1984)
5. L. Reggiani, Topics in Appl. Phys. 58 (Springer, Berlin, Heidelberg 1985)
6. R.W. Hockney, J.W. Eastwood, Comput. Simul. Using Particles (McGraw-Hill, New-York 1981)
7. J. Zimmermann, E. Constant, J. Appl. Phys. 825 (1981)
8. P. Hesto, J-F. Pône, M. Mouis, J-L. Pelouard, R. Castagné, Nasecode IV Conference Proceedings Dublin (1985)
9. K. Tomizawa, IEEE Proc. 129-1 (1982) 131.
10. J-P. Nougier, and col., J. Appl. Phys. 52, n° 2 (1981).
11. K. Blotekjaer, IEEE Trans ED-17, 38 (1970)
12. M. Shur, Electr. Lett. 12 615 (1976)
13. E. Constant, Topics in Appl. Phys. 58, (Springer, Berlin, Heidelberg 1985)
14. M. Reiser, IEEE Trans ED-20, n° 1 (1973)
15. J-F. Pône, Thèse de doctorat, Université d'Orsay (1986)
16. A. Cappy, IEEE Trans ED 27, n° 11 (1980)
17. A. Ghis, Thèse de doctorat d'ingénieur, CHS Université de Lille (1984)
18. P. Hesto, and col., Appl. Phys. Lett. 45 641 (1984) Lille (1984)
19. J-L. Pelouard, and col. IEEE Cornell Univ. Conf. (1985)

Ballistic Transport and Electron Spectroscopy in Tunnelling Hot Electron Transfer Amplifiers (THETA)

M. Heiblum

IBM Thomas J. Watson Research Center,
Yorktown Heights, NY 10598, USA

Recently, the great improvement in material quality and processing techniques led to a revival of the field of hot electron devices. Devices having current gains greater than 10 have been made successfully. Moreover, the devices proved to be extremely useful as electron spectrometers, establishing unambiguously, for the first time, the existence of ballistic electrons in heavily doped GaAs. I review briefly the history of the hot electron devices, then describe their usefulness in studying device physics: for example, in the determination of the ballistic mean free path, in identifying the major scattering mechanisms influencing the transport, in observing the transfer of hot electrons into upper satellite valleys, and by exhibiting quantization effects that take place in small structures.

1. Some History

The field started in 1960 with the proposal of MEAD to build a ***cold cathode transistor*** from metals and insulators. [1] In his device electrons tunnel between two metal layers and through a thin oxide. Upon emerging into the second metal the electrons have high kinetic energy and velocity, enabling them to traverse the metal quickly. If the metal is thinner than the mean free path (mfp) of hot electrons in it, most of the electrons will maintain their high energy and the direction of motion and surmount a second potential barrier (an oxide) to be collected eventually by a metal collector. The thin metal layer in the center, the base, has a very high conductivity, however, unfortunately, the mfp of hot electrons there is very short. Indeed the transfer coefficient, α, was found by MEAD to be very small (about 2%).

In an attempt to increase the mfp in the metal base SPRATT et al. [2] have replaced the collector oxide - metal combination by a semiconductor - metal combination. The collector Schottky diode thus has formed a barrier with height lower than in the oxide - metal case, hence allowing the hot electrons to be 'cooler' in the base and consequently have a longer mfp. This device was followed by a similar device, the ***metal base transistor***, proposed by GEPPERT and KHANG. [3] In this device, both oxides are replaced by a Schottky barriers. Electrons are injected into the metal base from a negatively biased semiconductor with kinetic energy equal to the Schottky barrier height. Subsequently they are collected by a reversed biased collector Schottky barrier. In all metal base devices pinholes in the thin metal base prevented the expected operation.

More recently SHANNON fabricated the ***camel transistor***. [4] Potential barriers were created via the implantation of a p type dopant into otherwise heavily doped n - type Si. Current gains as high as 19 were measured for very high energy injected electrons. A similar idea [5] was implemented in GaAs in the ***planar doped barrier transistor*** by MALIK and HOLLIS and coworkers. [6,7] These devices were grown by molecular beam epitaxy (MBE), and potential barriers were fabricated by the insertion of planar layers of negative charges in n - type GaAs. The highest reported transfer coefficient was about 0.75 in a

870Å wide base device. HOLLIS has proposed, for the first time, that coupled plasmon - phonon modes are responsible for the scattering of hot electrons in heavily doped GaAs, resulting thus in a mfp of about 300Å.

My first entry into the field was in 1977 with the revival of the fabrication of the ***cold cathode transistor.*** [8] The difficulties encountered in the fabrication process led me later to propose the ***tunnelling hot electron transfer amplifier*** (THETA) devices [9,10] that utilize heterojunctions to form barriers. For example, undoped AlGaAs layers replace the original oxides implemented by Mead. The good lattice match between the conductors and insulators enables the fabrication of single crystal material with perfect interfaces, however the lower potential barriers require a low temperature operation. The THETA device was first realized by YOKOYAMA et al. [11] in 1984. Using MBE and fabricating a 1000Å wide base device the group measured $\alpha = 0.56$ at 10 K. This was followed by our work, at IBM, on the THETA devices. [12 – 15] We have shown conclusively the existence of ballistic transport of up to about 75% of the injected electrons, and recent α's as high as 0.9 in devices with base width of 300 Å.

What follows will describe in some detail the operation principles of the THETA device, its use as an energy spectrometer, and some recently obtained results related to transport that might be applicable to other GaAs high speed devices.

2. Structure and Operation of the THETA Device

The structure is comprised of five epitaxial, MBE grown, layers. Three n - type heavily doped GaAs layers are separated by two undoped AlGaAs barriers. The first GaAs (emitter) - AlGaAs - GaAs (base) structure forms a tunnel injector for hot electrons into the base. The second GaAs (base) - AlGaAs - GaAs (collector) structure forms a barrier with height Φ_C enabling only hot electrons with $p_x^2/2m > \Phi_C$ to surmount it. Here, p_x is the component of the momentum which is perpendicular to the one dimensional collector barrier. The emitter barrier preselects for tunnelling predominantly those electrons in the emitter that move perpendicular to the barrier. Thus, a collimated hot electron beam, with an energy spread, associated with the perpendicular momentum, of about 60 meV is injected into the base. The kinetic energy of the injected electrons is approximately equal to $eV_{BE} + \zeta$, where e is the electron charge, V_{BE} is the applied voltage between base and emitter, and ζ is the Fermi level energy above the bottom of the conduction band in the base. If the base is made thinner than the mfp a substantial fraction of the ballistic electrons will maintain their energy and direction after traversing the base and be collected above the collector barrier. (Fig. 1)

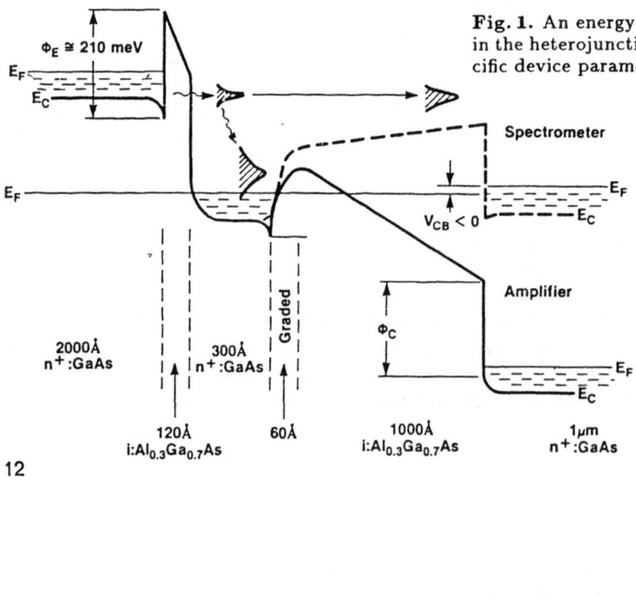

Fig. 1. An energy band diagram for Γ electrons in the heterojunction – THETA device. The specific device parameters are given

If hot electrons are injected in the (100) direction into a 300 A base with kinetic energy of 0.3 eV, they would traverse the base with a drift velocity of about 1×10^8 cm/s in a transit time of 30 fs. If the collector barrier is 500 Å wide and the electrons velocity there is about 5×10^7 cm/s, their transit time would be 100 fs. The short total transit time is the most attractive feature of the device. Maintaining this speed of operation in a practical device is difficult; I'll comment on it briefly at the end.

Before describing some of the recent results obtained with the hot electron devices, let me say a few words on the fabrication procedure. Since these devices are unipolar, contacting to the base can be done only by etching and exposing the base. Care must be taken in forming the ohmic contacts to the base to prevent a deep penetration and the shorting of the base and the collector layers. Etching to the base is done via a standard reactive ion etching process, ohmic contacts are formed by alloying thin NiAuGe overlayers, and isolation is achieved by boron implantation.

3. THETA Devices as Amplifiers

When the THETA devices are operated as amplifiers the emitter is negatively biased and the collector is positively biased with respect to the base. In the first devices fabricated by YOKOYAMA et al. [11] and in our initial devices [12], doping levels were at or below 5×10^{17} cm^{-3} ; consequently, a large injection voltage, $V_{BE} > 0.6$V, was needed in order to have a reasonable tunnelling current into the base. (This is because most of the applied voltage depleted the base and accumulated the emitter). Inevitable transfer into upper valleys (L and X valleys) then occurred, thus preventing ballistic motion. I'll discuss this effect in some more detail later. In devices with a 1000 Å thick base a maximum α of 0.7 was observed at temperatures of 100 K and below. The very weak temperature dependence suggested that optical phonon scattering was negligible. Since the hot electrons relaxed to the bottom of the available conduction bands (Γ, L and X) before arriving at the collector, increasing the collector voltage, V_{CB}, reduced the corresponding collector barrier heights (especially when barriers were graded), and aided in the electron collection.

Figure 2 shows the output characteristics of two similar devices: in one the AlGaAs of the collector barrier is graded toward GaAs in the base over 200 Å, and in the second the collector barrier is abrupt. The higher α in the graded device results in part from the lowering of the barrier Φ_C with the increased V_{CB}, and in part due to reduced quantum mechanical reflections. The disadvantage of grading is a lower output resistance, an effect that is reduced when the transport is predominantly ballistic.

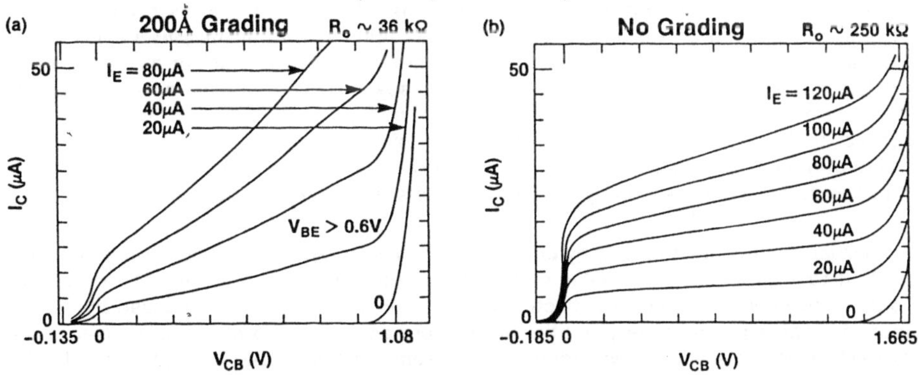

Fig. 2. The output characteristics of two devices with base doping of 2×10^{17} cm^{-3} and base width of 1000 Å. In one device the collector barrier is graded and in the other it is abrupt

When we increase the doping levels and thus are able to reduce the injection energies below 0.3 eV, ballistic transport begins to be evident for base width of about 800 Å.[13,15] It is clearly observed in Fig. 3, for a device with a 300 Å wide base and doping level of 1×10^{18} cm^{-3}, that hot electrons arrive at the collector with excess energy above the collector barrier. The collector current can be turned down only if the collector voltage is reversed. In devices with lower collector barrier heights, $\Phi_C = 140$ meV, and a doping level of 7×10^{17} cm^{-3}, maximum α's as high as 0.9 were observed with at least 75% of the current traversing the base ballistically.

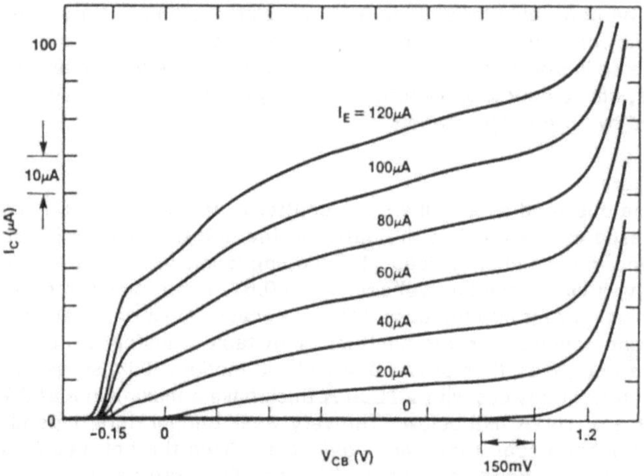

Fig. 3. Output characteristics, measured at 77 K, of the device that has a potential band diagram shown in Fig. 1. The maximum α is 0.75

4. THETA Devices as Energy Spectrometers

Spectroscopy in semiconductors was done initially in 1963 in Si [16] where the surface work function of Si used as a barrier spectrometer. In 1983, HESTO et al. [17] proposed to look for ballistic electrons in GaAs by fabricating a Schottky barrier as a spectrometer at the end of a short transport region. This was first realized in the ***planar doped barrier transistor*** by HAYES et al. [18] and in the THETA device by YOKOYAMA et al. [11] By this method energy distributions of hot electrons are scanned by a variable potential barrier; Φ_C in our case. Since the collected current is $e\int_{\Phi_C}^{\infty} nv dE$, the derivative of the current with respect to Φ_C is the momentum distribution, nv, of the collected carriers. Over a small range of energy (the distribution width $\simeq$60 meV) the electrons velocity is fairly constant, thus the derivative leads directly to the number distribution of the hot electrons.

Barrier heights are changed by changing V_{CB}; thus to do energy spectroscopy it is important to know in detail the value of Φ_C and its dependence on V_{BC}. Also, barriers have to be uniform without statistical fluctuations in their potential (can be caused by dopant atoms). The square - type potential barrier formed by a uniform AlGaAs layer is the easiest to fabricate and is a well-defined spectrometer (since $\Delta\Phi_C = \Delta V_{CB}$). Its only disadvantage is that electrons to be analyzed have to pass through the total thickness of the barrier without scattering. This requires an extremely high quality AlGaAs material. [19]

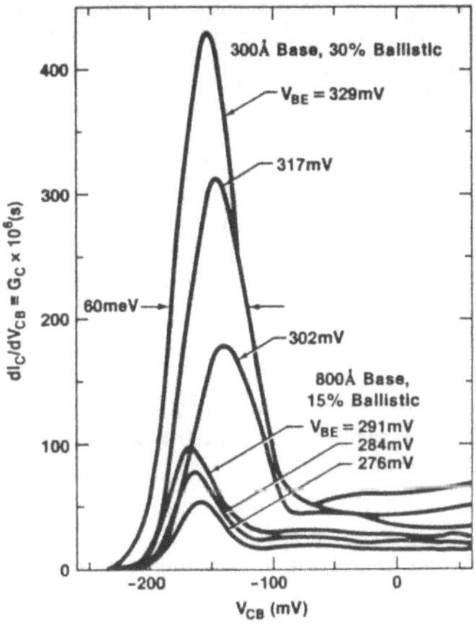

Fig. 4. Measured electron momentum distributions as a function of energy of the collected ballistic electrons for 300 Å and 800 Å base widths. The widths of the distributions are similar. The devices have slightly different collector barrier heights

When a derivative of the collector current, shown in Fig. 3, with respect to the collector voltage (in the range $V_{CB} < 0$) is taken, energy distributions like the ones shown in Fig. 4 are obtained. The peak positions coincide very well with the expected peak energy of the injected ballistic electrons, and the collected distribution widths are as narrow as the injected ones, indicating transfer without collisions. [13] When spectroscopy was done in the devices with a base width of 800 Å, similar energy distributions were found, but with a smaller fraction of collected ballistic electrons. In the 300 Å and 800 Å cases the ballistic fractions were 30% and 15%, respectively, for doping level of 1×10^{18} cm^{-3} and $\Phi_C \simeq 200$ meV.

To reduce electron-electron scattering the base doping was reduced to 7×10^{17} cm^{-3} in the 300 Å base. In addition, the available window for ballistic transfer (between the L valley minima and the top of the collector barrier, Φ_C, was increased by reducing Φ_C to about 140 meV (corresponding to an Al mole fraction of 16%). These changes resulted in a ballistic fraction of about 75%, leading to a mfp greater than a 1000 Å. However, I'd like to point out two complications that prevent an exact determination of the mfp: some of the ballistic electrons reflect from the base - collector boundary (due to the potential discontinuity) and the other have to traverse the AlGaAs barrier against an opposing field (scattering events will cause them to be collected by the base), thus leading to an underestimated ballistic fraction and mfp; on the other hand the base is partly depleted under injection conditions, hence, the conductive path is shorter than the base width, resulting in an overestimated mfp for a uniformly conductive base.

Note that the resolution of our spectrometer is limited by quantum mechanical reflections of electrons that are close above the barrier top, and tunnelling of electrons that are close below the barrier top. I estimate our spectrometer resolution to be about 20 meV. Due to the inherent energy width of the injected electron distribution, ruling out small-angle scattering events for some of the electrons in the ballistic part of the distribution seems incorrect. However, one should expect an eventual broadening of the distribution as the base width increases, an effect that we do not see. (see Fig. 4) Thus, it is reasonable to conclude that most scattering events are thermallizing or randomizing events, namely, each electron that scatters is 'plucked' out from the distribution and the remaining measured part represents the true ballistic electrons.

This 'traumatic' scattering mechanism is not yet understood. HOLLIS et al. [7] and later LEVI et al. [21] had proposed that coupled modes of optical phonons and plasmons are responsible for the main scattering events. This assumption results in a calculated mfp of 300 - 400 Å in GaAs doped to 1×10^{18} cm^{-3}. However, in their calculations quantization effects and the dominance of surface plasmon over the bulk plasmon modes in the thin quasi two-dimensional base were ignored; both are expected to give rise to a longer calculated mfp.

5. Physics Studied with THETA Devices

Having quasi monoenergetic ballistic electrons enables us to probe energetically, with a resolution of the order of the ballistic distribution width, properties of the transport region. Let me point out a few examples that we currently study:

5.1 Scattering Mechanism and mfp

By changing the base width and the doping level we can learn of the main factors that determine the ballistic mfp. The tunnel injector allows us to probe the transmission through the structure by varying the injection energy. This study may shed light on the dominant scattering mechanism that affect hot electrons in doped GaAs regions, and in particular, in narrow regions where quantization effects take place.

5.2 Quantization Effects in Narrow Structures

When electrons are injected via tunnelling into a very thin base, the resultant current depends strongly on the density of states in the base. When the base is narrow enough (less than 500 Å), the spacings between the allowed energy levels in the base - which lead to peaks in the density of states - are on the order of the distribution width. This results in a modulation of the tunnelling current as the injection energy scans through the states. When $eV_{BE} + \zeta < \Phi_C$, tunnelling occurs into bound states, as reflected clearly in the derivative of the emitter current shown in Fig. 5 (a). However, when $eV_{BE} + \zeta > \Phi_C$, the tunnelling current and α reflect the ***virtual levels*** in the base. [22] This can be seen in Fig. 5 (b) where α is plotted as a function of the injection voltage for different collector voltages.

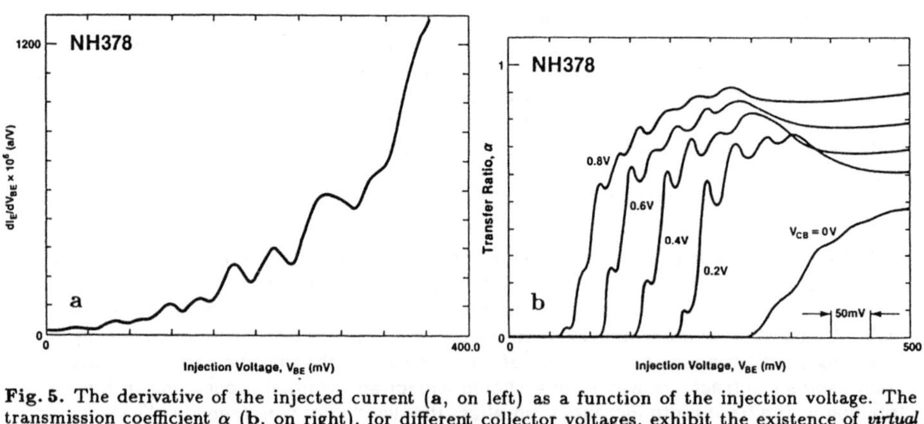

Fig. 5. The derivative of the injected current (**a**, on left) as a function of the injection voltage. The transmission coefficient α (**b**, on right), for different collector voltages, exhibit the existence of *virtual states* in the base. The device has a 300 Å wide base and doping of 1×10^{18} cm^{-3}

This is a beautiful demonstration of simple quantum mechanics ('particle in a box') in solid state devices, and a manifestation of the type of problems that small device structures will eventually exhibit.

5.3 Transfer into Upper Satellite Valleys

When the injection energy exceeds the energy splitting between the L and Γ valleys, a fraction of the ballistic electrons will scatter into the L valleys (via optical phonon or impurity scattering). The scattering will randomize the electrons that will diffuse toward the collector at the bottom of the L valleys. (scattering times back to the Γ valleys is on the order of 1 pS, much longer than the transit time). The 'L electrons' encounter a potential barrier at the collector, and thus, at low temperatures, they will not be collected, leading thus to a reduction in α. This phenomenon is observed clearly in devices that have a large fraction of the current carried by ballistic electrons.

We have proven the transfer by applying hydrostatic pressure on the devices and observing a transfer onset at lower injection energies due to the smaller separation between the L and Γ valleys. [14] One set of such results is shown in Fig. 6. Note that after transfer takes place the modulation in α (due to the *virtual states*) disappears because of a reduction in the fraction of ballistic electrons.

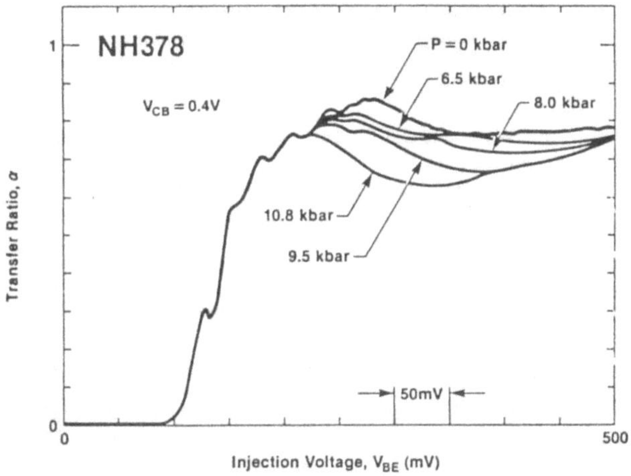

Fig. 6. $\alpha = dI_C/dI_E$ as a function of the injection voltage, at different applied hydrostatic pressures. The onset of transfer into the L valleys is indicated by the drop in α that is shifted to lower energies as the applied pressure increases

5.4 Determination of Barrier Heights

Since the collector current turns on abruptly, at low temperatures, when $eV_{BE} + \zeta \geq \Phi_C$, we can determine easily Φ_C, and from it the conduction band discontinuity, ΔE_C, between GaAs and AlGaAs. We find in some devices a greater barrier height than expected for the corresponding Al concentration. This is most probably due to negative charges that are found sometimes in the undoped AlGaAs. [23] In barriers with the least amount of negative charges, the conduction band discontinuity follows the approximate rule $\Delta E_C \simeq 9x$, where x is measured in [%] and ΔE_C in [meV]. This corresponds approximately to $\Delta E_C:\Delta E_V =$ 70:30, where ΔE_V is the valence band discontinuity.

6. Can Ballistic Transistors Operate in the Subpicosecond Regime ?

Because of the limited space I'll comment only briefly on the issue of the device speed. Obviously transit time is not a limiting factor. At this point it is not clear what is the highest

α we can achieve in the THETA or a similar hot electron device, however, let me assume that the gain is available. If resistances do not limit the current through the device, it can be shown that in a THETA device with an emitter area of 0.25 μm x 1 μm, emitter - base spacing of 0.25 μm, and a 500 Å base width, a current density of $3 \times 10^5 a/cm^2$ is needed to change the voltage of emitter and collector by 0.1 V and 0.2 V respectively in less than 1 pS. Since the emitter and collector capacitances are charged through the base current the base resistance has to be low enough, a difficult task for a very thin base. Using the ***modulation doping*** approach in the base (where the impurities are introduced in the AlGaAs barriers and electrons spill into the base, forming thus a quasi two-dimensional electron gas), [10] would allow shrinking of the base to 100 - 150 Å, while having $2 \times 10^{12} cm^{-2}$ electrons with a relatively high mobility at 77 K. If the formed electron gas could support a high enough current density through the base, [24] this structure (Modulation Doped THETA, or MODTHETA) would be a very high speed device.

Acknowledgments I wish to thank my collaborators in this work: D. C. Thomas, I. M. Anderson, C. M. Knoedler, J. W. Mitchell, L. Osterling, M. I. Nathan, W. P. Dumke, E. Calleja, P. J. Price, and M. V. Fischetti, and C. J. Kircher for his valuable comments on the manuscript.

References

1. C.A. Mead, Proc. IRE 48, 359 (1960); J. App. Phys. 32, 646 (1961).
2. J. P. Spratt, R. F. Schwarz, and W. M. Kane, Phys. Rev. Lett. 6, 341 (1961).
3. D. V. Geppert, Proc. IRE 50, 1527 (1961), D. Khang, Proc. IRE 50, 1534 (1961).
4. J.M. Shannon, IEE J. Solid-State & Electron Devices 3, 142 (1979).
5. M. Heiblum, IBM Technical Disclosure Y0881-0094, Feb. 1981, published in the IBM Technical Disclosure Magazine 24, 4507-9 (1982).
6. R.J. Malik, M.A. Hollis, L.F. Eastman, D.J. Woodard, C.E.C. Wood, and T.R. AuCoin, Proc. Conf. on Active Microwave Devices, Cornell Univ. (1981).
7. M.A. Hollis, S.C.Palmateer, L.F. Eastman, N.V. Dandekar, and P.M. Smith, Electron Device Lett. EDL-4, 440 (1983).
8. M. Heiblum, Thesis work, Univ. Cal. (Bekeley), 1978.
9. M. Heiblum, US patent no. 4,286,275, filed 1980.
10. M. Heiblum, Solid-State Electron. 24, 343 (1981).
11. N. Yokoyama, K Imamura, T. Ohshima, H. Nishi, S. Muto, K. Kondo, and S. Hiyamizu, IEEE Electron Device Meeting Proceedings, San Francisco (1984).
12. M. Heiblum, D. C. Thomas, C. M. Knoedler, and M. I. Nathan, Appl. Phys. Lett. 47, 1105 (1985).
13. M. Heiblum, M. I. Nathan, D. C. Thomas, and C. M. Knoedler, Phys. Rev. Lett. 55, 2200 (1985).
14. M. Heiblum, E. Calleja, I. M. Anderson, W. P. Dumke, C. M. Knoedler, and L. Osterling, Phys. Rev. Lett. 56, 2854 (1986).
15. M. Heiblum, I. M. Anderson, and C. M. Knoedler, Appl. Phys. Lett. 49, 207(1986).
16. D. J. Barterlink, J. L. Moll, and N. I. Meyer, Phys. Rev. 130, 972 (1963).
17. P. Hesto, J-F. Pone, and R. Castagne, Appl. Phys. Lett. 40, 405 (1982).
18. J.R. Hayes, A.F.J. Levi, and W. Wiegmann, Electron. Lett. 20, 851 (1984); Phys. Rev. Lett. 54, 1570 (1985).
19. The fabrication of high-quality barriers requires some care [20]; if poor quality AlGaAs is made spectroscopy is impossible because of severe scattering in the alloy. For spectroscopy, the preferred barrier is an asymmetric, compositionally graded AlGaAs triangular barrier with a peak close to the base, mimicking the planar doped barrier transistor. This barrier enables the spectroscopy to be done closer to the base. However, controling the exact barrier shape is very difficult,thus making accurate spectroscopy impossible.
20. M. Heiblum, E. Mendez, and L. Osterling, J. Appl. Phys. 54, 6982 (1983).
21. A. F. J. Levi, J. R. Hayes, P. M. Platzman, and W. Wiegmann, Phys. Rev. Lett. 55, 2071 (1985).
22. D. Bohm, ***Quantum Theory*** (Prentice - Hall, New York, 1951), Part 3.
23. T. Hickmott, P. Solomon, R. Fischer, and H. Morkoc, J. Appl. Phys. 57, 2844 (1985).
24. P. M. Solomon, private communications.

Hot Electron Transistors

J.R. Hayes[1], *A.F.J. Levi*[2], *A.C. Gossard*[2], *and J.H. English*[2]

[1]Bell Communication Research Inc., Murray Hill, NJ 07974, USA
[2]AT & T Bell Laboratories, Murray Hill, NJ 07974, USA

ABSTRACT

The transport properties of non-equilibrium electrons in GaAs Hot Electron Transistors has been successfully probed using Hot Electron Spectroscopy, and explained in terms of a fully coupled electron-phonon system. The mean free path, for a non-equilibrium electron is shown to be critically dependent on the electron concentration in the base region enabling us to conclude that GaAs is unsuitable for the fabrication of a useful Hot Electron Transistor. It is suggested that a semiconductor with a small effective mass or a two dimensional system in the base would show improved performance.

The advent of thin film semiconductor epitaxial growth techniques has led to a renewed interest in the fabrication of "Hot Electron Transistors" (HET's). Epitaxial crystal growth of the GaAs/AlGaAs alloy system is sufficiently well characterized that one may grow single crystal layers of almost arbitrary composition and doping. This unique degree of control allows one to translate metal-oxide-metal tunnel junctions into GaAs/AlGaAs/GaAs tunnel junctions and metal-semiconductor Schottky barriers into either GaAs/AlGaAs heterojunctions, camel diodes [1] or planar doped barriers [2]. Thus the HET concepts of the early 1960's are beginning to find their semiconductor analogues today with particular interest being shown in the development of a HET in the GaAs/AlGaAs semiconductor system [3]. In light of this fact we feel that an evaluation of the ability of fabricating a truly useful "ballistic" hot electron device is needed.

The hot electron spectra described in this paper were obtained using "Hot Electron Spectroscopy" by means of a planar GaAs structure grown by MBE at 650°C on a <100>-orientated semi-insulating GaAs substrate using cracked As_4. A schematic diagram of the energy band structure used for "Hot Electron Spectroscopy" is shown in Fig. 1. Electrical contact was made to the three degenerately doped n-type GaAs regions, that were separated by two bulk triangular potential barriers. When the emitter was biased negative electrons were injected into the base region with an excess energy $(\phi_{eb} - E_F)$ above the Fermi energy, E_F. With a small positive bias on the collector no electrons were collected as none had sufficient energy to transverse the collector barrier, ϕ_{bc}. With increasing base-collector bias, the barrier, ϕ_{bc} is decreased, as indicated by the broken lines in Fig. 1. Hence, by applying a base/collector bias a means of continously varying the analyzer barrier was obtained. Hot electrons arriving at the analyzer barrier will have a particular value of momentum normal to the barrier, $p_\perp$, and may be described in terms of a distribution of values $n(p_\perp)$ such that the total current flow, I_c, for a given barrier energy ϕ_{bc} is

$$I_c = -(e/m_e^*) \int_{p_\perp^0}^{\infty} p_\perp n(p_\perp) dp_\perp \qquad (1)$$

where m_e^* is the effective electron mass, e the electronic charge and $p_\perp^0 = (2m_e^* \phi_{bc})^{1/2}$.

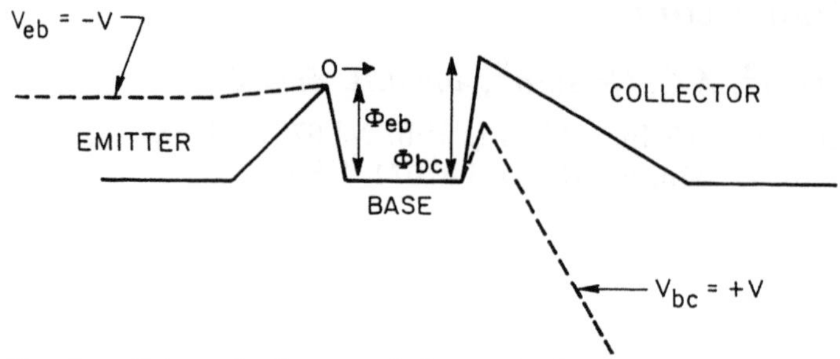

Fig. 1 Schematic diagram of the conduction band edge of the Hot Electron Transistor. The hot electron injector is at the emitter/base junction and the hot electron analyzer is at the base/collector junction. The broken lines indicate the conduction band edge under applied bias.

Taking the derivative of I_c, in (1), with respect to V_{bc} it is easy to show that

$$\frac{dI_c}{dV_{bc}} \propto n(p_{\perp}^{0}) \ . \tag{2}$$

Hence, by differentiating I_c with respect to V_{bc} we may obtain the electron momentum distribution at the base-collector junction [6]. Typical hot electron spectra are shown in Fig. 2 for samples having different transit region widths but the same carrier concentration ($1 \times 10^{18}\ cm^{-3}$). The spectra were obtained at 4.2K, to eliminate thermal smearing effects, and each had an injection energy $E_i \simeq 0.25$ eV. The spectra show two distinct peaks, for samples having narrow transit regions (< 900 Å), one at the injection energy (E_i) and the other close to the Fermi energy (E_F). The peak at the injection E_i is evidence of ballistic electrons an observation first presented by us in an invited presentation at the 1985 March meeting of the American Physical Society and subsequently published in *Physical Review Letters* [6]. This observation was later confirmed by Heiblum *el al.* [3] using a differently designed spectrometer. Unfortunately, a complete hot electron spectrum was not obtained, so that experimental results at low-analyzer barrier energies could not be compared.

To understand the electron scattering mechanisms giving rise to the hot electron spectra shown in Fig. 2 we have developed a theory of nonequilibrium electron transport [5]. The calculated scattering rates for coupled electron/phonon and ionized impurity scattering, as a

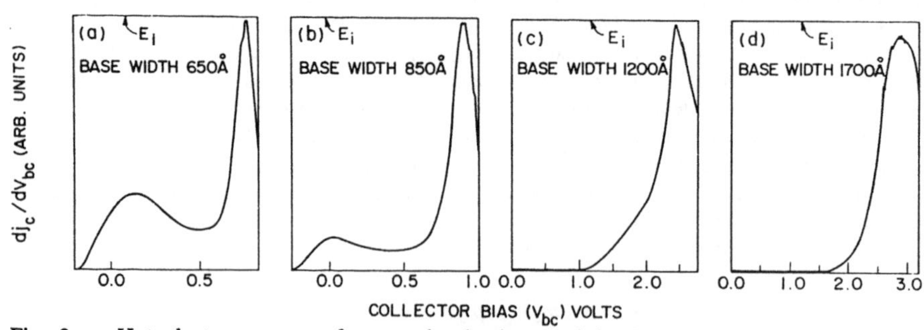

Fig. 2 Hot electron spectra for samples having an injection energy E_i ~0.25 eV and a transit region doping of $1 \times 10^{18}\ cm^{-3}$. The measurements were made at a current density of 12.7 A/cm^2

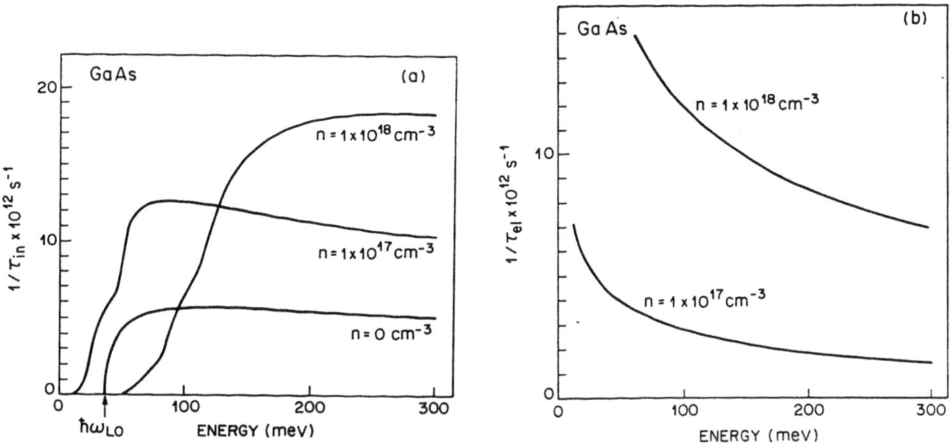

Fig. 3 Calculated inelastic (a) and elastic (b) scattering rates for GaAs.

function of energy, are presented in Fig. 3 for GaAs at the indicated carrier densities. Assuming that electrons injected at 0.25 eV have a velocity of 1×10^8 cm/s the mean free path, considering both mechanisms, would be ≃350Å. This theoretical mean free path is in excellent agreement with the experimental value obtained by studying the magnetic field dependence of the hot electron spectra. In addition the theory enables us to interpret the features of the hot electron spectra. The peak a low energies (high reverse bias) is predominately due to electrons actived from the Fermi sea in the short samples, but in the long samples it also has a contribution from electrons scattered down from high energies. The high energy peak (low or negative bias) is due to ballistic electrons.

Both our experimental and theoretical results indicate that GaAs is unsuitable for the fabrication of a useful, high-performance, "Ballistic" Electron Transistor (BET) because the injected hot electrons have such short mean free paths. There are two approaches to consider when designing a BET in a material system other than GaAs/AlGaAs. One possibility is to chose a semiconductor with a wide intervally separation, in order to take advantage of the decreased scattering rate with increasing energy. The other is to consider a material with a low effective mass and thereby lower density of states, giving a reduced electron scattering rate.

A semiconductor illustrating the first case could be CdTe which has a direct band gap of 1.3 eV and a subsidiary minimum 1.1 eV above the conduction band minimum. Such a material, however, with its large effective electron mass $m_e^* = 0.1m_0$ is unsuitable for "ballistic" devices. Alternatively, a semiconductor which satisfies the low mass condition would be InAs or InSb. InAs and InSb both have significantly lower inelastic scattering rates compared to GaAs. Because it may be possible to lattice match wider band gap alloys such as GaInAsSb to InAs we consider it in preference to InSb. As is the case with many low m_e semiconductor InAs has a small band gap energy. $E_g \sim 0.41$ eV which is less than the energy difference between the subsidiary and conduction band minimum. Consequently, whereas the maximum injection energy E_i^{max} in GaAs was determined by the subsidiary L minimum [8],for InAs E_i^{max} must satisfy $E_i^{max} \leqslant E_g + E_F$, to avoid the possibility of direct excitation of electrons from the valence band into the conduction band: as also occurs in InSb [9].

In Fig. 4(a) and (b) we plot the inelastic and elastic scattering rate respectively, for InAs as a function of E_i for various base impurity concentrations. The calculated scattering rates have features similar to those shown in Fig. 3 only now the rates for a given E_i are

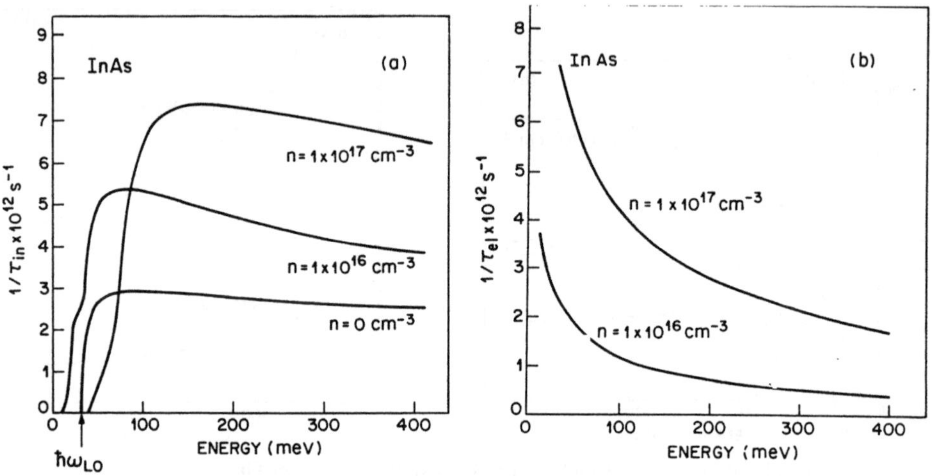

Fig. 4 Calculated inelastic (a) and elastic (b) scattering rates for InAs.

significantly reduced. Differences in electron mobility between InAs and GaAs mean that a HET device with an $n = 1 \times 10^{18}$ cm^{-3} doped GaAs base should be compared to an $n \sim 1 \times 10^{16}$ cm^{-3} doped InAs base. For $n = 1 \times 10^{16}$ cm^{-3} and $E_i = 0.4$ eV electrons in InAs have a total mean free path of around 3000Å giving, for a 500Å base width, a "ballistic" common base current gain $\alpha_B = 0.85$. This is a promising improvement when compared to GaAs.

Irrespective of whether the semiconductor used in GaAs or InAs to improve device performance still further we are forced to consider a different means of confining thermal electrons to the base region. An obvious approach is to create a uniform (fluctuation free) potential well such as occurs in a two-dimensional electron gas at a GaAs/AlGaAs interface. In this case, elastic scattering may be reduced to a minimum by use of "modulation doping" which spatially removes the donor ions from the confined electron gas [10]. The base region will only be ~100Å wide and inelastic electron scattering rates are reduced over those calculated for the three-dimensional case described above (because of the reduction in density of states) so that α_B could easily approach unity.

REFERENCES

[1] J. M. Shannon, Appl. Phys. Lett., Vol. 35, pp. 63-65, 1979.

[2] R. J. Malik, T. R. AuCoin, R. L. Ross, K. Board, C. E. C. Wood and L. F. Eastman, Elec. Lett., Vol. 16, pp. 836-838, 1980.

[3] S. Muto, K. Imamuwa, N. Yokoyama, S. Hiyanizu and H. Nishi, Elec. Letts., Vol. 21, pp. 555-556, 1985: J. M. Woodcock, J. J. Harris and J. M. Shannon, Physica, Vol. 134B, pp. 111-118, 1985: M. Heiblum, D. C. Thomas, C. M. Knoedler and M. I. Nathan, Appl. Phys. Letts., Vol. 47, pp. 1105-1107, 1985.

[4] J. R. Hayes, A. F. J. Levi and W. Weigmann, Electron Lett., Vol. 20, pp. 851-852, 1984. J. R. Hayes, A. F. J. Levi and W. Wiegmann, Phys. Rev. Lett., Vol. 54, pp. 1570-1572, 1985.

[5] A. F. J. Levi, J. R. Hayes, P. M. Platzman and W. Weigmann, Phys. Rev. Lett., Vol. 55, pp. 2071-2073, 1985.

[6] M. Heiblum, M. J. Nathan, D. C. Thomas and C. M. Knoedler, Phys. Rev. Lett., Vol. 55, pp. 2200-2203, 1985.

[7] A. F. J. Levi, J. R. Hayes, P. M. Platzman and W. Weigmann, Physica, Vol. 134B, pp. 480-486, 1985.
[8] J. S. Blakemore, J. Appl. Phys. Vol. 53, pp. R123-181, 1982.
[9] M. Glicksman and M. C. Steele, Phys. Rev., Vol. 110, pp. 1204-1205, 1958.
[10] R. Dingle, H. L. Störmer, A. C. Gossard and W. Wiegmann, Appl. Phys. Lett., Vol. 33, pp. 665-667, 1978.

Tunneling Through III–V Low-Barrier Heterostructures

P. Guéret [1], *A. Baratoff* [1], *S. Bending* [1], *H. Meier* [1], *E. Marclay* [2], *and M. Py* [2]

[1]IBM Zurich Research Laboratory, CH-8803 Rüschlikon, Switzerland
[2]Institute for Microelectronics, Swiss Federal Institute of Technoloy, CH-1015 Lausanne, Switzerland

This paper deals with tunnelling through very low and thick $Al_xGa_{1-x}As$ barriers. Experimental results are presented and compared with theory. Measurements in a transverse magnetic field are also described, and their relationship to the tunnelling time discussed. They provide further confirmation of direct, "ballistic" tunnelling through thick barriers.

The transmission coefficient D through a tunnel barrier depends exponentially on barrier height Φ and length ℓ, viz., $D \propto \exp\left[-A\sqrt{m^*/m_0 \Phi}\,.\ell\right]$, where m^*/m_0 is the ratio of effective to free electron mass, and A = 0.033 if Φ is in meV, and ℓ in Å . Accordingly, a small barrier and a low effective mass, in principle, make it possible to tunnel through macroscopic distances. Thus, the exponent in D has the same value for a barrier with $m^* = m_0$, Φ = 4 eV, ℓ = 15 Å, and one with $m^* = 0.07\, m_0$, Φ = 13 meV, ℓ = 1000 Å.

There are several reasons for investigating tunnelling through such low and thick barriers. One being that the large thickness may make it feasible to modulate the tunnel current with a control gate, leading to potentially interesting device applications. Low barriers are also well suited to study small barrier-height corrections as may be due to space-charge and image potential, or to undesirable process asymmetries. A further motivation is that low barriers provide a better lattice match at the $GaAs/Al_xGa_{1-x}As$ interface as well as greater accuracy of the effective mass approximation. Thick barriers are also less sensitive to thickness uncertainties.

We have therefore started a systematic study of tunnelling through GaAs-based heterostructures with barrier heights below 100 meV, Guéret et al. [1]. Vertical structures are grown by MBE on a highly-doped GaAs substrate. Their main part consists of a layer of nominally undoped $Al_xGa_{1-x}As$ between two identical layers of Si-doped GaAs. The donor concentration is $N_D = 10^{17}\, cm^{-3}$ which, at 4.2° K, places the Fermi level at about 12 meV above the conduction-band minimum in both GaAs layers. Mesas are etched on the chip surface, and ohmic contacts using Ni/Au-Ge metallization are formed on the top and bottom areas. The experimental set-up allows electrical measurements down to 1.5° *K*.

Figure 1 shows data taken at 4.2° K for barriers with x = 0.06 to 0.02, and ℓ = 430 to 1000 Å. Barrier heights are obtained from thermionic conductance data, and used to compute the tunnel current [1]. The barrier transmission coefficient is calculated in the effective mass approximation for plane waves transmitted through a rectangular barrier tilted by the applied voltage. The effective

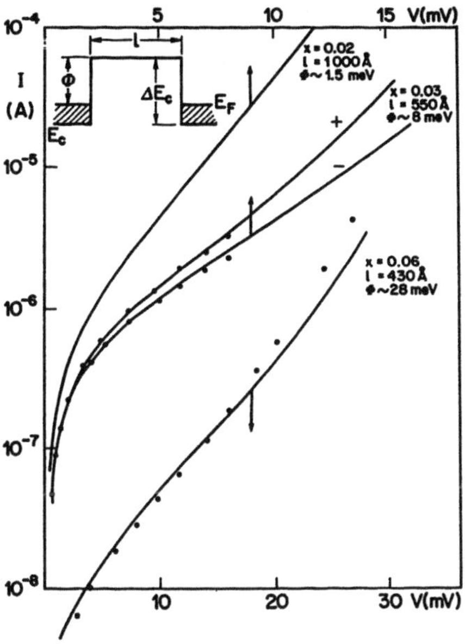

Fig. 1. Current-voltage characteristics of several $n - GaAs/Al_xGa_{1-x}As/n - GaAs$ barriers, measured (full lines) and calculated below the Fowler-Nordheim regime (dots), at $T = 4.2°K$. Diode areas $\sim 1.6\ 10^{-4}\ cm^2$

barrier thickness is used as a parameter to fit the computed to the experimental data below the Fowler-Nordheim regime. Fitted barrier thicknesses are about 50 Å smaller than those deduced from process calibration. The theoretical model includes corrections for band-bending in the electrodes and for the presence of quantum-mechanical space-charge in the barrier. This last correction arises from the fact, Büttiker [2], that the electronic wave function has an exponential tail in the barrier, creating space-charge and a corresponding increase $\Delta\Phi_{SC}$ in the effective barrier height. A first-order estimate from Poisson's equation yields

$$\Delta\Phi_{SC}/\ \Phi = \frac{1}{8}\ (\hbar\omega_{p}/\ \Phi)^2, \tag{1}$$

where $\omega_p^2 = q^2 N_D\ /\ \varepsilon m^*$ is the plasma frequency in the electrodes, and $\hbar\,\omega_p = 12.65$ meV for $N_D = 10^{17}\ cm^{-3}$, $m^*/m_0 - 0.067$. The quantum-mechanical space-charge correction should therefore become significant for barrier heights $\Phi\ \sim$ 10 meV or lower. A fit to the data for x = 0.02 has not yet been made since a self-consistent theory appears to be required in this extreme case. Also shown in Fig. 1 for the case x = 0.03 is a fit to the I-V data for positive and negative bias, obtained by assuming a process-related asymmetry of 0.5 meV. Such an asymmetry may be due to a slight decrease of the mole-fraction x during barrier growth, or to a minor difference in the donor concentrations in the GaAs electrodes. It appears quite systematically and always in the same direction, the barrier being tilted down, from the substrate side towards the top of the structure.

Figure 2 summarizes our data in a logarithmic plot of the specific zero-bias conductance as a function of the tunnelling exponent in the range x = 0.03 to 0.12 and ℓ = 250 to 550 Å, space-charge correction included. The exponential dependence is seen to be rather well-confirmed over about three orders of magnitude.

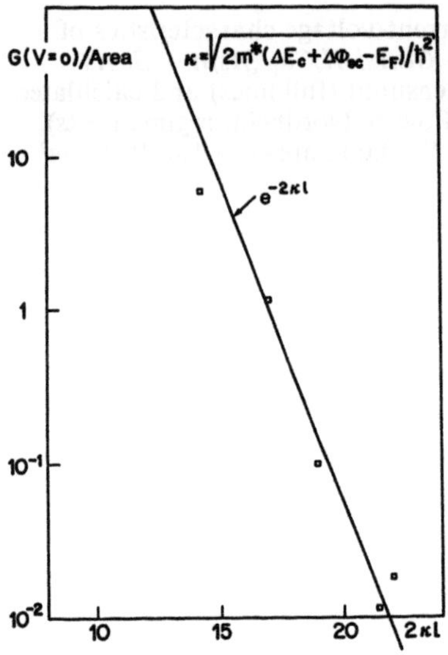

Fig. 2. Measured specific zero-voltage conductance (squares) versus tunnelling exponent, in the range x = 0.03 to 0.12, ℓ = 250 to 550 Å

Finally, we note that if we assume a $GaAs/Al_xGa_{1-x}As$ gap discontinuity $\Delta E_g = 1.247$ x, our data yield a conduction-band discontinuity $\Delta E_C \sim 0.54 \Delta E_g$ in the range investigated. This ratio compares well with the value 0.59 (± 0.03) obtained by Meynadier et al. [3] for x = 0.13.

We have also made a set of measurements in a magnetic field B transverse to the current. As a simple calculation shows, the transverse momentum imparted to an electron by this field increases its classical path length in the barrier by

$$\Delta\ell = \ell/6\ (\omega_C\ \tau_T/\ 2)^2, \tag{2}$$

where $\omega_C = q\ B/m^*$, and τ_T is the traversal time. Solving Schrödinger's equation in the WKB approximation yields exactly the same result, i.e., a reduced tunnelling probability by a factor exp($-2\kappa\Delta\ell$), thus

$$D(B) = D(B=0) \bullet \exp(-2\kappa\Delta\ell) \tag{3}$$

where $\Delta\ell$ is given by (2), $\kappa = \sqrt{2m^*\ \Phi\ /\ \hbar^2}$ and $\tau_T = \ell/(\hbar\kappa/m^*)$ is the tunnelling time introduced by Stevens [4] and Büttiker and Landauer [5].

It can be verified that low and thick barriers, like those we are dealing with, will exhibit a substantial current decrease even with the moderate fields (up to 4 Teslas) used for our experiment. This is quite a favorable situation since under these conditions, the Landau levels in the electrodes remain smeared out and the effect of the magnetic field remains entirely confined to the barrier. Results are shown in Fig. 3 for two distinct cases, and show quite good agreement between experimental and calculated data. This provides further confirmation that we are indeed

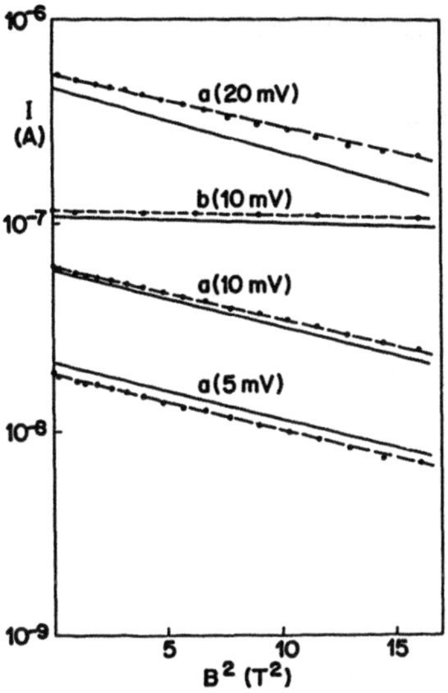

Fig. 3. Theoretical (full lines) and experimental (dots) magnetic-field dependence of the tunnel current at 4.2°K for different applied biases and for two barriers with $\Delta E_c = 40$ meV, $\ell = 430$ Å (a), and $\Delta E_c = 83$ meV, $\ell = 250$ Å (b)

observing direct, "ballistic" tunnelling through the barriers. In addition, this experiment provides a simple and accurate way to obtain the tunnelling time. Several thought experiments have in fact been proposed for this purpose [5,6], all rather difficult to implement. One such static thought experiment analyzed by Büttiker [6] makes use of the spin precession of an electron subject to a transverse field confined to the barrier. The transmission coefficient is reduced in that case by a factor exp(- $\omega_L \tau_T$), where ω_L is the Larmor frequency, so that the magnetic-field dependence of the tunnel current can then be used to obtain the tunnelling time. The present experiment is the orbital analog of the former, using the orbital motion rather than the spin precession as a clock. For case a in Fig. 3, the slope of the data line together with (2) and (3) yields $\tau_T = 0.1$ psec for the tunnelling time.

References

1. P. Guéret, U. Kaufmann, and E. Marclay: Electron. Lett., 21, 344 (1985)
2. M. Büttiker: private communication
3. M.H. Meynadier, C. Delalande, G. Bastard, M. Voos, F. Alexandre, and J.L. Liévin: Phys. Rev. B, Rapid Communications, 31, 5539 (1985)
4. K.W.H. Stevens: J. Phys. C 16, 3694 (1983)
5. M. Büttiker and R. Landauer: Phys. Rev. Lett., 49, 1739 (1982)
6. M. Büttiker: Phys. Rev. B 27, 6178 (1983)

Velocity Overshoot and Suppression of Diffusivity and Microwave Noise in Short n^+-n-n^+ Structures of GaAs

V. Bareikis, K. Kibickas, J. Liberis, A. Matulionis, R. Miliušytė, J. Paršeliūnas, J. Požela, and P. Sakalas

Semiconductor Physics Institute, Lithuanian Academy of Sciences, K. Poželos 52, Vilnius, SU-232600, USSR

1. Introduction

Fast devices usually require semiconductors with high drift velocity of electrons and prefer the conditions of operation favourable for low noise and diffusivity.

The maximum steady-state drift velocity in bulk GaAs is accompanied by intense diffusivity and noise /1,2/. Less unfavourable conditions are possible during a transient process to the steady state, because the transient behaviour is different for drift velocity, temperature and diffusion coefficient.

The drift velocity reaches the steady state through the overshoot: a very fast acceleration is followed by a slower relaxation down to the steady value /3/. The transient behaviour of the diffusion coefficient depends on many circumstances /4,5,6/, and it is important for us now, that the velocity overshoot coexists with a reduced diffusivity /4/. Microwave noise is related to diffusion /7,8/, therefore a low noise and a high velocity are expected simultaneously in a limited interval of time or space. Our task is to confirm this possibility experimentally through microwave investigations of short n^+-n-n^+-structure of GaAs performed at room temperature.

2. High Drift Velocity in Short Structures

The 37 GHz microwave integral mobility technique /9/ is applied to measure the deviation from the Ohm's law and to determine the drift velocity averaged over the n-region of the structure. Some additional details are described elsewhere /10/.

The results /10/ are presented in Fig.1 together with those for long samples /1/ for comparison. A velocity-field dependence for the 7.5 micron structure does not differ essentially from that measured for long samples. The length of the n-region being reduced from 7.5 to 1.0 micron the drift velocity at the threshold field for the negative differential mobility is found to increase 3 times. The minimum drift time is less than 2 ps and 5 ps for the 1.0 and 1.5 micron structures correspondingly.

3. The Hot Electron Microwave Noise and Diffusion

The noise temperature measured at 10 GHz frequency is presented in Fig.2. The data for the bulk samples /2/ are close to those

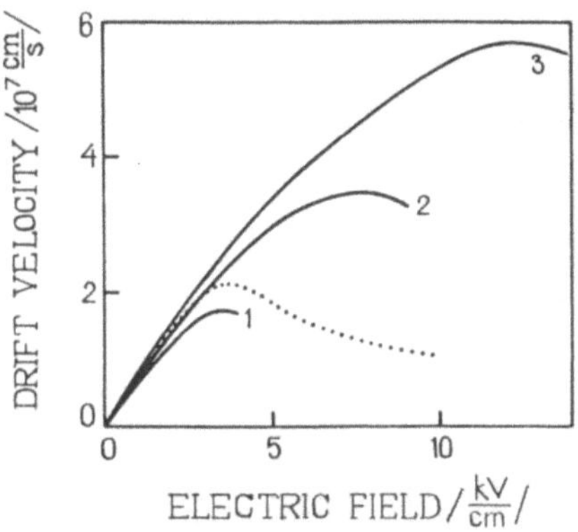

Fig.1. Velocity-field dependence for the n^+-n-n^+-structures with the room-temperature mobility 7500...7600 $cm^2/(V\cdot s)$, $(1...2.8)\times 10^{15} cm^{-3}$ donor density and various length of the n-region (in microns): 1-7.5, 2-1.5, 3-1.0 /10/, together with the data of /1/ (dots)

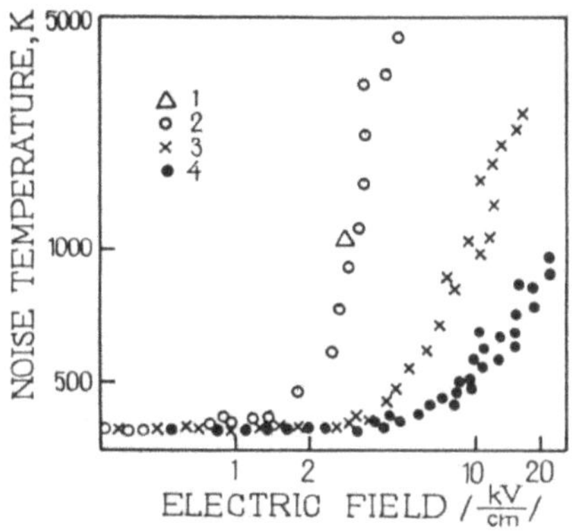

Fig.2. Noise temperature vs effective electric field (E= U/L). The structure length (in microns): 1-1000 /2/; 2- 7.5, 3- 1.5, 4- 1.0 /11/

for the 7.5 micron structure /11/: the steep increase due to intervalley transfer takes place at approximately 2 kV/cm field. At 4 kV/cm the temperature reaches 5000 K in the 7.5 micron structure, however it falls down to less than 350 K in the shortest structures /11/. The maximum suppression of the noise temperature exceeds 10 dB.

The results of Fig.2 show that the distance of 1 micron is too short for an onset of the intervalley transfer and the related noise at fields below 8 kV/cm. Thus we have a high velocity and a low noise simultaneously (Fig.1 and Fig.2).

Figure 3 shows the diffusion coefficient deduced from the microwave noise and the differential conductance measurements performed at 10 GHz /2,11,12/. The 11 micron structure /12/ and the 7.5 micron one /11/ show a slight minimum at 1.8 and 2.4 kV/cm field, respectively. This behaviour is not observed in the bulk sample /2/. The decrease of the diffusion coefficient well below the thermal equilibrium value is typical for the 1.0 and 1.5 micron structures. We see that an essentially reduced diffusion does not require submicron structures.

4. The Zones at the Cathode

Now we would like to discuss briefly different types of zones at the cathode, which are of interest for fast devices.

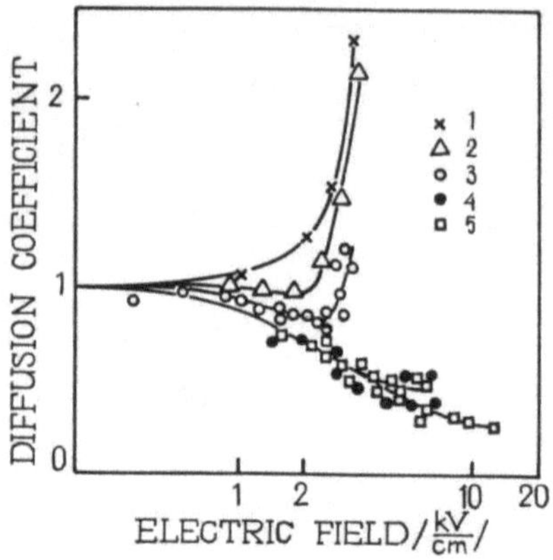

Fig.3. Relative diffusion coefficient for the structures of various length (in microns): 1- 1000 /2/; 2- 11.0 /12/; 3- 7.5, 4- 1.5, 5- 1.0 /11/

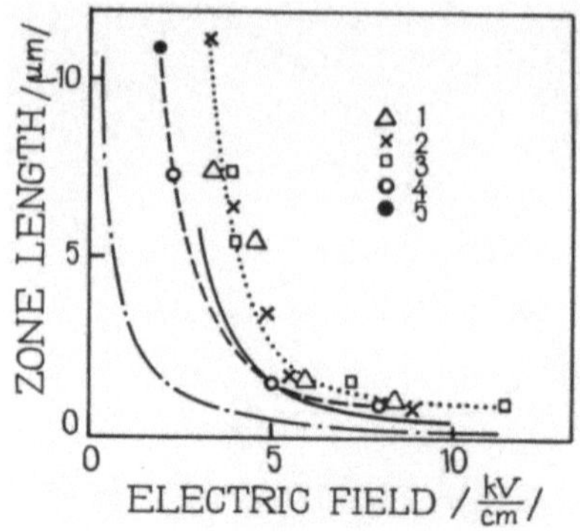

Fig.4. Field dependence of the zone length: for incubation of intervalley processes (solid line) /14/, ballistic (dashed-dotted), noiseless (dashed), speeding (dotted). Experimental points: dc data 1- /10/, 2- /13/; microwave data 3- /10/, 4- /11/, 5- /12/

The speeding zone length is determined from the maximum electron drift velocity measured for the structures of various length in the microwave experiment /10/. Essentially the same results are obtained from direct current measurements /10,13/. The noise temperature data /11,12/ are used to determine the noiseless zone length; the suggestion that the intervalley nois onsets when the noise temperature reaches 400 K /2/ is used. The results are plotted in Fig.4 together with those for the speeding zone. Monte Carlo calculations /14/ give us the length of the zone necessary for incubation of the intervalley processes. The noiseless zone is found to be shorter than the speeding zone. The results of Monte Carlo calculations are quite close to the noiseless zone length. The distance of ballistic acceleration of lucky electrons up to the intervalley energy is too short to be of importance for the discussion of the experimental data.

1. J.G.Ruch, G.S.Kino: Phys. Rev. 174, 3 (1968)
2. V.Bareikis, V.Viktoravičius, A.Galdikas, R.Miliušytė: Fiz. Techn. Poluprovodn. 14, 1427 (1980)
3. J.Ruch: IEEE Trans. ED-19, 652 (1972)
4. J.Požela, A.Reklaitis: Solid State Comm. 27, 1073 (1978)
5. A.Matulionis: Lietuvos Fizikos Rinkinys 20, 65 (1980)
6. V.Bareikis, A.Matulionis, J.Požela, S.Ašmontas, A.Reklaitis, A.Galdikas, R.Miliušytė, E.Starikovas: In Hot Electron Diffusion, ed. by J.Požela, Electrons in Semiconductors, Vol. 3 (Mokslas, Vilnius, 1981)
7. A.Matulionis, J.Požela, E.Starikov: Lietuvos Fizikos Rinkinys 21, 45 (1981)

8. Y.K.Pozhela: In Hot-Electron Transport in Semiconductors, ed. by L.Reggiani, Topics Appl. Phys., Vol. 58 (Springer, Berlin, Heidelberg 1985) p. 113
9. V.Dienys, A.Dargis: J. de Physique 42, C7-33 (1981)
10. K.Kibickas, J.Paršeliūnas, S.Bočkarev: Lietuvos Fizikos Rinkinys 26, 195 (1986)
11. V.Bareikis, K.Kibickas, J.Liberis, A.Matulionis, R.Miliušytė, P.Sakalas: In Fluctuation Phenomena in Physical Systems, Abstr. 4th Soviet Conf. Pushchino (1985) p.42
12. D.Gasquet, M. de Murcia, J.P.Nougier, C.Gontrand: Physica 134 B, 264 (1985)
13. S.Kačiulis, K.Kibickas, J.Paršeliūnas: phys. stat. solidi (a) 85, K 179 (1984)
14. A.Matulionis, R.Miliušytė: Lietuvos Fizikos Rinkinys 26, 293 (1986)

Mobility Overshoot of Hot Electrons

V.N. Freire, A.R. Vasconcellos, and R. Luzzi
Instituto de Física, DFESCM, UNICAMP, 13100-Campinas, SP, Brasil

We consider the question of how the evolution of the nonequilibrium state of a highly photoexcited plasma in polar semiconductors (HEPS) affects the mobility of the hot carriers (photoinjected electrons and holes). For that purpose we resort to the use of the nonequilibrium statistical operator method (NSOM) in Zubarev's approach [1], in the form already applied to the study of ultrafast relaxation phenomena in HEPS [2]. A nonlinear transport theory follows from the NSOM, which allows to write down the equations of evolution for a basis set of nonequilibrium thermodynamic variables appropriate for the description of the macroscopic state of physical systems arbitrarily away from equilibrium. For the case of a HEPS in the presence of an electric field $\mathcal{E}$, we choose the quasi-temperatures of carriers, $T_c^*(t)$, and of the different phonon branches, $T_{LO}^*(t)$, $T_{TO}^*(t)$, $T_A^*(t)$, the quasi-chemical potentials of electrons, $\mu_e(t)$, and of holes, $\mu_h(t)$, as well as the drift velocities $v_e(t)$ and $v_h(t)$. The concentration of photoinjected carriers, n, is taken as a constant, since recombination is negligible in the ps time scale we are considering.

In the NSOM-linear theory of relaxation [2], defining the current densities $I_\alpha(t) = \mp$ en $v_\alpha(t)$, where α = e or h, we obtain the transport equation

$$\frac{d}{dt} I_\alpha(t) = (ne^2/m_\alpha) - \varphi_\alpha(t)\, |I_\alpha(t)| \quad , \tag{1}$$

where $\varphi_\alpha(t)$ is the reciprocal of a relaxation time, resulting from the collision of carriers with phonons, and depending on time through the instantaneous values of the nonequilibrium thermodynamic variables on which the macroscopic state of the HEPS depends. Eq. (1) has the formal solution

$$I_\alpha(t) = \sigma_\alpha(t)\,\mathcal{E} = (ne^2/m_\alpha)\,\tau(t)\,\mathcal{E} \quad , \tag{2}$$

which has a Drude-type form, but with an instantaneous transport relaxation time

$$\tau_\alpha(t) = \exp\{-\psi_\alpha(t)\} \int_0^t dt' \exp\{\psi_\alpha(t')\} \quad , \tag{3}$$

with $\psi_\alpha(t) = \int_0^t dt' \varphi(t')$.

Clearly a complete solution requires to solve the set of coupled highly nonlinear integro-differential equations for the seven nonequilibrium thermodynamics variables. Using the fact that if at time t_M there occurs overshoot then the first derivative of the current is null and the second must be negative, and thus (2) together with (3) allow to establish the following criterion: *mobility overshoot occurs when there exists a cross-over of the evolution curves for transport and current relaxation times, and the latter is decreasing at the cross-over.*

Such conditions strongly depend on the time evolution of the nonequilibrium thermodynamic state of the HEPS. In the case of typical pump-probe experiments performed at room temperature using ultrafast time-resolved laser spectroscopy, there is no noticeable heating of the phonon system [2]. Thus a necessary condition for overshoot to occur at t_M is that

$$\left.\frac{d\varphi^{-1}}{dt}\right)_{t_M} \simeq \frac{\partial\varphi^{-1}}{\partial T_c^*} \cdot \left.\frac{dT_c^*}{dt}\right)_{t_M} < 0 \qquad , \qquad (4)$$

i.e. (a) if the carriers system is cooling down then the current relaxation time must decrease along with T_c^*, and (b) φ^{-1} must decrease with increasing T_c^* if the carriers system is warming up. This implies in case (a) that the electric field intensity should verify that $e^2\mathcal{E}^2[(\varphi_e^{-1}/m_e) + (\varphi_h^{-1}/m_h)]$ is smaller than the rate of energy transfer from carriers to the phonon field, and therefore overshoot cannot occur above an upper limit of $\mathcal{E}$. The inequality must be reversed in case (b), for example the normal case of overshoot starting with the HEPS in equilibrium, and there is a lower limit of $\mathcal{E}$ for the effect to follow.

Next we consider the specific case of GaAs samples with a photoinjected carriers concentration of 2×10^{18} cm^{-3}, and a kinetic excess energy of 2.4 eV per pair. At low electric field intensities ($\mathcal{E} \lesssim 3$ kV/cm) the quasi-temperature of the hot carriers rapidly decreases (ps scale) to values slightly above the reservoir temperature. The current density increases in time, passes through a maximum, showing a sharp overshoot at delay times of a fraction of ps (at a point in accordance with the criterion above), and next attains a stationary value in a ps time-scale; (ii) At intermediate electric fields ($\mathcal{E}$ roughly between 3 and 8 kV/cm) the behavior of the system is similar to case (i), except that the hot carriers quasi-temperature show a near-plateau in time with values well above the reservoir temperature, and the current overshoot is less and less sharp with increasing field intensity; (iii) For moderately high electric field intensities ($\mathcal{E} \gtrsim 8$ kV/cm to 10 kV/cm) the hot carriers system begins to cool down during a couple of ps but heating effects resulting from the transference of energy from the field are strong and try to keep or even increase the quasi-temperature of carriers, which eliminates overshoot effects once the criterion above cannot be satisfied since $\varphi^{-1}(t)$ becomes a monotonicly increasing function of time. In all three cases the phonon system does not show appreciable heating.

Comparison with experimental data is not quite feasible at present because of the lack of detailed observations and necessity to improve the theoretical model. However, there seems to exist a good qualitative agreement with existing experimental results [3].

1. D.N. Zubarev, Neravnovesnaia Termodinamika Statischekaia (Izd. Nauka, Moskwa, 1970)[English Transl.: Nonequilibrium Statistical Thermodynamics (Consultants Bureau, New York, 1974)].
2. A.C. Algarte and R. Luzzi, Phys. Rev. B27, 7563 (1983); R. Luzzi and A.R. Vasconcellos, in Semiconductors Probed by Ultrafast Laser Spectroscopy, Vol. I, ed. by R.R. Alfano (Academic, New York, 1984).
3. C.V. Shank, R.L. Fork, B.I. Greene, F.K. Reinhardt, and R.A. Logan, Appl. Phys. Lett. 38, 104 (1981).

Monte-Carlo Simulation of the Effects Induced by Real-Space Transfer in a HEMT

M. Mouis, P. Dolfus, B. Mougel, J.-F. Pône, and R. Castagné

Institut d'Electronique Fondamentale, UA 22, Bât. 220, Université Paris XI, F-91405 Orsay Cedex, France

1. Introduction

In a HEMT (High Electron Mobility Transistor), electron conduction occurs within the accumulated layer confined against the interfacial hetero-barrier. So, the device behavior is similar to MOST behavior if the large gap semiconductor upper layer is considered as an insulating layer. Nevertheless, the technology used to grow this layer allows for a greater interface quality than in the classical MOS transistor. Furthermore, non-stationary transport phenomena are favoured by the small channel length of the HEMT's. When heated by the drain electric field, the carriers can gain enough energy to step over the low (about 0.3 eV) hetero-barrier. This is called real-space transfer/1/.

Since this phenomenon is highly dependent on both electron heating and wave-vector scattering in the direction normal to the interface, a detailed description of each semiconductor band structure and of electron-phonon interactions is necessary to understand the behavior of the device in the saturation regime. In order to provide such a detailed description in a single system, we studied real-space transfer and its consequences in a GaAs/GaAlAs HEMT using a self-consistent ensemble Monte-Carlo technique. This algorithm uses a two-dimensional geometrical space (for the Poisson solver and particle motion) and a three-dimensional k-space. Phonon scattering (intervalley and intravalley) and ionised impurity scattering are included. Γ, L and X valleys are taken as spherical, non-parabolic/2/.

The simulated GaAs/GaAlAs HEMT has standard geometry (fig.1). We should point out that source and drain contacts are ohmic both on GaAs and GaAlAs. This is the situation when contacts are taken on GaAs through the GaAlAs layer. Real - space transfer and saturation mechanisms have been studied by varying some significant geometrical and structural parameters.

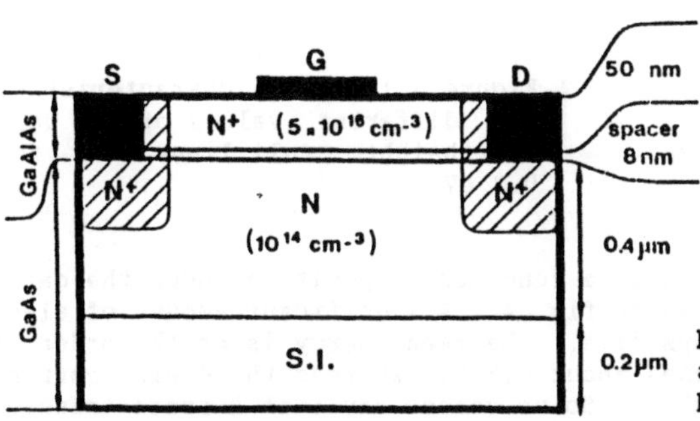

Figure 1 : Geometrical parameters and doping levels of simulated HEMT's

2. Saturation Mechanism of Short Gate HEMT

In 1μm gate length device, previous simulations/3/ have shown that channel pinch-off was the reason for the saturation of drain current. The situation in the 0.5μm device is somewhat different : both k-space tranfer and real-space transfer occur near the drain when the device current is saturated. So the problem is to determine what is the key mechanism in current saturation phenomenon. Different possible mechanisms can be responsible for current saturation of the device : (i) classical pinch-off of the channel; (ii) k-space transfer, which creates a space-charge region screening the drain potential; and (iii) real-space transfer in the upper layer, where electron conduction is less effective.

In order to understand if this real-space transfer can be taken as responsible for the saturation, we performed simulations using different heterojunction barrier heights. Carrier distributions for barrier heights of 0.25, 0.3 and 0.5 eV are shown on fig. 2. Although these distributions are very different, the electron concentration being higher in GaAlAs when the barrier is lower, the drain currents are the same in the three cases (fig.3). So, real-space transfer is not directly responsible for the current saturation.

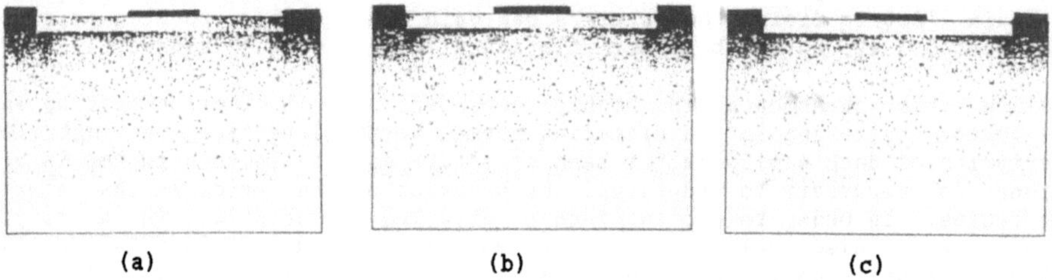

Figure 2 : Spatial distribution of electrons in a 0.5μm gate length HEMT for different values of the heterojunction barrier height ; (a) 0.25 eV, (b) 0.3 eV, (c) 0.5 eV

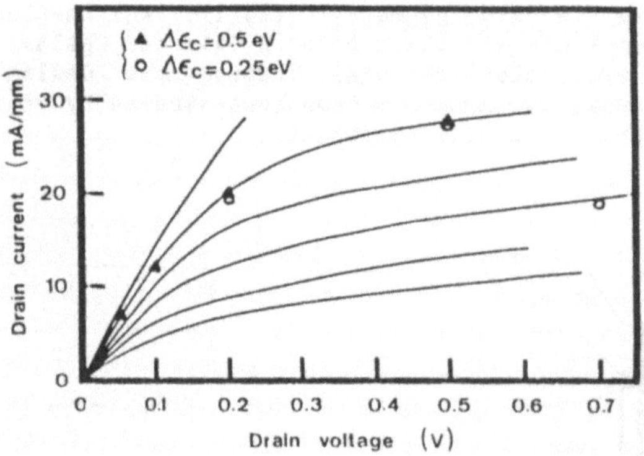

Figure 3 : Drain characteristics for different values of the barrier height : full lines are for 0.3 eV

The mean energy of the carriers as a function of position under the gate, for different drain biases, is shown on fig. 4. A significant amount of electrons is transferred in upper valleys (i.e. the mean energy is of the order of the intervalley gap) when V_{DS} reaches about 0.5 V, whereas the drain current is already saturated since V_{DS}=0.25 V. So we cannot say that k-space transfer induces the saturation. When k-space tranfer becomes effective, the isotropy of the electron-phonon interaction in L valleys can favour the orientation of the wave-vector in the direction normal to the interface, enhancing the real space - transfer probability. This can be observed on fig. 4, where the mean energy saturates at the barrier height value. The real - space transfer then prevents the formation of a domain by spreading the electrons in the GaAlAs layer.

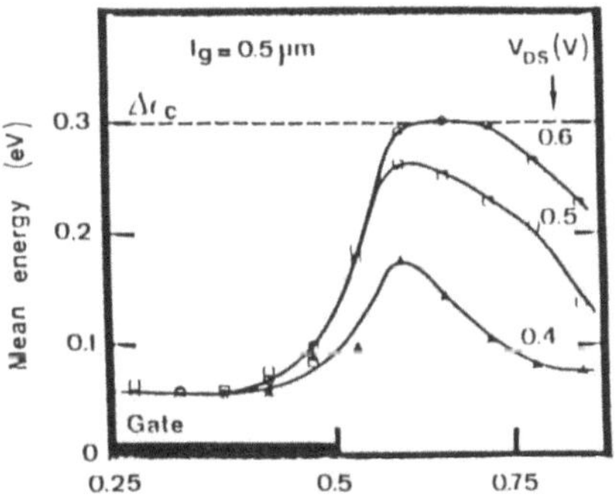

Figure 4 : Mean energy of the carriers in the channel under the gate as a function of position, for different drain voltages

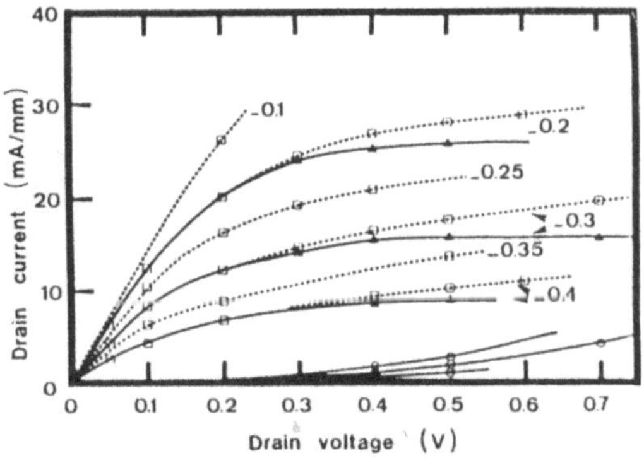

Figure 5 : Drain characteristics of the 0.5μm gate length HEMT;
triangles : total drain current
squares : current collected on GaAs
circles : current collected on GaAlAs

So, several electron-spreading mechanisms prevent current control by a dipolar domain. the drain current is then conditioned by the channel narrowing under the gate, and screening of the drain potential is low. This leads to a high value of the drain conductance g_D.

3. Influence of the Drain Contact

In the previous case, real-space transfer leads to a leakage current through the GaAlAs layer. Drain current collected in each layer is drawn on fig.5. If drain contact is taken only on the GaAs layer (using a mesa for example), the charge stored in GaAlAs is greater, but the drain current stays the same (the electrons transfer back to GaAs and contribute to the total current). This is consistent with previous results, the carrier distribution under the gate being not disturbed.

Let us note, however, that if current flowing through GaAlas could be collected by another electrode, drain current flowing through GaAs can be very well saturated, and even show a negative differential resistance.

4. Utilisation of Real-Space Transfer in Negative Differential Resistance Device

In order to evaluate the possible negative differential resistance (NDR)/4/ that can be obtained by this way, we simulated a device with two different drain contacts, one on the GaAs layer (D_1) and the other on the GaAlAs layer (D_2)(fig.6). Its drain characteristics, shown on fig.7, together with the carrier spatial distributions, show a negative differential resistance for ordi-

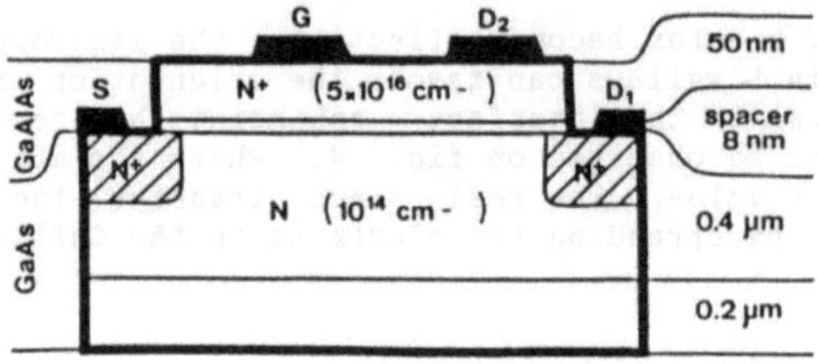

Figure 6 : Device with two drain contacts

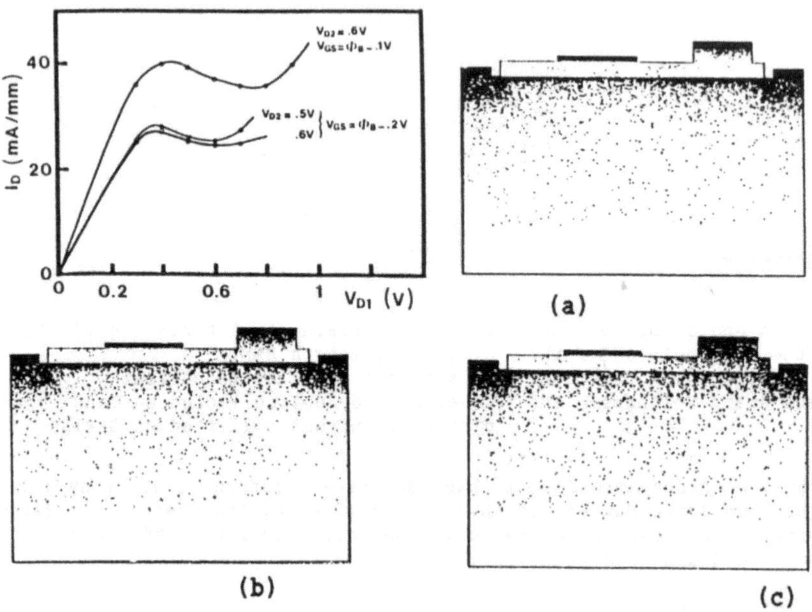

Figure 7 : Drain characteristics and particle distributions of the two-drains device for $V_{GS}=\phi_B-.1$ V, $V_{D2}=0.6$ V and (a) $V_{D1}=0.3$ V, (b) $V_{D1}=0.6$ V, (c) $V_{D1}=0.9$ V

nary drain current, when the real-space transfer **of electrons from GaAs to GaAlAs appears. However, the part of the characteristic presenting this NDR cannot be very important, the real-space** transfer **concerning only a part of the population, which can be** considered **as determined by scattering probabilities.**

5. Conclusion

The drain current saturation of the short gate device studied here is controled by the channel narrowing under the gate. The real-space transfer prevents the formation of a dipolar domain by allowing the electrons to spread in the GaAlAs layer, and is then responsible for a higher g_D value. It can also change the distribution of the drain current between the different conducting layers.

<u>References</u>

1. See for example : K. Hess, in <u>proceedings of ICPS 16</u>, Montpellier, (North-Holland, 1982) p.723
2. P. Hesto, J.-F. Pône, M. Mouis, J.-L. Pelouard, R. Castagné , in <u>proceedings of NASECODE IV Conference</u>, Dublin (Boole Press, Dublin,1985)
3. M. Mouis, J.-F. Pône, P. Hesto, R. Castagné, <u>Cornell Conference on advanced concepts in high - speed semiconductor devices and circuits</u>, July 1985
4. S. Luryi, A. Kastalsky, in <u>proceedings of 4th International Conference on Hot Electrons in Semiconductors</u>, Innsbruck, (North-Holland, 1985) p.453

Application of the Integral Boltzmann Equation to the Hot Electron Problem in an Inhomogeneous Submicron Structure

P. Kuivalainen [1,2] *and K. Lindberg* [1]

[1] Electron Physics Laboratory, Department of Electrical Engineering, Helsinki University of Technology, Otakaari 7 A, SF-02150 Espoo, Finland

[2] Semiconductor Laboratory, Technical Research Centre of Finland, Otakaari 7 B, SF-02150 Espoo, Finland

Charge transport in a GaAs n^+-n-n^+ submicron device is studied by applying the path variable method for solving the Boltzmann equation self-consistently with the Poisson equation. This allows the study of ballistic transport in the submicron structures where the charge density and thermal gradients are large.

1. Introduction

Modern microelectronics has now reached the submicrometer scale of miniaturization. In submicron devices large non-uniform electric fields and charge density inhomogeneities are typically present. In addition, a substantial number of charge carriers may cross the active region of the device without scattering. The Boltzmann transport equation (BTE) is expected to be valid in the submicron devices at least if the length of the device is larger than the mean free path of the carriers. Reliable solutions of BTE in the submicron regime have been obtained thus far by Monte-Carlo simulations /1/ and by numerical solution of BTE expressed as partial differential equation /2/.

In this paper we present the self-consistent calculation of the electrical transport properties of a GaAs n^+-n-n^+ submicron device using the path variable method and the integral formulation of BTE. This is a well-known method first introduced by BUDD /3/, but it has not been applied for inhomogeneous semiconductor structures. We found that especially in the relaxation time approximation (RTA) integral formulation of BTE provides a simple and accurate method for calculating the charge carrier distribution function in the submicron structrures. As pointed out by BARANGER /2/ RTA is expected to be adequate for the study of the transport in the submicron structures, since for ballistic electrons the details of the scattering processes are irrelevant. The main effect is determined by the magnitude of the scattering rate which can be correctly included within RTA.

2. Method

In RTA the time and position dependent velocity distribution for electrons is /3/:

$$f(x,v,t) = f_0(x,v)\ \exp\{ -\int_0^t \frac{ds}{\tau[x(t,s)v(t,s)]}\}$$
$$+ \int_0^t ds\ \frac{f_0[x(t,s),v(t,s)]}{\tau[x(t,s),v(t,s)]}\ \exp\{-\int_0^s \frac{ds'}{\tau[x(s'),v(s')]}\}\quad , \qquad (1)$$

where $f_0(x,v) = n(x)\exp(-v^2/2v_{th}^2)/\sqrt{2\pi}\, v_{th}$ is the local equilibrium distribution, $n(x)$ the electron concentration and $v_{th} = (kT/m*)^{1/2}$. $[x(s),v(s)]$ is the collision-free trajectory of a carrier in phase space and $\tau(x,v)$ is the position and velocity-dependent relaxation time. Using the Green's function $G(x;x') = (x - x')\theta(x - x') - x'(L - x)/L$ for a one-dimensional structure whose length is L, we readily obtain the solution of the Poisson's equation for the electrostatic potential

$$\phi(x) = - \int_0^L dx'\, G(x;x')\rho(x') + xV/L\ , \qquad (2)$$

where $\rho(x) = |e|\,[N_d(x) - n(x)]$ is the charge density with the donor concentration $N_d(x)$ and V is the applied voltage.

When in the numerical estimation the time integrals (1) are changed into summations, the free electron trajectory $[x(s),v(s)]$ can be calculated by assuming that within the small region where the electron moves during each small time step Δs, the electric field obtainable from (2), $\varepsilon(x) = - d\phi(x)/dx$, is constant. Then the classical equations of motion for x(s) and v(s) can be easily solved. The error due to this approximation can be made arbitrarily small by choosing Δs small enough.

3. Results and Discussion

The above method was applied to a quasi-one-dimensional GaAs n^+-n-n^+ submicron structure (Fig. 1) in a regime where the charge density inhomogeneities and ballistic effects are large. The results for $n(x)$, $\phi(x)$ and $f(x,v)$ obtained from a self-consistent solution of (1) and (2), are shown in Figs. 1, 2 and 3, respectively. Here a simple parabolic conduction band and a constant relaxation time $\tau_0 = 2.9\cdot10^{-13}$ s were assumed. The peak in the distribution function in Fig. 3 corresponds to those electrons which travel in the

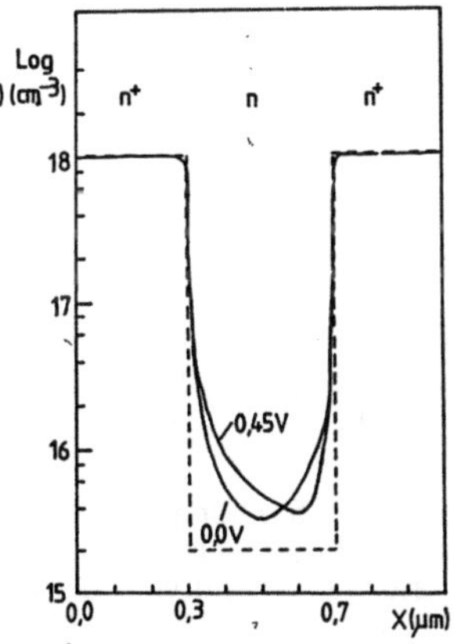

Fig. 1
Electron density vs. distance in a GaAs n^+-n-n^+ device (doping profile shown as a dashed line)

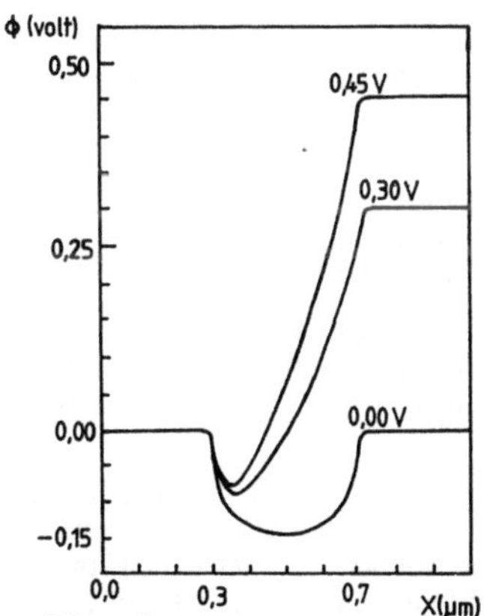

Fig. 2
Electrostatic potential vs. distance for various applied voltages

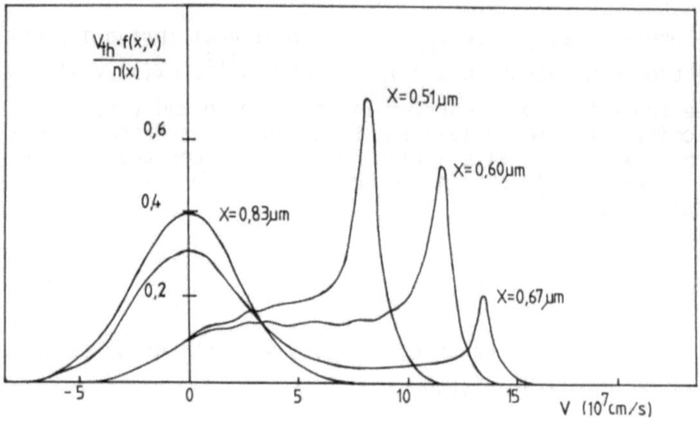

Fig. 3
Normalized velocity distribution of electrons at various spatial points in the device of Fig. 2

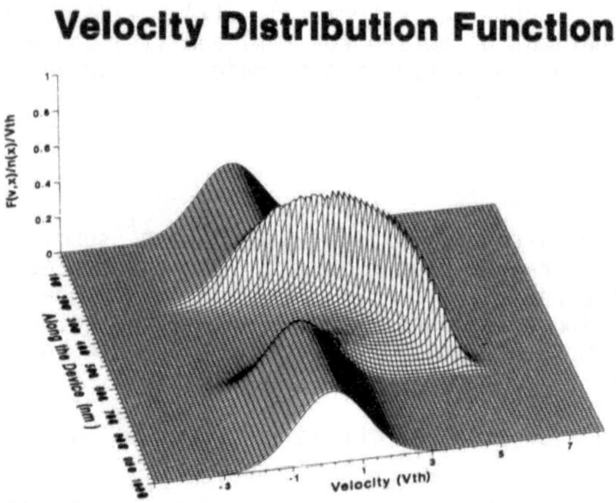

Fig. 4
Distribution function f(x,v) at various spatial points in the device of Fig. 1

n-region without scattering, i.e. the ballistic electrons. The shift of this peak to higher velocities due to the acceleration in the potential slope of Fig. 2 is shown clearly in Fig. 4.

The effect of the energy dependence of the relaxation time $\tau(E)$ to f(x,v) is shown in Fig. 5. Here the polar optical phonon scattering was assumed with the approximation $\tau(E) \sim \sqrt{E}$ and taking only the phonon absorption part into account.

The result of Figs. 1 to 4 are similar to those of Ref. 5 where a different method for solving the Boltzmann equation had been chosen. The advantages of our method are the following: the complete energy dependence of the relaxation time can be taken into account, the time evolution of the distribution function (1) can be easily studied and, finally, the coupled BTE and Poisson's equation are solved self-consistently from (1) and (2) both in the heavily doped contact region and in lightly doped active region of the device without any further approximations.

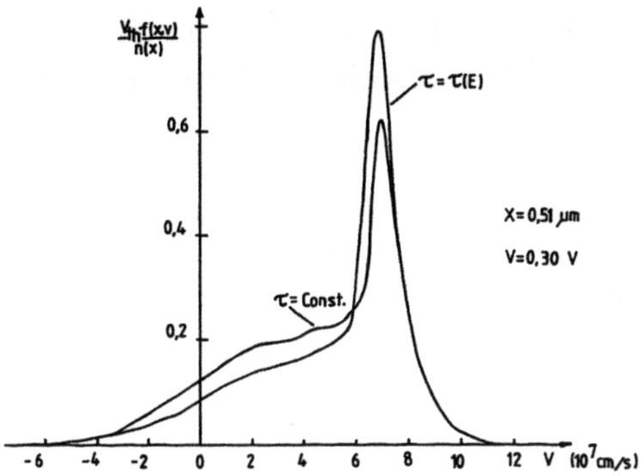

Fig. 5
f(x = 0.51 µm, v) for constant and energy-dependent relaxation times

References

1. P. Hesto, J.F. Pone, R. Castagné and J.L. Pelourd: In The Physics of Submicron Devices, ed. by H.L. Grubin, K. Hess, G.J. Iafrate and D.K. Ferry. (Plenum Press, New York 1984)
2. H.U. Baranger: Doctoral thesis, Cornell University (1986); H.U. Baranger and J.W. Wilkins: Phys. Rev. B30, 7349 (1984) and Physica 134 B, 470 (1985)
3. H. Budd: Phys. Rev. 158, 798 (1967)

Three Picosecond Oscillations in Avalanche Electron-Hole Plasma Induced by Energy Relaxation Phenomena

D. Lippens and J.L. Nieruchalski

Centre Hyperfréquences et Semiconducteurs, UA. CNRS. 287, Bât. P4, Unversité des Sciences et Techniques de Lille I, Flandres-Artois, F-59655 Villeneuve d'Ascq Cedex, France

1. Introduction

Although there are extensive theoretical and experimental investigations of transient carrier transport at moderate field /1/, relatively little attention has been paid to high-field range (typically 500 kv/cm), for which Impact Ionization is observed /2/ - /4/. We have recently found from numerical simulation of non-stationary high-field carrier transport that submicron GaAs p-i-n diodes show high-frequency coherent instabilities in the particle current under Impact Ionization conditions /5/. In this paper, this finding is further investigated by concentrating on the conditions by which the transient behaviour of the multiplication process is expected to become significant.

The numerical procedure we used to describe the carrier transport is based on a many-particle simulation program of avalanche devices /6/. The most important features of this program may be summarized as follows : Impact Ionization is described by a statistical process. Each carrier has a probability of undergoing an ionizing collision. Random trials such as those performed in a Monte Carlo procedure permit us to find out whether an ionizing collision has occurred or not. Drift in the electric field and diffusion are modeled using the diffusive model of HOCKNEY /7/. Ionizing probabilities and transport parameters are assumed to be functions of the carrier energy. With this treatment eventual delays in the drift velocity and in the multiplication of carriers by impact ionization are simulated. The self- consistent field according to Poisson's equation is also considered.

2. Results

Figure 1 shows the variations of the total multiplied particle current (electrons and holes) versus time for a p-i-n device with an intrinsic region length of 0.2µm. The doping concentration is 10^{15} cm^{-3}. The diode is reverse biased at a dc voltage of 10v. The simulation time and sampling time are 150 and 510^{-2} ps respectively. Periodic fluctuations, undamped during the interval of observation, can be identified in the time evolution of particle currents. In order to test the possibility that these self-oscillations are due to mathematical instability, the magnitude of the mesh interval in space and time was varied. No numerically induced effects were observed, confirming that these oscillations have a physical basis. Besides, it can be shown that these self-oscillations affect both types of carrier and thus are not due to the negative electron differential mobility.

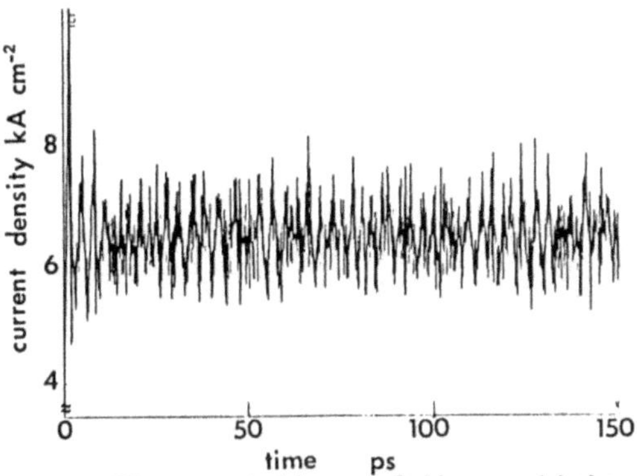

Fig.1 Time variations of the particle current

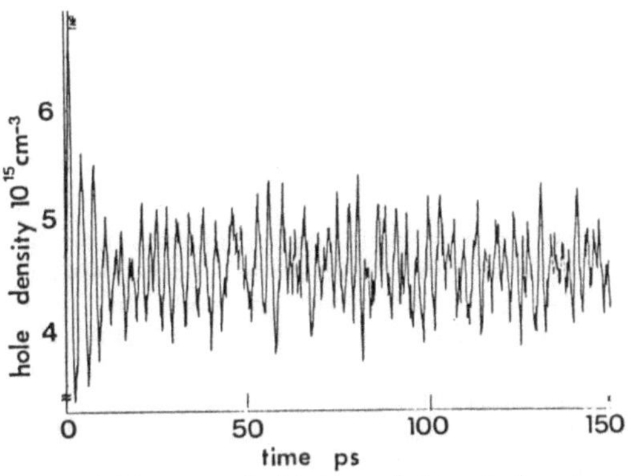

Fig.2 Time variations of the hole densities

Figure 2 shows the variations of the hole densities versus time. Similar coherent instabilities are also revealed. With regard to this result, one is led to conclude that the oscillating nature of the time variations of particle currents results from coherent fluctuations in the carrier densities.

Besides, it was shown before that no marked oscillating forms are detected on the time variations of velocities /5/. The fact that the ionization probabilities are very sensitive functions of carrier energy explains why a variation in carrier energy produces a significant variation in the impact ionized carrier densities compared to eventual fluctuations in the velocities.

Confirmation of the first interpretation may be obtained from analysis of correlation functions of fluctuating quantities. Figures 3 a and b show the auto-correlation functions of the fluctuations in hole current and in hole density. Comparable variations are obtained corroborating the dominant effect of carrier density instabilities. Moreover, we can note, as it could be expected, periodic correlations of the fluctuations in hole current and density. The time interval between two maxima is close to 3.3pS which corresponds to a 300GHz oscillation. This frequency value may be compared to the transit

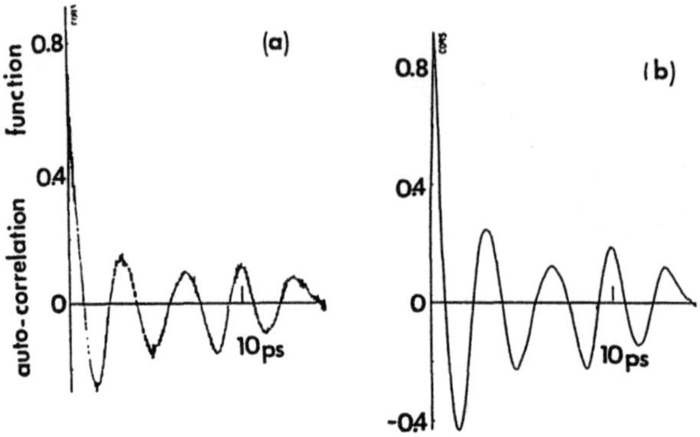

Fig.3 Auto-correlation functions of hole currents (a) and hole densities (b)

frequency. This can be explained in the sense that all the excess carriers inside the region can disappear because of the absorption by the two highly p^+ and n^+ doped layers after a period corresponding to a transit time.

Insight into the fundamental origin of the self-oscillations can be obtained by examining the cross-correlation functions. Figure 4(a) reports the cross-correlation of the terminal voltage and the hole ensemble energy. The main point to note in this figure is that the correlation between these two quantities occurs later than those of auto-correlation functions. This is particularly revealed at time t=10pS. The time delay observed in this figure is roughly 0.4pS. This time constant is close to the energy relaxation time. Therefore it is expected that this retardation effect results from the imbalance between the electric field and the energy. In order to connect this result to the generated carrier densities, the cross-correlation function of the terminal voltage and of the hole concentrations are reported in figure 4(b). Here, it is shown that the correlations between the impact-ionized carriers and the driving voltage occur with a time delay of 1.3pS.

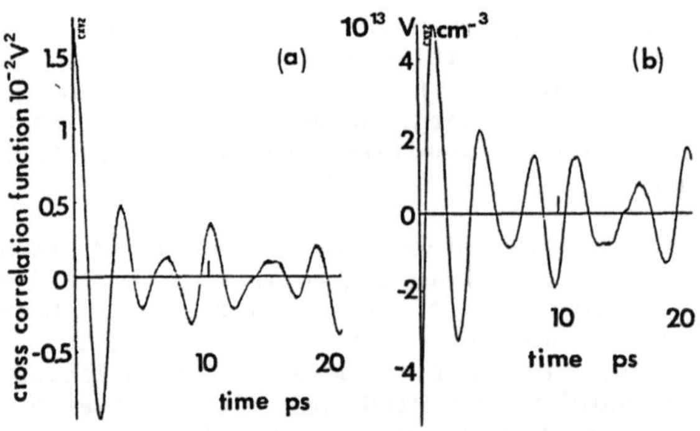

Fig.4 Cross- correlation functions of voltage and hole energy (a) of voltage and hole densities (b)

3.Discussion

Fundamentally it takes a finite time for the carriers to gain energy from the electric field. Thus energy relaxation effects occur and the carrier energy lags the terminal voltage by nearly $\pi/4$ rad at 300GHz. The energy variation produces a strong variation in the ionization probabilities.Therefore an excess charge is building up rapidly with time. The important feature to note here is that the peak of the ionized carrier density waveform is delayed by $\pi/2$ with respect to the driving force (the carrier energy for the frequencies we are interested in). This is inherent to the avalanche process. The net effect of retardation in the particle energy and in the avalanche process is that the current lags the temporal field gradient with a phase shift of nearly 1,5 $\pi/2$ rad. This corresponds to a negative component of the conductivity and therefore to the possibility of the appearance of spontaneous oscillations.

Since the importance of these relaxation effects is strongly dependent on energy relaxation times values, it is reasonable to question whether the calculation of these terms is good enough for predicting accurately the occurrence of a negative conductance. From the energy balance equation, we deduce a simple relation for energy relaxation time $\tau_\varepsilon=\varepsilon-\varepsilon_{th}/qE_v$, where ε is the mean energy, E the field amplitude, v the mean carrier velocity, ε_{th} the mean energy at thermal equilibrium. In practice, the calculation of τ_ε is complicated by the fact that little is known about high electric field transport. To illustrate this fact, we shall consider two evolutions of the mean energy versus electric field. First, let us assume that carriers ionize as soon as they reach threshold energy ε_i. In this case, it has been shown,from Monte Carlo simulations,/2/, a slow rise of electron energy with electric field. This can be understood since it is expected that ε_i constitutes an asymptotic limit for the evolution of $\varepsilon(E)$. Therefore, when the effective ionization energy lies very close to threshold, a typical energy relaxation time value is about $4 10^{-13}$ s. Let us now suppose that the ionization cross-section is slowly rising with energy. In this case, since electron energy must be widely in excess of ε_i for impact ionization, the net mean carrier energy may increase with electric field. This yields energy relaxation time values which could be higher by a factor two. From the above considerations, it is obvious that although transient mechanisms may be clearly identified, the ultimate frequency limitations depend on the amount of physics included in the simulations.

With the afore mentioned restriction in mind, it is believed that energy relaxation mode GaAs devices could be extremely useful as oscillators up to submillimeter wavelengths. Besides, it is worth noting that this possibly coherent emission being not due to the negative electron differential mobility, material such as Si could be used.

1 See for example:In Hot Electrons in Semiconductors, Physica Volume.134 B+C (1985) Nos 1-3
2 H.Shichijo and K.Hess: Physical Review B,Vol.23, 8 (1981)
3 B.K.Ridley: J.Phys.C: Solid State Phys. 16 (1983)
4 S.Mc.Kenzie and M.G.Burt: J.Phys.C: Solid State.Phys. 19 (1986)
5 D.Lippens and J.L.Nieruchalski: Appl.Phys.Lett. 48, 21 (1986)
6 D.Lippens, J.L.Nieruchalski and E.Constant: IEEE.Trans.Electron Devices. Vol.ED-32, 11 (1985)
7 R.Hockney, R.A.Warriner and M.Reiser: Electron.Lett. Vol.10, 485 (1974)

Part II

Heterostructures, Superlattices and Quantum Wells

Resonant Tunneling Transistors, Tunneling Superlattice Devices and New Quantum Well Avalanche Photodiodes

F. Capasso

AT & T Bell Laboratories, Murray Hill, NJ 07974, USA

1. Introduction

Recent advances in quantum well (QW) optoelectronic and electron devices are discussed. These include the first realization of a resonant tunneling (RT) bipolar transistor; infrared lasers and detectors based on sequential RT; avalanche photodiodes (APDs) and solid-state photomultipliers based on the recently discovered impact ionization across the band-edge discontinuity. The last section discusses the high speed operation of 1.3-1.6 μm QW APDs.

2. Resonant Tunneling Heterostructures.

2.1 Resonant Tunneling through Double Barrier Structures.

Recently, SOLLNER et al. [1],[2] have demonstrated negative differential resistance (NDR), microwave and millimeter wave operation in double-barrier (DB) QW structures. Since the pioneering work of TSU, ESAKI, and CHANG [3] on RT in DBQW structures, the material quality has improved to the point that negative differential resistance (NDR) can be observed at room temperature [4],[5]. Recently peak-to-valley ratios in current of nearly 3:1 were obtained at room temperature [4],[5] and 9:1 at 77 K [6]. As shown by LURYI [7], the NDR in DBQW diodes is a consequence of the dimensional confinement of states in a QW, and the conservation of energy and lateral momentum in tunneling. In addition to that, the operation of these structures has often been discussed in connection with a RT effect analogous to that in a Fabry-Perot resonator. This *coherent* effect is presumed to occur when the energies of incident electrons in the emitter match those of unoccupied states in the QW. Under such conditions, the amplitude of the resonant modes builds up in the QW to the extent that the electron waves leaking out in both directions cancel the reflected waves and enhance the transmitted ones. This can lead to much higher peak currents than in the case when phase coherence of the electrons waves is distroyed by scattering. In the latter case collisions in the double barrier region randomize the phase of the electron waves and prevent the build-up of the probability density in the well. No resonant enhancement of the transmission is then possible and the electrons must be viewed as tunneling into and out of the well *sequentially* without preserving the phase coherence of the incident wave. One should distinguish therefore coherent (Fabry-Perot like) RT from incoherent (sequential) RT. It is the ratio of the intrinsic resonance width to the total scattering width (collision broadening) that determines which of the two mechanisms controls RT, as recently shown by CAPASSO et al. [8]. If the latter ratio is $>>$ 1 the probability density can build-up in the well to the resonant value in a time much smaller than the scattering time. In the opposite case instead the build-up time of the wave in the Fabry-Perot ($\approx \hbar$/intrinsic resonance width) is long compared to the scattering time, so that collisions tend to randomize the phase of the wave function and prevent the resonant enhancement of the transmission.

2.2 Resonant Tunneling Bipolar Transistors

The physical picture of coherent RT has led to a design strategy intended to optimize the Fabry-Perot resonator conditions. In particular, DAVIS and HOSACK [9] and RICCO and AZBEL [10] pointed out that achievement of a near-unity resonant transmission requires equal transmission coefficients for both barriers at the operating point - a condition not fulfilled for barriers designed to be symmetric in the absence of an applied field. To counter that, a RT structure was proposed [11] in which a symmetric DBQW was built in the base of a bipolar transistor, and the Fabry-Perot conditions were maintained through the use of minority-carrier injection. (Figs. 1 and 2.) Thus, this novel geometry maintains the crucial, structural symmetry of the double barrier, allowing near unity transmission at all resonance peaks and higher peak-to-valley ratios and currents compared to conventional RT structures. Both tunneling and ballistic injection in the base are considered (Figs. 1 and 2).

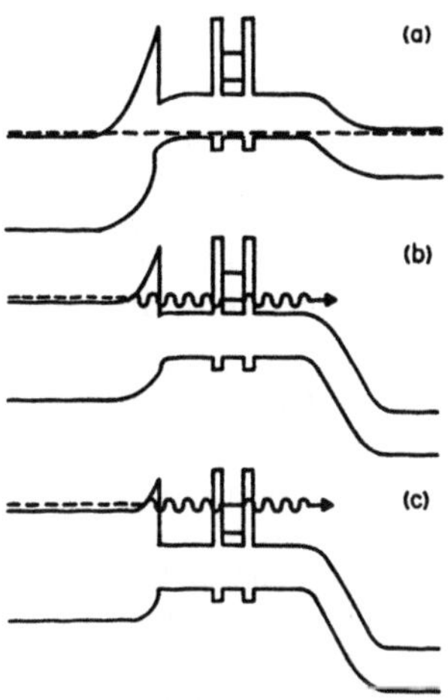

Fig. 1 Band-diagram of RBT with tunneling emitter and quantum well base.

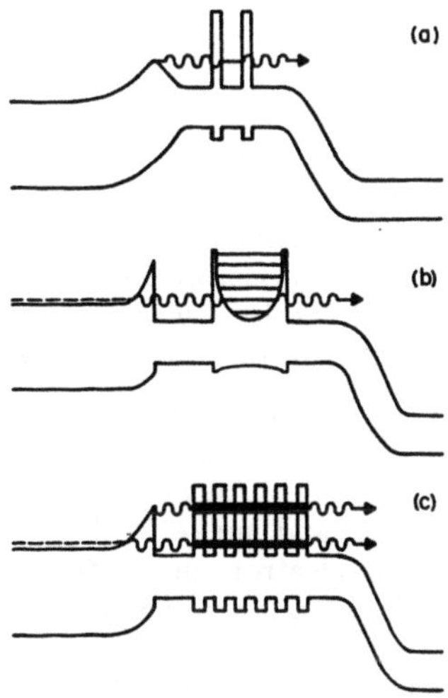

Fig. 2 Band diagram of RBTs with (a) ballistic injection, (b) parabolic QW, (c) superlattice base.

Shortly after this initial proposal, YOKAYAMA et al reported the low temperature operation (70 K) of a *unipolar* RT hot electron transistor (RHET) [12]. This structure contains a double barrier in the emitter and is discussed in detail by MUTO et al in this volume.

Recently, we have demonstrated the room temperature operation of the first RT bipolar transistor (RBT) (CAPASSO et al [13]). The band diagram of the transistor under operating conditions is sketched in Fig. 3, along with a schematic of the composition and doping profile of the structure (bottom). Details of the structure

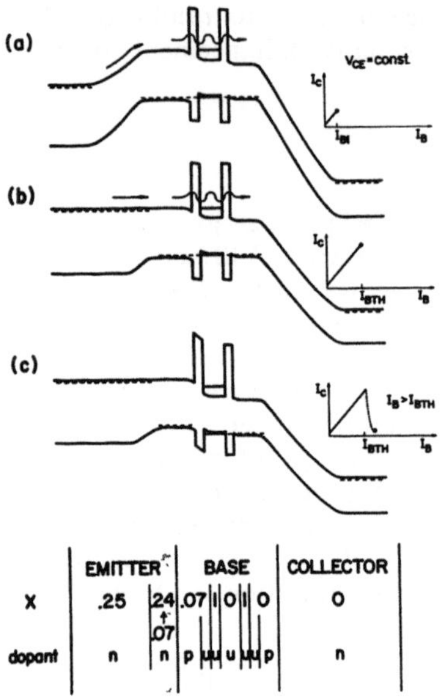

Fig. 3 Energy band diagrams of RBT with thermal injection and corresponding schematics of collector current I_C for different base currents at a fixed collector-emitter voltage V_{CE}. Shown is also the composition and doping profile of the structure.

are given elsewhere [13]; here it suffices to say that the DB consists of a 74 Å undoped GaAs QW sandwiched between the two undoped 21.5 Å AlAs barrier and that the AlGaAs graded emitter is doped to $\approx 3\text{x}10^{17}\text{cm}^{-3}$. The portion of the base ($Al_{0.07}Ga_{0.93}As$) *adjacent to the emitter* was anodically etched off, while the rest of the base was contacted using AuBe. These base processing steps, discussed in detail in [13], are essential for the operation of the device.

There is an essential difference with respect to the previous RBT [11] and the unipolar RT transistors [12]. These structures rely on *quasi-ballistic or hot-electron* transport through the base. These schemes place stringent constraints on the design and make it difficult to achieve room temperature operation due to the small electron mean free path ($\leq$500 Å at 300 K), since electrons that have suffered a few phonon collisions cannot reach the collector. The key to the present structure is that electrons are *thermally* injected into and transported through the base, thus making the device operation much less critical. This new approach has allowed us to achieve for the first time RT transistor action at room temperature. Thermal injection is achieved by adjusting the alloy composition of the portion of the base adjacent to the emitter in such a way that the conduction band in this region lines up with the bottom of the ground-state subband of the QW (Fig. 3a). For a 74 Å well and 21.5 Å AlAs barriers the first quantized energy level is $E_1 =$ 65 me V. Thus the Al mole fraction was chosen to be x = 0.07 (corresponding to $E_g = 1.521$ eV) so that $\Delta E_c \cong E_1$. This equality need not be rigorously satisfied for the device to operate in the desired mode, as long as E_1 does not exceed ΔE_c by more than a few kT. The QW is undoped; never-the-less it is easy to show that there is a high concentration ($\cong 7\text{x}10^{11}\text{cm}^{-2}$) two-dimensional hole gas in the well. These holes have transferred from the nearby $Al_{0.07}Ga_{0.93}As$ region, by tunneling through the AlAs barrier, in order to achieve Fermi level line-up in the base. Consider a common emitter bias configuration. Initially the collector-emitter voltage V_{CE} and the base current I_B, are chosen in such a way that the base-emitter and the base-collector junctions are respectively forward and reversed-biased. If V_{CE} is kept constant and the base current I_B is increased, the base-emitter potential also increases until flat-band condition in the emitter region is reached (Fig. 3b left). In going from the band configuration of Fig. 3a to that of Fig. 3b the device behaves like a conventional transistor with the collector current increasing with the base current (Figs. 3a-3b right). The slope of this curve is, of course, the current gain β of the device. In this region of operation electrons in the emitter overcome, by thermionic injection, the barrier of the base-emitter junction and undergo RT through the double barrier. If now the base current is further increased above the value I_{BTH} corresponding to the flat band condition, the additional potential difference drops primarily across the first semi-insulating AlAs barrier (Fig. 3c), between the contacted and uncontacted portions of the base, since the highly doped emitter is now fully conducting. This pushes the conduction band edge in the $Al_{0.07}Ga_{0.93}As$ above the first energy level of the well, thus quenching the RT. The net effect is that the base transport factor and the current gain are greatly reduced. This causes an abrupt drop of the collector current as the base current exceeds a certain threshold value I_{BTH} (Fig. 3c, right). The devices were biased in a common emitter configuration at 300 K and the I-V characteristics were displayed on a curve tracer. For base currents $\leq$ 2.5 mA the transistor exhibits normal characteristics, while for $I_B \geq$ 2.5 mA the behavior previously discussed was observed. Fig. 4 shows the collector current versus base current at V_{CE} = 12 V, as obtained from the common emitter characteristics. The collector current increases with the base current and there is clear evidence of current gain ($\beta = 7$ for $I_C > 4$ mA). As the base current exceeds 2.5 mA, there is a drop in I_C because the current gain is quenched by the suppression of RT. Figure 5 illustrates the common emitter characteristics at 100 K. It is apparent that in addition to the previously discussed dependence of I_C on I_B (see Fig. 4), there is also a large negative conductance in the I_C vs V_{CE} curve for base currents in excess of the threshold value (= 4 mA at 100 K). This is easy to understand by noting that in order to reach the band configuration of Fig. 3c and quench RT (at a fixed $I_B > I_{BTH}$) the collector-emitter voltage V_{CE} must be large enough for the collector-base

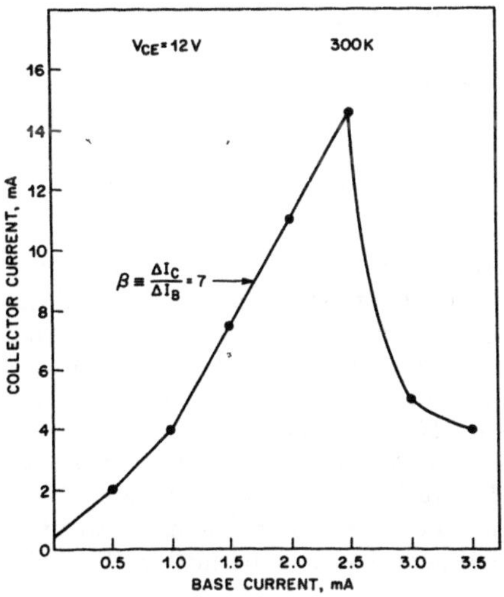

Fig. 4

I_C vs I_B in the common emitter configuration, at room temperature, with the collector-emitter voltage held constant. The line connecting the data points is drawn only to guide the eye.

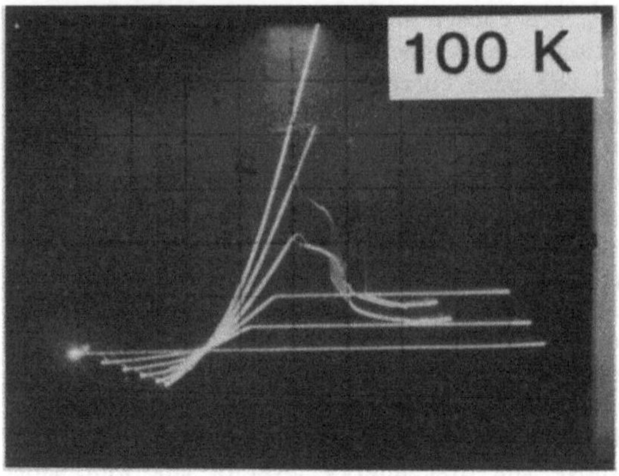

Fig. 5
Common emitter characteristics of the RBT bipolar at 100 K. The vertical scale is 10 mA/div. while the horizontal scale is 2 V/div. Each step in the base current is 2 mA.

junction to be reverse biased and draw a significant collector current. Single frequency oscillations (at 25 MHz, limited by the probe stage) have been observed in these devices when biased in the negative conductance region of the characteristics [13]. One important advantage of this device [13] compared to the RHET [12] is that it has much higher current gain in its "on" state. Other RT transistors have also been proposed ranging from Stark effect [14] to quantum wire unipolar RTs [15] and new RT triodes [16].

RT transistors are potentially very valuable for many logic and signal processing applications. These include new ultrahigh-speed analog-to-digital converters, parity generator circuits, high density memories using multiple valued logic and have been discussed by the author in a recent volume [17].

2.3 Sequential Resonant Tunneling in Superlattices and its Device Applications.

In a strong electric field in a superlattice the miniband picture breaksdown when the potential drop across the superlattice period exceeds the miniband width. When this condition is satisfied the quantum states become localized in the individual wells. In this limit an enhanced electron current will flow at sharply defined values of the external field, when the ground state in the n-th well is degenerate with the first or second excited state in the $(n+1)$-th well, as illustrated in Fig. 6a. Under such conditions, the current is due to electron tunneling between the adjacent wells with a subsequent de-excitation in the $(n+1)$-th well, by emission of phonons. In other words, electron propagation through the entire superlattice involves sequential RT.

Experimental difficulties in studying this phenomenon are usually associated with the non-uniformity of the electric field across the superlattice and the instabilities generated by negative differential conductivity. To ensure a strictly controlled and spatially uniform electric field, CAPASSO et al. [18] placed the superlattice in the i region of a reverse-biased p^+-i-n^+ junction. This structure allowed for the first time to observe the sequential RT predicted in [19]. Two NDR peaks observed in the photocurrent characteristics, Fig. 7, correspond to the resonances shown schematically in Fig. 6. For the sequential RT regime, KAZARINOV and SURIS [19] had predicted the possibility of a laser action at the inter-subband transition frequency - an effect not yet observed experimentally in superlattice (Fig. 6b).

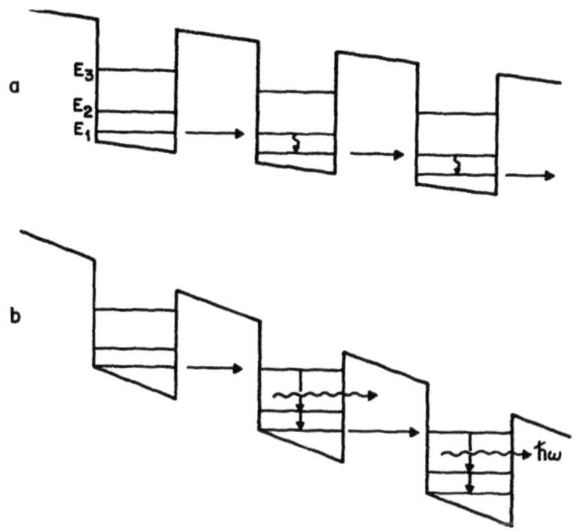

Fig. 6

a. Band diagram of sequential RT of electrons through a superlattice (SL).

b. Band-diagram of far ir four level laser based on sequential RT.

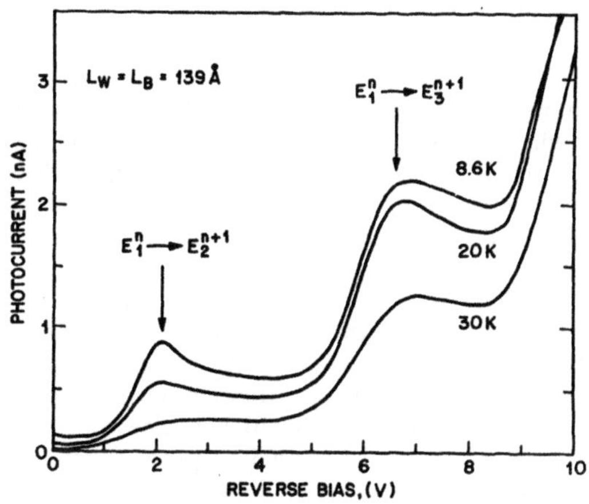

Fig. 7

Photo I-V for a SL with 139 Å thick wells and barriers and 35 periods. The two peaks correspond to RT between the ground and first two excited states of adjacent wells. (See Fig. 6a and 6b).

Another interesting possible application of sequential RT is a narrow band (a few meV) ir detector which uses absorption between the first two subbands of the wells [8]. The applied voltage is such the first and second excited states of adjacent wells are resonant. Thus the photoexcited electrons tunnel to the nearby well and give rise to a photocurrent.

3. Superlattice Photodetectors

3.1 Effective Mass Filtering: New High Gain, Low Voltage Photoconductors

Capasso et al. [20], recently reported the observation of a new extremely large photocurrent amplification phenomenon at very low voltages in a superlattice of $Al_{.48}In_{0.52}As/Ga_{0.47}In_{0.53}As$ in the quantum coupling regime (35 Å wells, 35 Å barriers). Room temperature responsivities at $\lambda = 1.3$ μm are typically $2x10^3$ and 300 Amps/Watt, at 0.3 Volt and 0.08 Volt bias respectively, while the highest measured value is $4x10^3$ A/W, corresponding to a current gain of $2x10^4$. This

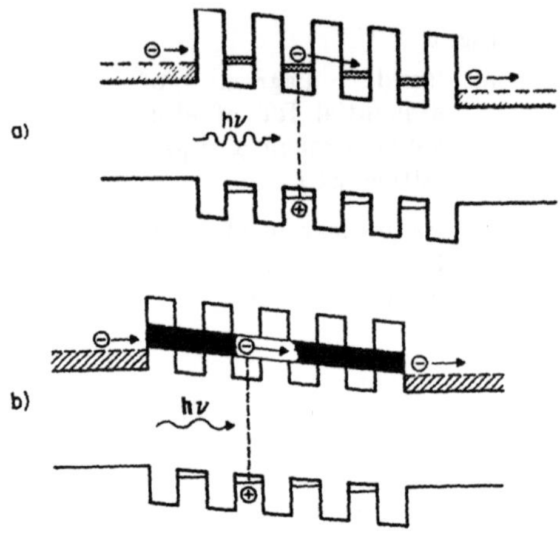

Fig. 8 Band diagram of effective mass filtering in the case of

a. phonon assisted tunneling and

b. miniband conduction

effect, which represents a new quantum type photoconductivity, is caused by the extremely large difference in the tunneling rates of electrons and heavy holes through the superlattice layers (*effective mass filtering;* Fig. 8) [20],[21]. When thickness and compositional fluctuations cause fluctuations in the subband energies of the order or greater than the miniband width ΔE, miniband conduction cannot be sustained and hence conduction will proceed by phonon-assisted tunneling between adjacent wells, (hopping conduction). Since electrons have a much smaller mass than holes, their tunneling rate between adjacent wells is much larger (*effective mass filtering*). Photogenerated holes therefore remain relatively localized in the wells (their hopping probability is negligible) while electrons propagate through the superlattice (Fig. 8a). This effective mass filtering effect produces a photocurrent gain, given by the ratio of the lifetime to the electron transit time. The gain strongly decreases with increasing $Al_{0.48}In_{0.52}As$ barrier layer thickness and becomes unity when this exceeds 100 Å . This confirms effective mass filtering as the origin of the large gain, since, as the barriers are made thicker, electrons also eventually tend to become localized, thus decreasing the tunneling probability and increasing the recombination rate. The temperature dependence of the responsivity conclusively confirmed hopping conduction.

For superlattices made of the same two materials with wider electron minibands (achieved by using thinner barriers) the electron transport is by miniband conduction, while holes are still localized (Fig. 8b). Such superlattice effective mass filters will have a much greater gain-bandwidth product than the other kind (Fig. 8a) due to the much shorter electron transit time.

3.2 Impact Ionization Across the Band-discontinuity: A New Solid-state Photomultiplier.

The difference between the conduction and valence band discontinuities is known to enhance the band-to-band impact ionization rates ratio α/β in a superlattice [22,23]. This effect led to the concept of the staircase solid-state photomultiplier (PMT) [22]. Recently we have observed a new avalanche phenomenon in superlattices, namely the impact ionization across band discontinuities of carriers confined in the wells, [24]. This phenomenon, independently predicted by CHWANG and HESS [25], could lead to a new type of solid-state PMT. This effect

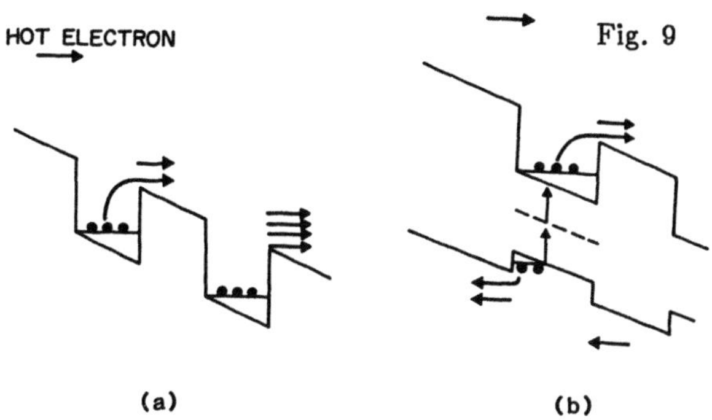

Fig. 9 Impact ionization across the band-edge discontinuity.

a. Quantum wells are doped *n* type.

b. Wells are undoped.

Shown is the ionization across band discontinuities of carriers dynamically stored in the wells. These carriers originate from thermal generation processes via midgap centers.

is illustrated in Fig. 9a. Consider a multiple QW structure with *n*-type doped wells and undoped barriers. Electrons from the parent donors can remain confined in the wells even in the presence of a relatively strong electric field, provided the barriers are thick enough to minimize tunneling. Consider now a hot electron in a barrier layer. When it enters the well with sufficient energy, it can impact ionize one of the bound carriers out of the well. In this ionization effect only one type of carrier is created so that the positive feedback of impact ionizing holes is eliminated, leading to the possibility of a quiet avalanche with small excess noise. Of course in this case, one must constantly supply the electrons in the wells by applying suitable selective contacts to the well regions. From a conceptual point of view this effect has some similarities with the impact ionization of deep levels in the sense that the QW may be treated as an artificial trap. It is important to point out that doped wells are not required for the observation of the effect. Due to the thermal generation of electrons and holes in the well layers (which gives rise to bulk dark current), relatively large electron and hole densities can be *dynamically* stored in the wells if the band discontinuities are appreciably larger than the average energies of the carriers in the wells and the dark current is relatively large. This situation may occur in the high field region of certain QW *pin* photodiodes such as the ones investigated by us [24] and is illustrated in Fig. 9b. These structures contain an $Al_{0.48}In_{0.52}As/Ga_{0.47}In_{0.53}As$ superlattice in the i region, with barriers and well thicknesses in the 100-500 Å range. A large ratio of the multiplications for holes and electrons was observed (M_h/M_e), implying that holes ionize at a significantly higher rate than the electrons in these structures. A similar effect has been found in pin diodes containing AlSb/GaSb [24] and $InP/Ga_{0.47}In_{0.53}As$ superlattices [26]. A systematic study of the temperature and chopping frequency dependence of the multiplication showed conclusively that the observed effect is not a band-to-band process. Also deep level ionization could be ruled out [24], since it would require unrealistically large densities of such centers ($\geq 10^{17} cm^{-3}$). Such densities are also in contrast with DLTS data, which in the case of $InP/Ga_{0.47}In_{0.53}As$ superlattice diodes, give an upper limit of $\approx 10^{14} cm^{-3}$.

By appropriately grading the interface of the wells, the storage of electrons can be eliminated, while holes are still confined (Fig. 10). This should maximize the ionization rates ratio, by minimizing electron initiated multiplication. This structure was grown by MBE in the AlInAs/GaInAs system [27]. The structures

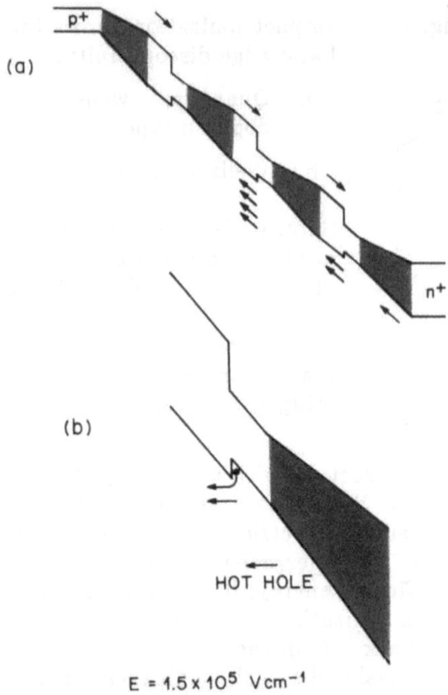

Fig. 10

a. Band structure of a multiple graded well photomultiplier (the graded regions are shaded) showing multiplication of holes. The small multiplication of electrons is not shown.

b. Mechanism of hole multiplication by impact ionization across the band-edge discontinuity.

consisted of three period superlattice, placed in the i region of a pin photodiode, with 501 Å $Al_{0.48}In_{0.52}As$ barriers, 292 Å $Ga_{0.47}In_{0.53}As$ wells and 1022 Å AlInGaAs graded regions. These regions were grown linearly graded and lattice matched to InP by computer-controlled MBE. For hole-initiated multiplication, avalanche gain occurs at a reverse bias of 7 V and reaches $\approx$ 20 at -12 V, at a temperature $\approx$ 100 K. For electrons, the multiplication is dominated by ionization across the band-gaps and is less than 1.4, resulting in a ionization rate ratio β/α ($\approx M_h - 1/M_e - 1$) in excess of 50 (Fig. 11). This is the highest value measured in a III-V material. Thus we have observed near single carrier-type multiplication of holes by ionization over the discontinuity, with feedback provided by band-to-band ionization of electrons.

3.3 Ultrahigh-Speed Quantum Well Avalanche Photodiodes

The speed of response of superlattice avalanche photodiodes can be affected by the pile-up of photocarriers in the wells (we are here considering APDs based on ionization across the band-gap). In InP / $Ga_{0.47}In_{0.53}As$ avalanche photodiodes with separated multiplications and absorption regions (SAM APDs) the pile-up of photogenerated holes is particularly severe, due to the large valence band discontinuity. This limits the response time to > 10 ns causing long tails in the pulse response. Such tails have been reduced or eliminated by the insertion of grading layers or chirped-superlattices between the multiplication and the absorption layer [28]. In $Al_{0.48}In_{0.52}As$ / $Ga_{0.47}In_{0.53}As$ SAM APDs the valence band offsets is instead significantly smaller (0.2 eV), so that the hole-pile effects, at the high fields typical of APDs ($\gtrsim 3 \times 10^5$ V/cm), is small and high speed of response

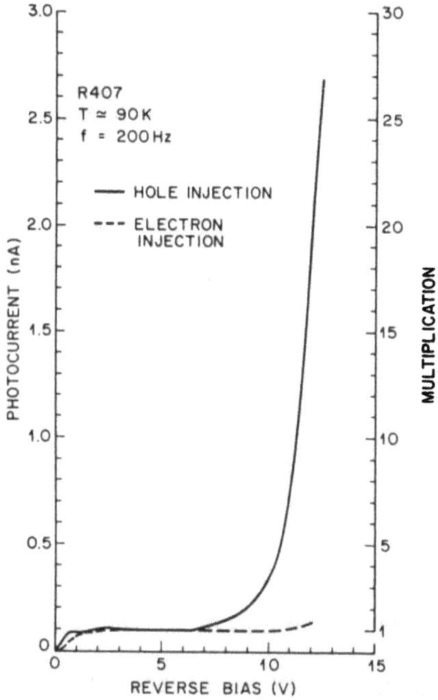

Fig. 11 Reverse bias photocurrent for the sample of Figure 10, under conditions of pure hole and pure electron injection.

can be achieved (FWHM ≈ 200 ps), [29]. In multiquantum well APDs of the latter material system, nevertheless, electron pile up in the wells can be a potential problem since $\Delta E_c = 0.5$ eV. These structure were recently investigated [30].

The p^+in^+ structure, grown by molecular beam epitaxy, consists of a 35 period $Al_{0.48}In_{0.52}As$ (139 Å)/$Ga_{0.47}In_{0.53}As$ (139 Å) QW i region sandwiched between p^+ and n^+ $Al_{0.48}In_{0.52}As$ transparent layers. DC and high-frequency multiplications of 32 and 12, respectively, have been measured; the dark current at unity gain is 70 nA. High speed of response (Fig. 12) with full width at half-maximum of 220 ps at

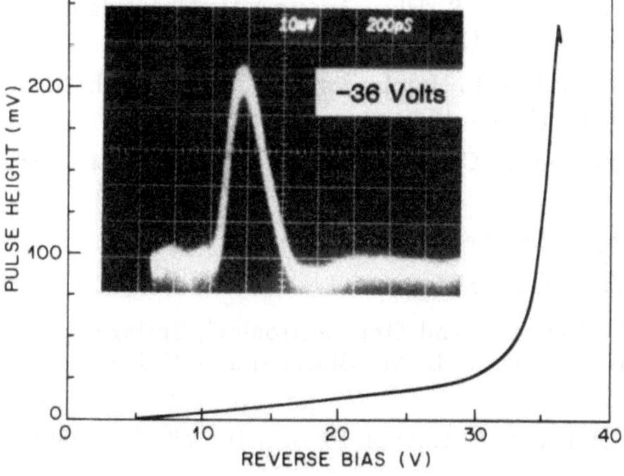

Fig. 12

Pulse height vs reverse bias of 1.3 μm QW APD showing microwave gain. The inset shows the pulse response at an ac gain of 10 and $\lambda = 1.3$ μm.

a microwave gain of 12 and the absence of tails are demonstrated, indicating that carrier pile-up in the wells is negligible. This result is important, considering that the conduction band discontinuity is large ($\Delta E_c = 0.5$ eV). Thus grading of the interface at the exit of the wells (for electrons) is not always necessary to achieve high speed. It should be noted that in these structures, the carrier densities dynamically stored in the wells are negligible, due to the small dark currents. Thus only ionization across the gap and not across the discontinuity could be observed, unlike the AlInAs/GaInAs devices considered in section 3.2. The results of Fig. 12 imply that electron can get out of the deep wells, possibly due to fast thermionic or tunneling emission or combinations thereof.

REFERENCES

1. T. C. L. G. Sollner, P. E. Tannenwald, D. D. Peck, and W. D. Goodhue, Appl. Phys. Lett. *45,* 1319, (1984).
2. E. R. Brown, T. C. L. G. Sollner, W. D. Goodhue, B. J. Clifton, and P. E. Tannenwald, Technical Digest of the 1986 Device Research Conference. Paper IV-B1.
3. L. L. Chang, L. Esaki, and R. Tsu, Appl. Phys. Lett. *24,* 593 (1974).
4. T. T. Shewchuk, P. C. Chapin, P. D. Coleman, W. Kopp, R. Fischer and H. Morkoc, Appl. Phys. Lett. *46,* 508 (9185).
5. M. Tsuchiya, H. Sakaki and J. Yoshino, Jap. J. Appl. Phys. *24,* L466 (1985).
6. H. Morkoc, J. Chen, U. K. Reddy, T. Henderson, P. D. Coleman, and S. Luryi, Appl. Phys. Lett. *42,* 70 (1986).
7. S. Luryi, Appl. Phys. Lett. *47,* 490 (1985).
8. F. Capasso, K. Mohammed and A. Y. Cho, IEEE J. Quantum Electronics, *QE-22,* 1853 (1986).
9. R. H. Davis and H. H. Hosack, J. Appl. Phys. *34,* 864 (1963).
10. B. Ricco and M. Ya. Azbel, Phys. Rev. *B 29,* 1970 (1984).
11. F. Capasso and R. A. Kiehl, J. Appl. Phys. *58,* 1366 (1985).
12. N. Yokoyama, K. Imamura, S. Muto, S. Hiyamizu and H. Nishi, Japan J. Appl. Phys. part 2, *24,* pp. L583-L584, (1985).
13. F. Capasso, S. Sen, A. C. Gossard, A. L. Hutchinson and J. E. English, IEEE Electron Device Lett., October 1 (1986). In press.
14. A. R. Bonnefoi, D. H. Chow and T. C. McGill, Appl. Phys. Lett. *47,* 888, (1985).
15. S. Luryi and F. Capasso, Appl. Phys. Lett. *47,* 1347, (1985).
16. Y. Nakata, M. Asada and Y. Suematsu, Electron. Lett. *22,* 58, (1986).
17. F. Capasso in "Picosecond Electronics and Optoelectronics", Springer Series in Electrophysics Vol. 21. G. A. Mourou, D. M. Bloom and C. H. Lee, Eds. p. 112, (1985).
18. F. Capasso, K. Mohammed and A. Y. Cho, Appl. Phys. Lett. *48,* 474 (1986).

19. R. F. Kazarinov and R. A. Suris, Sov. Phys. Semicond. *6,* 120 (1972).

20. F. Capasso, K. Mohammed, A. Y. Cho, R. Hull and A. L. Hutchinson, Appl. Phys. Lett. *47,* 420 (1985).

21. F. Capasso, K. Mohammed, A. Y. Cho, R. Hull and A. L. Hutchinson, Phys. Rev. Lett. *55,* 1152 (1985).

22. F. Capasso in Semiconductors and Semimetals, R. K. Willardson and A. C. Beer eds. (Academic, New York, 1986) Vol. 22, part D, p. 2

23. F. Y. Juang, U. Das, Y. Nashimoto and P. K. Bhattacharya, Appl. Phys. Lett. *47,* 972 (1985).

24. F. Capasso, J. Allam, A. Y. Cho, K. Mohammed, R. J. Malik, A. L. Hutchinson, and D. Sivco, Appl. Phys. Lett. *48,* 1294 (1986).

25. J. L. Chwang and K. Hess, J.. Appl. Phys. *59,* 2885 (1986).

26. J. Allam, F. Capasso, M. B. Panish and A. L. Hutchinson, (unpublished).

27. J. Allam, F. Capasso, K. Alavi and A. Y. Cho. To be published in Technical Digest of the 1986 International Electron Devices Meeting, Los Angeles, December 7-10, 1986.

28. F. Capasso, H. M. Cox, A. L. Hutchinson, N. A. Olsson and S. G. Hummel, Appl. Phys. Lett. *45,* 1193 (1984).

29. F. Capasso, B. Kasper, K. Alavi, A. Y. Cho and J. M. Parsey, Appl. Phys. Lett. *44,* 1027 (1984).

30. K. Mohammed, F. Capasso, J. Allam, A. Y. Cho and A. L. Hutchinson, Appl. Phys. Lett. *47,* 597 (1985).

Novel Real-Space Transfer Devices

A. Kastalsky

Bell Communications Research, Murray Hill, NJ 07974, USA

Recent results of studying novel, three-terminal, high-speed device based on the real-space hot electron transfer are presented. Two modes of operation of this device: the NERFET and the CHINT — for which the most impressive new data have been obtained, are discussed emphasizing the physics of their operation.

1. Introduction

We review here our results of studying novel, three-terminal, high-speed devices employing a real-space hot electron transfer between two conducting layers separated by a potential barrier. Actually, we always deal with only one device, which for different circuit arrangements reveals a big variety of possible high-speed applications [1-5]. Below we shall discuss two modes of this device: the NERFET and the CHINT — for which the most impressive results have been recently obtained [6,7], emphasizing the physics of operation in each mode.

The effect of real-space hot-electron transfer in multilayer semiconductor structure was first suggested by Hess et al. [8]. They proposed that in layered heterostructures electrons heated by a lateral electric field can move between adjacent layers. When the layers have different mobilities, the real-space transfer was predicted to result on the NDR in the two-terminal circuit. This effect was used for microwave generation [9].

In spite of the fact that the main physical process governing the operations of our device is also the real-space transfer, the mechanism of NDR in our case is completely different. As we will see later, the NERFET action results from a depletion of the one conducting layer due to electron divertion to the second layer, rather than from the different mobilities in adjacent layers. Furthermore, existing in our device separate electrical access to each conducting layer vastly widens the area of important applications of the hot electron transfer effect. We begin with the discussion of basic physical processes underlying the operation of our device and important requirements to be met for its successful experimental realization.

2. Physical Model

Figure 1 illustrates, very schematically, the device structure and contact layout. A high-mobility channel with source and drain contacts to it represents the first conducting layer — the source of hot electrons. The second conducting layer-collector of hot electrons — is separated from the channel by the insulator, which provides a potential barrier for hot electrons. At low positive drain V_D and collector V_C voltages (relative to the source) the main current flows to the drain, except for exponentially small amount of electrons capable to overcome the barrier and reach the collector. At

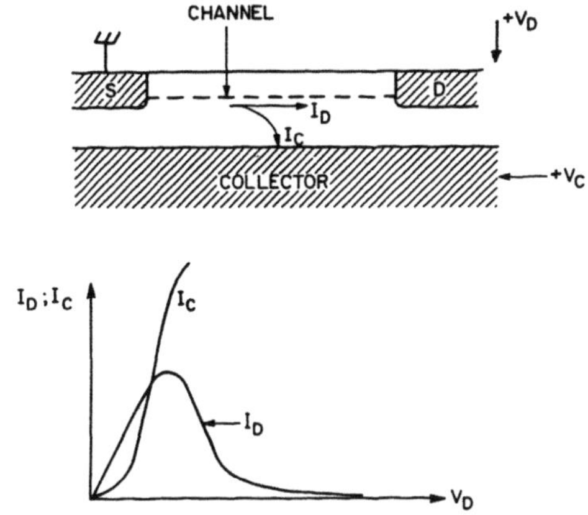

Fig. 1. Device cross-section and major I-V characteristics

higher V_D, however, electron temperature increases giving rise to the exponential enhancement of the thermionic emission over the barrier to the collector. At special bias conditions, namely, at a relatively high heating voltage V_D and even higher collector voltage V_C the majority of electrons will be pulled to the collector instead of going to the drain, thus causing the effect of NDR in the source-drain circuit, which we call the NERFET action [1,2]. As a result of such a diversion of electrons from the drain to the collector the reduction of the drain current I_D will be accompanied by a rapid rise of the collector current I_C, which eventually becomes dominant. The exponential dependence of thermionic emission on the electron temperature offers a new possibility for transistor action: a relatively small modulation of the drain (input) voltage causes a significant variation of the collector (output) current. We call this mode of our device a Charge Injection Transistor (CHINT) [3].

There are several important factors which control the performance of our device. First, electrons in the channel should possess high mobility and low effective mass, to provide an efficient electron heating. To this end, a two-dimensional, high-mobility channel at the AlGaAs/GaAs heterointerface, seems to be the most appropriate way of making a source of hot electrons. Second, source and drain device contacts to the channel must be shallow enough to prevent a contact punch-through to the collector. Finally, since electric field across the barrier can be very high, the undoped AlGaAs layer must provide a high quality insulation between conducting layers. Last requirement is critical for both above-mentioned modes of device operation and actually determines the choice of structure and method of its growth.

In earlier experiments [2,3,5] we used modulation doping to create the channel in AlGaAs/GaAs heterostructure grown by molecular beam epitaxy (MBE). Because of current leakage through the insulator the NDR effect at room temperature was small. Better results were obtained when we used the undoped heterostructure where the channel electrons were induced by a positively biased collector [6]. However, in both cases the CHINT action could not be demonstrated even at 77K. The leakage current has been vastly reduced in devices fabricated with materials grown by organo-metallic chemical vapor deposition. As a result, a strongly pronounced NDR and CHINT action have been achieved at room temperature [7]. Below we shall discuss the results obtained in OMCVD grown devices utilizing the undoped heterostructure approach for creation of a two-dimensional channel.

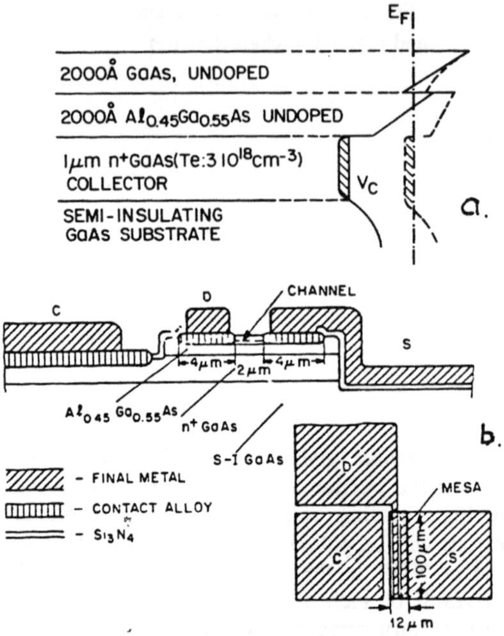

Fig. 2. a. Layer structure and energy band diagram. b. contact layout.

3. Experimental Results and Discussion

The layer structure grown by OMCVD and its energy diagram are shown in Fig. 2a. The undoped GaAs and AlGaAs layers (2000Å each) are depleted by the surface potential at equilibrium. When a positive voltage V_C is applied to the 1 μm thick second conducting layer, the GaAs/AlGaAs interface is pulled below the Fermi level, giving rise to a two-dimensional electron gas at the interface. A voltage V_D applied between source and drain contacts heats the electrons at the interface and activates the process of hot electron injection. The device cross-section and contact lay-out used in our experiments are shown in Fig. 2b. To reduce a parasitic capacitance between source and drain bonding pads and the second conducting layer we ion mill away this layer everywhere except under the 12 μm × 100 μm active region and the collector contact. Au/Ge/Ag source and drain contacts, each 4 μm long and 1-2 μm apart, are evaporated and carefully alloyed at T = 430°C for 1 sec to avoid a short to the collector (C). Solid line in Fig. 3 shows a typical source-to-collector I-V curve at T = 300K. We see a high-quality blocking characteristic for both polarities. For polarity of interest (plus on C) the turn-on voltage is about 6.5V, which is significantly higher than that for the structure grown by MBE (dashed line). Furthermore, as we shall see later, our best results were obtained at collector voltages as high as $V_C \approx 11V$ (corresponding to the electric field of $\sim 5.5 \times 10^5$ V/cm). The measured in this case parasitic leakage was only a small fraction of a total collector current. One can conclude here that the OMCVD grown AlGaAs layer has excellent insulating properties and provides a strong reduction of parasitic leakage compared to the MBE grown AlGaAs insulators.

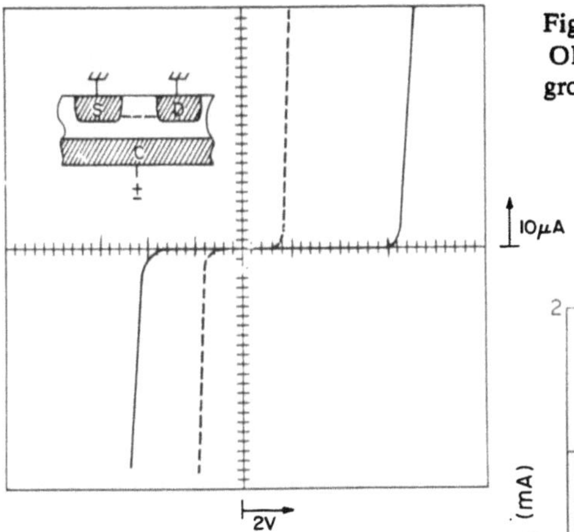

Fig. 3. Source-Collector I-V characteristics for OMCVD-(solid line) and MBE-(dashed line) grown devices. T = 300K.

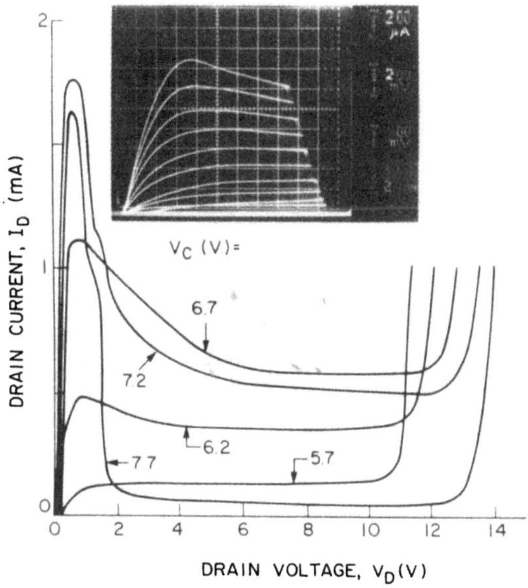

Fig.4. I_C-V_D characteristics for different V_C; characteristics on top oscillogram begin from V_C=5V.

Figure 4 displays the source-drain $I_D - V_D$ characteristics taken at different fixed V_C. At the relatively low $V_C \lesssim 6V$ the device operates as a buried gate FET with a transconductance $g_m^{FET} \sim 20 - 30 \frac{mS}{mm}$ (see top oscillogram). There are several reasons for such "poor" values of g_m^{FET}. First, the insulator thickness (2000Å) is larger than that in conventional (AlGa)As/GaAs modulation doped FET structure (~350 − 500Å). Second, the source and drain contact resistances seem to be rather high. At low temperatures (77K) the transconductance becomes even smaller and the $I_D - V_D$ characteristics exhibit diode-like superlinearities at low V_D, indicating the presence of contact barriers. Finally, hot electron injection from the channel to the collector prevents (as we will show later) the realization of a high drain current, and therefore, keeps the g_m^{FET} at low levels.

At higher $V_C(\gtrsim 6V)$ we observe a development of the NDR effect. At $V_C > 8V$ we normally obtain a very strong NDR with peak-to-valley ratio of ~20−30. The highest peak-to-valley ratio observed in our experiments was ~160.

An interesting feature is seen beyond the NDR region: the drain current exhibits a saturation within large region of drain voltages, the saturation current value decreasing with increasing V_C, which corresponds to the effect of negative transconductance. We shall return later to the discussions of these peculiarities after analysis of drain and

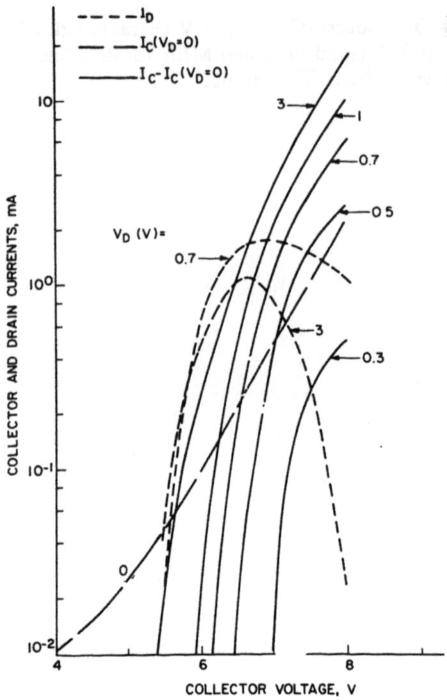

Fig. 5. Collector and drain currents versus

collector currents as a function of the V_C. They are presented in semilog plot in Fig. 5 for different fixed V_D.

One can see a high sensitivity of the net charge injection current $I_C - I_C(V_D = 0)$, (where $I_C(V_D=0)$ is the leakage current) to the heating voltage V_D. The injection current begins to exceed the leakage one at drain voltages as low as $\sim 0.5V$ and very rapidly becomes dominant. The slopes of these dependences are too weak to attribute them to the process of tunneling, but the effect of the barrier lowering for thermionic emission due to an increase of the Fermi level in the channel seems to be in reasonable agreement with the experiment.

The drain current (dashed curves in Fig. 5) exhibits a sharp rise at $V_C \gtrsim 5V$. Then it has a tendency to saturation and finally goes down for both drain voltages presented. $V_D = 0.75V$ represents a peak of $I_D - V_D$ characteristic, whereas $V_D = 3V$ corresponds to a current saturation region. In the latter case the drain current can be almost 1000 times less than the hot-electron transfer current ($V_C = 8V$, Fig. 4), but even for $V_D = 0.75V$ we observe a decrease of $I_D(V_C)$ at high V_C, obviously due to the real-space transfer to the collector. A comparison of drain and collector current behavior shown in Fig. 5 suggest the basic reason for the existence of the NDR: as V_C *increases beyond the value, corresponding to appearance of the 2-D channel the further linear rise of the channel electron density is overwhelmed by the sharp (almost exponential) increase of hot electron emission out of the channel resulting in a channel depletion and therefore in a drop of the drain current.*

The presence of NDR assumes the development of a strong nonuniformity of electric field along the channel and creation of a high field domain. *The p ecularity of such a domain in our case is its depletion due to real-space electron transfer to the collector.* Formally, this situation is close to that existing in a conventional MOSFET, where high field region arises near the drain due to pinch-off effect (see, for instance [10]). By analogy with the MOSFET, we attribute the observed(at high V_D) saturation of the drain current beyond the NDR region (Fig. 4) to the appearance of the high field domain which results in our case from an electric field instability along the channel as consequence of the NDR effect. It means that the drain current in saturation is controlled by the electron flux across the high field domain depleted by the real space transfer. Since the effect of domain depletion enhances with V_C the saturated drain current should decrease as V_C increases, thus, in our opinion, giving rise to the effect of negative transconductance shown in Fig. 4.

The observed rapid rise of the collector current as a function of V_D and relatively low level of the parasitic leakage current (shown in Fig. 5) encourages us to investigate experimentally a possibility for the discussed above novel transistor action, based on the real-space hot-electron transfer. As mentioned before we have failed to achieve a real transistor effect on MBE grown devices. In new experiments [7] we used OMCVD grown structure similar to that we discussed above (Fig. 3-5) with the only difference that leakage current was measured to be even lower than in previous case.

The results of DC characterization of our device are presented in Fig. 6 and 7. The input, I_D, and the output, I_C, currents in the CHINT versus the input bias V_D are shown in Fig. 6 for two different output voltages V_C: 10.4 and 10.8V. In both cases the NDR effect is clearly observed. A relatively small increase of the V_C (from 10.4 to 10.8V) brings about a strong enhancement in the I_C, whereas the input current

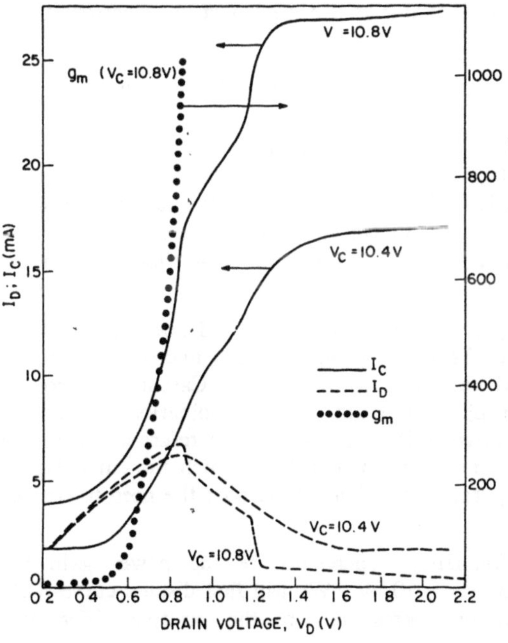

Fig. 6. I_D-V_D, I_C-V_D and g_m-V_D characteristics for two fixed values of V_C.

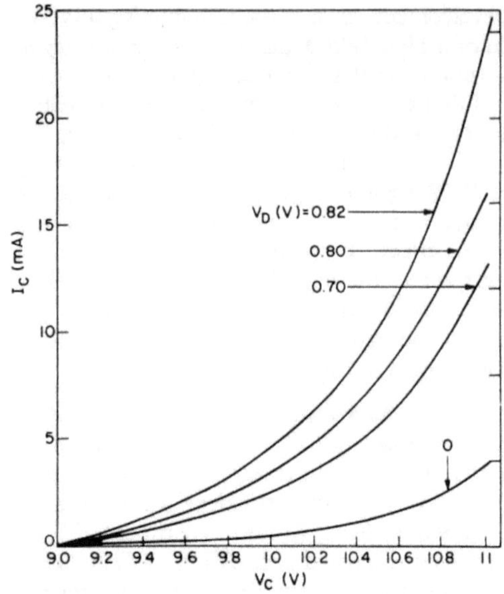

Fig. 7.
$I_C - V_C$ characteristics for different V_D.

changes insignificantly. We also see a high sensitivity of the output current to the input voltage, especially for $V_C = 10.8V$. Moreover, in the region of the peak of I_D ($V_D \approx 0.7-0.8V$) the largest increment of I_C corresponds to the lowest increment of the input current I_D. This range of the input voltages is expected to be the most promising with respect to gain in the CHINT. The fine features on the $I_D(V_D)$ curves have mirror reflections on the $I_C(V_D)$ curves. Such a correlation in the currents supports once more the idea that the NDR is due to a diversion of hot electrons from the drain to the collector.

The output current-voltage ($I_C - V_C$) characteristics of CHINT are shown in Fig. 7 for different fixed values of input voltages V_D. The curve at $V_D = 0$ gives a pure leakage to the buried gate (mainly due to the leakage through the S and D contacts). The rapid increase of the I_C with V_C begins at $V_D \gtrsim 0.7V$ and obviously results from the enhancement of hot-electron injection. One can notice that the latter dominates the leakage current for the entire range of V_C shown in this figure, in spite of the fact that the device area where the parasitic leakage occurs is at least 5-6 times larger than the area of hot electron injection.

The rf results are presented in Fig. 8. The rf current gain and power gain of the CHINT were obtained as a function of frequency using HP 8510 microwave network analyzer. The CHINTs were probed in wafer form using a Cascade Microtech model 42 probe station at frequencies up to 18 GHz. This combination of network analyzer and prober brings fully calibrated 50Ω transmission line references plane to the CHINT, allowing the network scattering parameters, from which the current gain and power gain are calculated by computer, to be obtained without the need to correct for the effects of packaging.

Figure 8 illustrates the typical frequency dependence of the power gain for different input bias voltages V_D and for $V_C = 10.8V$. We will first discuss curves (a)-(d), for which $V_D \leq 0.82V$ and which corresponds to the positive differential

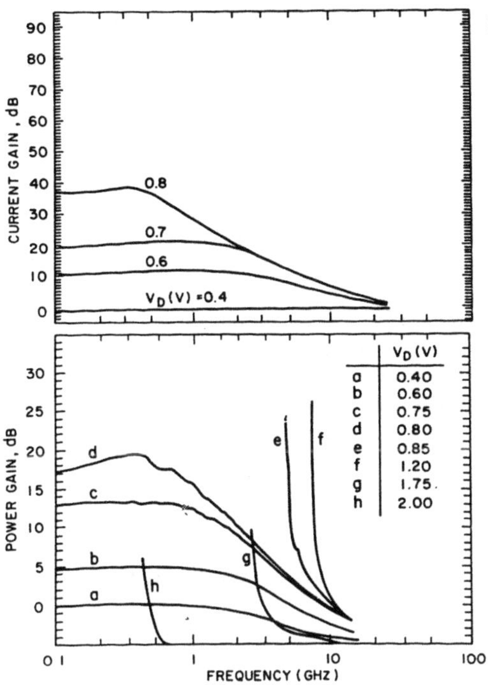

Fig. 8. Power and current gains in CHINT as a function of frequency for different V_D

resistance region in the input I-V characteristics (Fig. 6). The maximum power gain of 20 dB occurs near $V_D = 0.8V$ and corresponds to the peak of the input current I_D. At frequencies above 1 GHz, the power gain decreases for all values of V_D, and the maximum frequency of unity power gain achieved is $f_{max} = 9.8$ GHz. In curves (e)-(h), for which $V_D > 0.82V$ (negative differential resistance region of the input I-V characteristics) the power gain becomes infinite at low frequencies. This infinite power gain arises because the NDR of the input circuit provides the potential for oscillation at the frequency of the input signal under appropriate source impedance conditions. For each of curves (e)-(h), there is a maximum frequency of NDR f_{max}^{NDR} above which the power gain becomes finite, with the highest $f_{max}^{NDR} = 7.7$ GHz occurring for $V_D = 1.2V$ and corresponding to the second current drop in the input I-V characteristics of Fig. 6 (dashed curve for $V_{sub} = 10.8V$). The variation of f_{max}^{NDR} with V_D correlates with the slope of the NDR for the same V_D. The best f_{max}^{NDR} are obtained for the two values of V_D, namely, 0.85 and 1.2V, at which sharp drops in I_D occur. Thus, the rf performance of the NERFET is very sensitive to V_D in the NDR region.

The rf current gain of the CHINT is shown in Fig. 8 for four values of V_D, with $V_C = 10.8V$. We see a marked rise of current gain as V_D increases with the highest observed gain of 40 dB measured at $V_D = 0.8V$. The cut-off frequency (frequency of unity current gain) f_T is significantly higher then f_{max}, reaching 32 GHz at $V_D = 0.8V$. In the region of NDR the current gain decreases, finally disappearing for $V_D \geq 1.5V$.

An important parameter for characterizing the performance of our device is the transconductance, $g_m = \left.\frac{\partial I_C}{\partial V_D}\right|_{V_C}$. The latter as a function of V_D is shown in Fig. 6 for $V_C = 10.8V$. We see a steep rise of g_m with input voltage and for optimal bias conditons ($V_C = 10.8V$ and $V_D = 0.80V$) it reaches ~ 1000 mS/mm.

The 20 dB/decade RC falloff of current gain with frequency is determined by the excess resistance and capacitance rather than by the intrinsic device properties. Fundamental limitations on the intrinsic speed of our hot-electron devices arise due to the time-of-flight delays characteristic of a space-charge-limited current across the regions of the domain and the AlGaAs insulator and because of a finite time required for the establishment of an electron temperature [3]. Both limitations give the delay time of about 1-2 psec. Therefore, improvement of the rf response of the CHINT and the NERFET is expected from further reduction of RC in the output circuit, where R is controlled by source and channel resistances and C is dominated by the drain to buried gate capacitance. Simple estimates for charge density induced in the channel by the buried gate indicate that for the range of V_C we apply (10-11V) the drain current should be much higher. In earlier experiments on modulation-doped structure [2,4] we were able to obtain an order of magnitude higher currents at V_C lower by a factor of 2. We believe that the reason for such low drain currents in our case originates from the high source and drain contact resistances that limit the current.

4. Conclusion

In conclusion, we emphasize that there has recently been much consideration given to the concept of real-space hot-electron transfer [1,3,8]. This phenomenon shows particular promise for high-frequency operation because of the short transit distances of 1000Å or less between adjacent layers. Until recently, the only demonstration of real-space hot-electron transfer has been the observation of negative differential resistance and microwave generation in both two-terminal [8] and three-terminal [4] configurations. In our work, because of dramatic improvement of the insulating properties in AlGaAs layers grown by OMCVD real-space transfer has been used to achieve true transistor action through the modulation of electron temperature in the high-mobility channel. The obtained high-frequency characteristics in our devices reveal substantial power and current gains with a cut-off frequency for the latter of 32 GHz. Simultaneously, we evaluated improved negative differential resistance properties in the same structure. We believe that with the reduction of both channel resistance and parasitic capacitance our device can approach the intrinsic high-frequency limits associated with the real-space transfer and extend frequency response by about one order of magnitude. This work is in progress now.

References

1. A. Kastalsky and S. Luryi IEEE Electron Device Letters, EDL-4 (1983) p. 334.
2. A. Kastalsky, S. Luryi, A. C. Gossard and R. Hendel IEEE-Electron Device Letters, EDL-5 (1984) p. 57.
3. S. Luryi, A. Kastalsky, A. C. Gossard and R. Hendel IEEE Trans. Electron Device ED-31 (1984) p. 32.
4. A. Kastalsky, R A. Kiehl, S. Luryi, A. C. Gossard and R. Hendel IEEE Electron Device Letters, EDL-5 (1984) p. 321.

5. A. Kastalsky, S. Luryi, A. C. Gossard and W. K. Chan IEEE Electron Device Lett. EDL-6 (1985) p. 347.
6. A. Kastalsky, R. Bhat, W. K. Chan and M. Koza Solid-State Electron. 1986, to be published.
7. A. Kastalsky, J. H. Abeles, R. Bhat, W. K. Chan and M. Koza Appl. Phys. Lett. 43 (1986) p. 71.
8. K. Hess, H. Morkoc, H. Shichijo and B. G. Streetman Appl. Phys. Lett. 35, (1979) p. 469. See also K. Hess, Physica 117B (1983) p. 723.
9. P. D. Coleman, J. Freeman, H. Morcoc, K. Hess, B. Streetman and M. Keever Appl. Phys. Lett. 40, (1982) p. 493.
10. S. M. Sze, Physics of Semiconductor Devices, 2nd edition (Wiley, New-York, 1981).

Transport Characteristics in Heterostructure Devices

S. Muto, S. Hiyamizu, and N. Yokoyama

Fujitsu Limited, 10-1 Morinosato-Wakamiya, Atsugi 243-01, Japan

This paper describes some of the heterostructure devices, including the high electron mobility transistor (HEMT), the hot electron transistor (HET), and the resonant-tunneling hot electron transistor (RHET). This is followed by a discussion of transport characteristics in the RHET, with special attention given to the resonant tunneling barrier (RTB) used as the emitter barrier of the RHET.

1. Introduction

Recently, a lot of heterostructure devices have been developed. Each of these new devices presents new areas of physics, and studying its physics is not only interesting, but also indespensable for further improvement of the device. This paper describes the transport characteristics of the RHET, the first resonant tunneling transistor which we developed recently.

First, we outline the characteristics of heterostructure devices, including the high electron mobility transistor (HEMT), the hot electron transistor (HET), and the newly developed resonant-tunneling hot electron transistor (RHET). Next, we will introduce the transport characteristics of the RHET, emphasizing the resonant tunneling barrier used as the emitter barrier of the transistor.

2. Characteristics of Heterostructure Devices

The HEMT[1] seems to have gained commercial acceptance, especially in the field of microwave applications. Its further improvement is now dependent on the fabrication technology. For discrete HEMTs, the highest obtained cuttoff frequency is 80 GHz [2] and the lowest noise figure is 0.66 dB at 12 GHz[3]. For HEMT ICs, the shortest switching delay of 5.8 ps was reported[4] for a ring oscillator and, aiming towards LSI, a 4 K static RAM[5] and 1.5 K gate array[6] were developed.

The HEMT is a device which uses electron transport parallel to the heterointerface. However, we feel that more capable devices could be developed by using electron transport perpendicular to the heterointerface. A hot electron transistor (HET) has the potential to be an ultra-high-speed device, due to its extremely short base transit time. In 1984, we realized the first AlGaAs/GaAs HET[7]. Using the magnetic field dependence of the collector current, we found the base transit time of the HET to be subpicosecond[8]. However, the common-emitter current gain of the transistor was only 1.3 at 40 K. Since then, we have been unable to achieve a significant increase in the current gain.

The current gain of a HET is determined by various scattering mechanisms intrinsically present in the base layer. This point will be discussed later in the section on the transport characteristics of the RHET, which unfortunately inherits this problem.

Giving a transistor a new function and thus reducing the number of transistors required for an integrated circuit seems to be a very promising way of achieving a high-speed IC. For this purpose, we recently developed a new-functional transistor, which we christened the resonant-tunneling hot electron transistor, or the RHET[9]. Figure 1 shows a schematic cross section of the RHET. Here, we introduced an AlGaAs/GaAs resonant tunneling barrier as the emitter barrier of a HET. The base layer was 100 nm wide. The collector barrier was a single AlGaAs barrier with a smaller Al mole fraction. This is essentially a HET. However, due to the negative differential resistance (NDR) of the resonant tunneling barrier placed between emitter and base layers, the collector current of the transistor shows an unusual peak when plotted as a function of base-emitter voltage. This is the first transistor ever to use resonant tunneling.

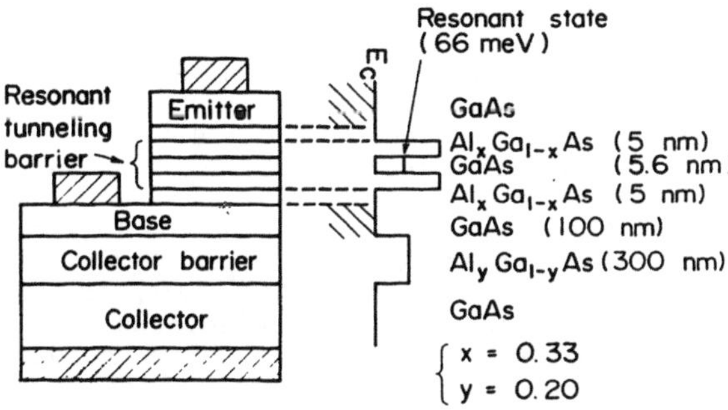

Figure 1. Schematic cross-section and band diagram of the RHET

Figure 2 shows the RHET collector current observed at 77K. The sudden decrease of the collector current is due to the negative differential resistance of the resonant tunneling barrier. This anomalous characteristic can never be realized with conventional transistors and this endows the RHET with a wide variety of new functions. We have already demonstrated a few of them: the frequency multiplication and Exclusive-NOR logic function, both at 77K[9] using simple circuits with only one RHET. Considering that 7 switching MESFETs are needed to achieve the Exclusive-NOR logic function, we see that the RHET has the potential for use in very high-speed and high-density integrated circuits.

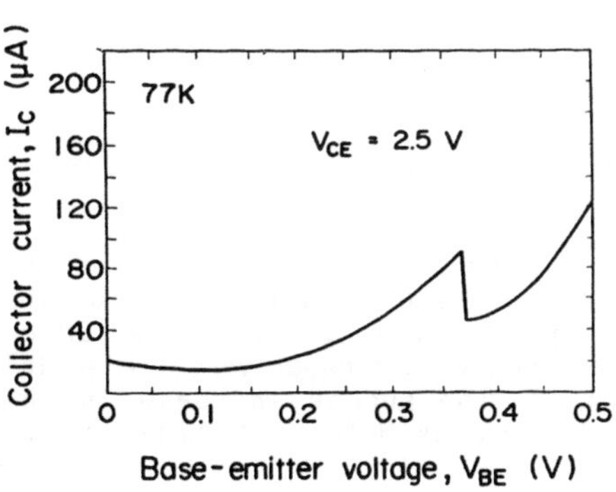

Figure 2. Collector current of the RHET as a function of base-emitter voltage

3. Transport Characteristics in RHET

3.1 Transport in the Base Layer

The transport characteristics of the RHET related to the current gain are essentially the same as those of the HET. One important difference is the uniformity of the electron energy injected through the resonant tunneling barrier. In fact, the half-width of the injection energy by the resonant tunneling barrier used for the RHET is calculated to be as small as 0.2 meV. This is in contrast to the HET, where the width of the injection energy is some 50 meV.

Taking this into account, we performed particle Monte Carlo simulation to study the electron scattering occurring in the base layer of the RHET. In the simulation we used a 3-valley band model, including Gamma, L, and X valleys. As the scattering mechanisms in the base layer, we considered 1) ionized impurity scattering, caused by ionized silicon impurities in the base layer, 2) intravalley and intervalley scattering, which is essentially LO phonon scattering in this case, 3) plasmon scattering, which results from the collective motion of electrons in the base layer, and finally 4) the quantum mechanical scattering or reflection from the collector barrier.

Figure 3 shows the result of the Monte Carlo simulation of the forward velocity distribution of electrons at the end of the 100-nm RHET base. The initial velocity of the injected electrons is about 8×10^7 cm/s, so the sharp peak with the highest velocity is caused by those electrons which reach the end of the base layer without any scattering. The broader peak with lower velocity corresponds to those electrons which have experienced inelastic scattering by LO phonons or plasmons, and the shoulders are caused by the rotational motion of electrons due to elastic scattering by ionized silicon impurities.

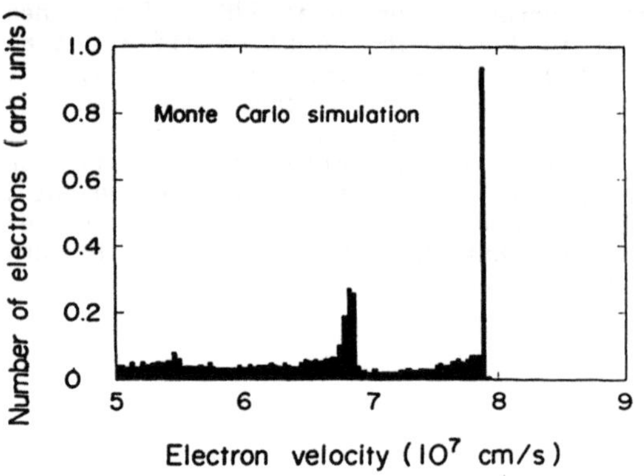

Figure 3. Result of the Monte Carlo simulation on the forward velocity distribution of electrons at the end of the 100-nm RHET base

It should be noted that there are ballistic electrons still present at the end of the 100-nm RHET base.

We can actually see evidence of the ballistic electrons in the transistor characteristics. Figure 4 shows the collector current of the RHET, when the collector-emitter voltage was kept to 1 V. The sudden decrease of the collector current is again due to the negative differential resistance of the RTB as seen in Fig. 2. The sudden increase in the collector current has also never been observed in HETs. We believe that

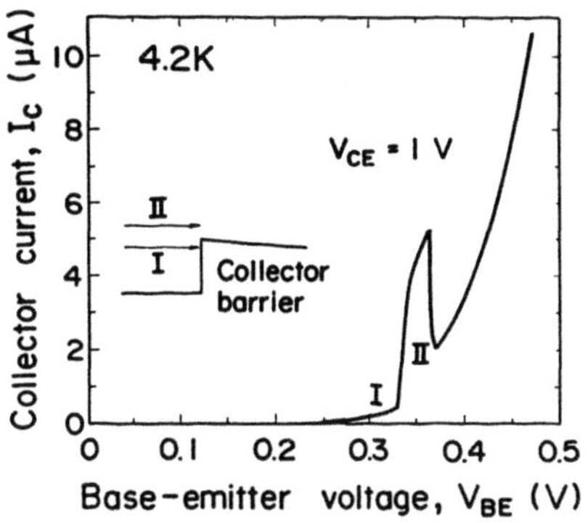

Figure 4. Collector current of the RHET as a function of base-emitter voltage at collector-emitter voltage of 1 V

this indicates the appearance of monoenergetic and ballistic electrons just above the collector barrier. As shown in the inset, when the base-emitter voltage is small, most electrons are reflected by the collector barrier and cannot reach the collector. When the voltage is increased, however, injection energy also increases and ballistic electrons start to pass the collector barrier. This explains the sudden increase.

If we introduce the RHET a transverse magnetic field, we can observe the effect of the cyclotron motion of the hot electrons in the base layer of the RHET. While travelling through the base layer, hot electrons transform their longitudinal kinetic energy into transverse kinetic energy by the cyclotron motion, and thus lose their longitudinal kinetic energy, and finally they cannot surmount the collector barrier[8]. Figure 5 shows the magnetic field dependence of the RHET collector current observed at 4.2K together with the result of the Monte Carlo simulation.

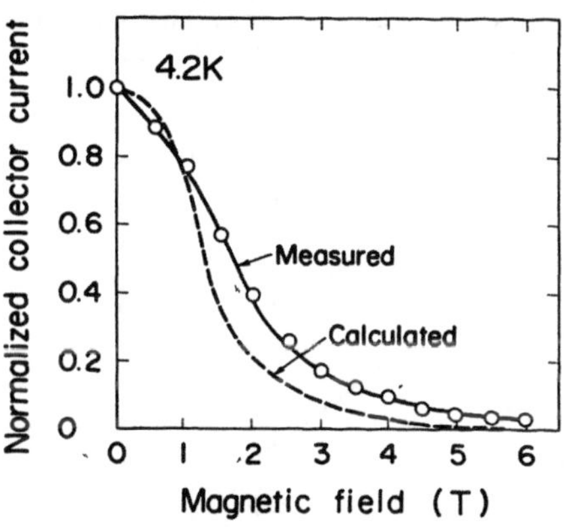

Figure 5. The RHET collector current measured at the peak of the NDR as a function of the transverse magnetic field

3.2 Transport in the Emitter Barrier

Many things are yet to be learned about the resonant tunneling barrier (RTB), used as the emitter barrier of the RHET. As far as the static properties are concerned, the negative differential resistance at low temperatures is the most important. In fact, the peak current density, J_p of the negative differential resistance region limits the maximum collector current available from the transistor and the valley current density J_v results in leakage current when the transistor is in the "off" state[10].

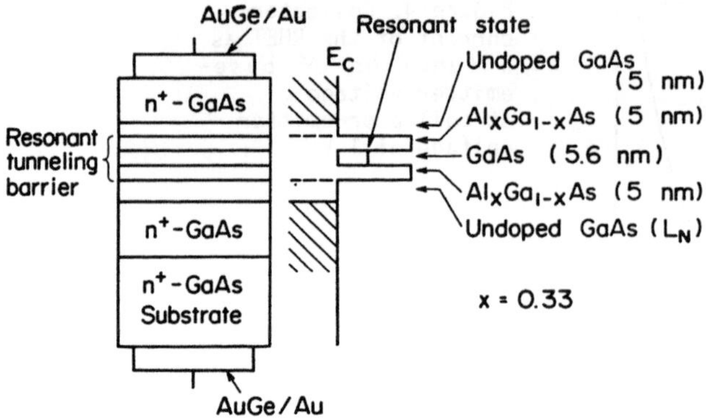

Figure 6. Schematic cross-section and band diagram of the RTB diode

Therefore, we hope to simultaneously have a high J_p and a high peak-to-valley ratio, J_p/J_v.

To study the negative differential resistance characteristics at low temperatures, we made a diode with a resonant tunneling barrier structure as shown in Fig. 6[10]. Here, a resonant tunneling barrier sandwiched by n^+-GaAs layers doped with Si to $1\times10^{18}/cm^3$ was grown on a n^+-GaAs substrate by MBE. At low temperatures, silicon impurities may play an important role in the quality of the barrier. Therefore, we introduced thin, undoped GaAs layers, which we call spacer layers, on both sides of the barrier structure. The thickness of the lower spacer layer was varied from 0 to 15 nm, and the thickness of the upper spacer layer was fixed at 5 nm.

Figure 7a shows a plot of the peak-to-valley ratio of diodes measured at 77 K as a function of the spacer layer thickness, L_N. Here, the surface electrode of the diode is positively biased with respect to the substrate

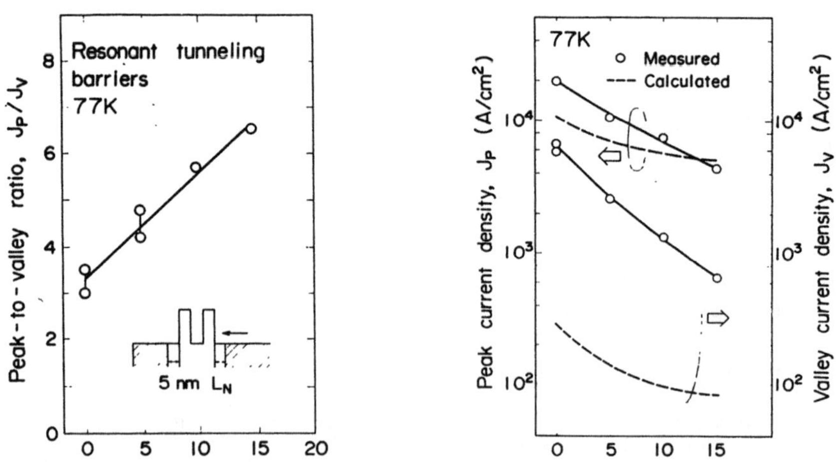

Figure 7. (a) Peak-to-valley ratio of RTB diodes at 77K as a function of spacer layer thickness, L_N, (b) peak and valley current densities of the diodes as a function of L_N

electrode and electrons enter the barrier from the spacer layer. During growth, silicon impurities slightly diffuse into the barrier structure, which could cause an unwanted increase in the valley current. Therefore, drastic improvement of the peak-to-valley ratio, when the thickness, L_N is increased, is thought to be due to a reduction in the ionized Si impurities in the barrier structure.

Figure 7b shows the effect of L_N on the peak current density and the valley current density. The open circles correspond to measured values of the diodes and the broken curves are values calculated from the tunneling theory[11]. In the calculation, we took into account the band bending in and around the barrier structure. Note the good agreement between the measured and calculated values of the peak current density. For the valley current densities, however, the agreement is rather poor, and the measured values are about an order of magnitude higher than the calculated values. The reason for this discrepancy is not yet known, but it should be related to electron scattering other than the quantum mechanical reflection at the heterointerfaces, since we assumed no other scattering mechanism in the tunneling calculation.

In fact, the resonant state only refers to the kinetic energy of electrons in the longitudinal direction, and so even elastic scattering such as ionized impurity scattering by silicon can interfere with the simple resonant tunneling mechanism.

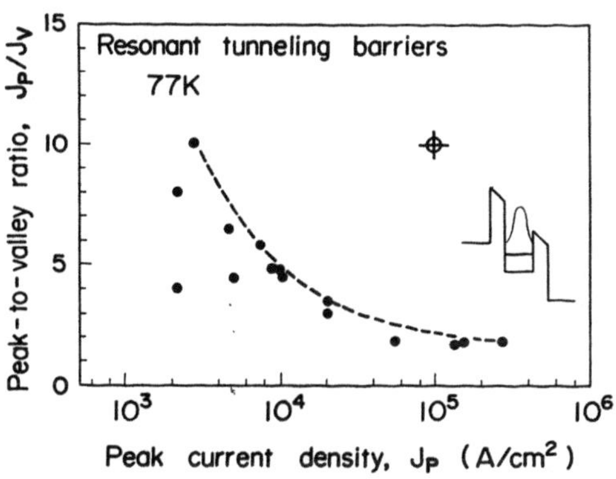

Figure 8. Peak-to-valley ratio as a function of peak current densities[10,12-14]

Replacing the AlGaAs emitter barrier of a HET with the resonant tunneling barrier results in a high current density. In this way, we have already achieved a high peak current density of 2.7×10^5 A/cm^2 for a resonant tunneling barrier structure[10], the highest to date. The peak-to-valley ratio of the barrier structure was only 1.8 at 77 K. Figure 8 shows some of the values of the peak-to-valley ratio as a function of the peak current density[10,12-14]. The apparent negative correlation between the peak-to-valley ratio and the peak current density is not yet understood. However, calculations show that, as the current density increases, the density of electrons bound in the quantum well of the barrier structure increases. Therefore, it is natural to attribute this decrease in the peak-to-valley ratio to the scattering of tunneling electrons by the accumulated electrons.

The negative differential resistance expected for a resonant tunneling barrier for the RHET is a peak current density of 1×10^5 A/cm^2 and a peak-to-valley ratio of more than 10. Therefore, further improvement of the resonant tunneling barriers is of prime importance, and this, in turn, calls for investigation of the origin of the valley current.

In summary, some of the heterostructure devices we developed are described, including the HEMT, the HET, and the RHET. Although in different phases of development, all of them show great promise as high-speed devices. Concerning the transport characteristics of the RHET, the presence of monoenergetic and ballistic electrons was pointed out at the end of the 100-nm base of the RHET we developed. As for the resonant tunneling barrier, which is the characteristic emitter of the RHET, the significance of electron scattering by the ionized silicon impurities, and also by the electrons accumulated in the quantum well of the barrier structure, was emphasized.

The authors wish to express their thanks to T. Yamaoka, A. Shibatomi, M. Kobayashi, O. Ryuzan, and T. Misugi for their continuous encouragement of this line of work. A part of this work was performed under the management of the R & D Association for Future Electron Devices as a part of the R & D Project of Basic Technology for Future Industries sponsored by Agency of Industrial Science and Technology, MITI.

References

1. T.Mimura, S.Hiyamizu, T.Fujii, and K.Nanbu, Jpn. J. Appl. Phys. 19 (1980) L225.
2. P.C. Chao, S.C. Palmateer, P.M. Smith, U.K. Mishura, K.H.G. Duh, and J.C.M. Hwang, IEEE Electron Device Letters, EDL-6 (1985) 531.
3. K.Kamei, H. kawasaki, S. Hori, K. Shibata, M. Higashiura, M.O. Watanabe, and Y. Ashizawa, in: Proceedings 12 th Intern. Symp. on GaAs and Related Compounds, Karuizawa, 1985, p.541.
4. N.J.Shah, S.S. Pei, C.W. Tu, and R.C. Tiberio, IEEE Trans. Electron Devices ED-33 (1986) 543.
5. S. Kuroda, T. Mimura, M. Suzuki, N. Kobayashi, K. Nishiuchi, A. Shibatomi, and M. Abe, in: GaAs IC Symp. Tech. Dig. (1984) p.125.
6. Y. Watanabe, K. Kajii, K. Nishiuchi, M. Suzuki, I. Hanyu, M. Kosugi, K. Odani, A. Shibatomi, T. Mimura, M. Abe, and M. Kobayashi, in: ISSCC Tech. Dig. (1986) p.80.
7. N.Yokoyama, K.Imamura, T.Ohshima, H.Nishi, S.Muto, K.Kondo, and S.Hiyamizu, IEDM 84 Tech. Dig. (1984) p.532.
8. S.Muto, K.Imamura, N.Yokoyama, S.Hiyamizu, and H.Nishi, Electron. Lett. 21 (1985) 555.
9. N.Yokoyama, I.Imamura, S.Muto, S.Hiyamizu, and H.Nishi, Jpn. J. Appl. Phys. 24 (1985) L853.
10. S.Muto, T.Inata, H.Ohnishi, N.Yokoyama, and S.Hiyamizu, Jpn. J. Appl. Phys. 25 (1986) L577.
11. H.Ohnishi, T.Inata, S.Muto, N.Yokoyama, and A. Shibatomi, to be published.
12. T.J. Shewchuk, P.C. Chapin, P.D. Coleman, W. Kopp, R. Fischer, H. Morkoc, Appl. Phys. Lett. 45 (1985) 508.
13. M. Tsuchiya, H. Sakaki, and J. Yoshino, Jap. J. Appl. Phys. 24 (1985) L466.
14. M. Tsuchiya and H. Sakaki, in: IEDM, Tech. Dig. (1985) p.662.

Technical Issues of High-Speed Heterostructure Devices

T. Ikegami, K. Yamasaki, T. Mizutani, T. Ishibashi, and M. Fujimoto

NTT Electrical Communications Laboratories, Morinosato Wakamiya, Atsugi-shi, Kanagawa, 243-01, Japan

1. INTRODUCTION

Recently, heterostructure devices have been extensively investigated because of their superior properties and new applications. In addition, the development of new epitaxial techniques, such as molecular beam epitaxy (MBE) and metal organic chemical vapor deposition (MOCVD), have enabled precise control of thickness and doping in the fabrication of those devices. In this paper, relatively orthodox heterostructure devices are discussed from the technical and application points of view. First of all, current technical issues of heterostructure FETs (HFETs) /1/ directed at integrated circuits are described. Next, heterojunction bipolar transistors (HBTs) /2/ are discussed from the same point of view. Finally, current topics of semiconductor laser diodes concerning fast direct modulation are presented.

2. HETEROSTRUCTURE FET (HFET)

2.1 Scaling-down of FETs

Recently, high-speed GaAs MESFET ICs, such as an 11 GHz frequency divider /3/ have been successfully fabricated with a half-micron gate length. In order to achieve higher speed operation, further reduction of gate length is required. Figure 1 shows the relationship of propagation delay time versus gate length measured with ring oscillators at room temperature. Several types of MESFETs and HFETs are included. Delay times obtained decrease in approximate proportion to gate length reduction. By using rough extrapolation, 5 ps/gate at 0.2 μm gate length and 2.5 ps/gate at 0.1 μm gate length are predicted. With gate length shortening, shallower and higher-doped channel layers are required in order to alleviate the short

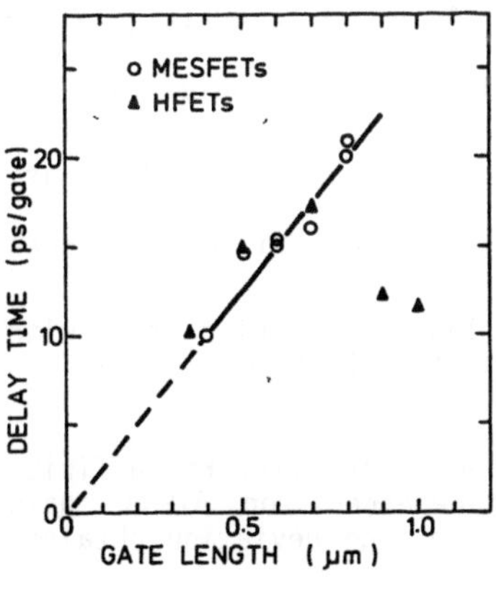

Fig. 1 Propagation delay time vs. gate length for GaAs MESFETs and AlGaAs heterostructure FETs

channel effects. For instance, a 30 nm channel thickness with over 1.0 x 10^{18} cm^{-3} doping is necessary for 0.1 μm gate length MESFETs. However, it is very difficult to obtain these scaled-down thin channel layers by present ion implantation technology used in IC fabrication. Moreover, heavily doped channel layers reduce electron mobility due to ionized impurity scattering.

2.2 Modulation-Doped Heterostructure FET (MODFET)

These scaling-down related problems should be resolved by a heterostructure FET (HFET). This is because very thin layers are obtained by current epitaxial technologies (MBE and MOCVD) and carriers induced in the heterointerface drift without impurity scattering. Therefore, an HFET is one of the most promising candidates for very high-speed devices with short gate lengths. A record propagation delay time of 5.8 ps/gate was measured at 77 K with a ring oscillator integrated with 0.35 um gate length MODFETs /4/.

However, one of the serious problems with MODFETs is difficulty in threshold voltage reproducibility. The threshold voltage V_{th} is determined by

$$V_{th} = \Phi_b + \Delta E_c/q - (q/2\varepsilon_s)N_d t_d^2, \qquad (1)$$

where Φ_b is Schottky barrier height, ΔE_c is the conduction band discontinuity, and N_d and t_d are doping density and thickness of AlGaAs /5/. It is a difficult task to consistently reproduce the doping density and thickness of the AlGaAs layer with present MBE or MOCVD technology.

Another problem is related to deep levels (D-X centers) in n-AlGaAs, which are induced by donors. These defects result in the persistent photoconductivity effect /6/ and drain I-V collapse /7/ at low temperature. They also cause low-frequency generation recombination noise /8/.

2.3 MISFET

The most simple and effective method to overcome the above problems is to use undoped-AlGaAs in place of a doped-layer. From this point of view, MIS-like HFETs (MIS-HFETs) have been proposed /9-13/. Among these, an n^+-Ge gate MISFET, proposed by ARAI et al. /13/, is very attractive for application to LSIs. The schematic cross-section of the device, fabricated using MBE, is shown in the left part of Fig. 2. An n^+-Ge layer is used for gate material for the following reasons.

(1) Expected threshold voltage of about 0.1 V is suitable for LSI application.
(2) Heavy n^+ doping (5 x 10^{19} cm^{-3}) permits non-alloy contact with a gate metal electrode (WSi).
(3) Selective etching of Ge against GaAs by RIE results in small process-induced deviation of FET characteristics.

Source and drain n^+-layers are formed by self-aligned ion implantation with gate materials as a mask. The activation of the implanted layers is performed by rapid thermal annealing.

The threshold voltage is essentially determined by the electron affinity difference between Ge and GaAs, and is not dependent on AlGaAs gate insulator thickness. As a result, the threshold voltage deviation obtained

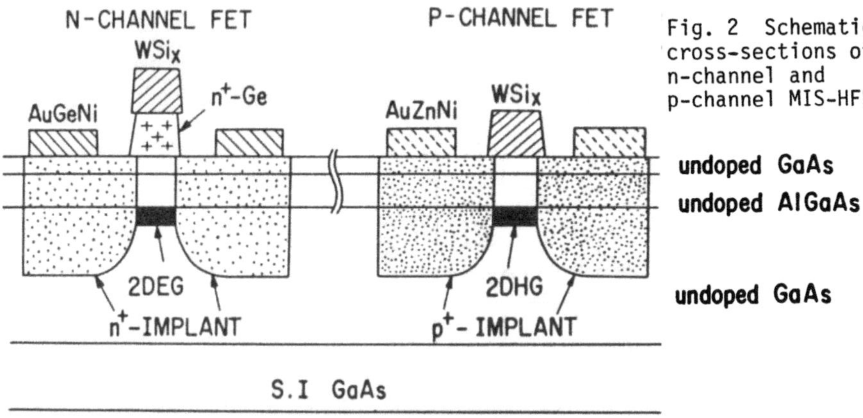

Fig. 2 Schematic cross-sections of n-channel and p-channel MIS-HFETs

was very small; standard deviation in a quarter chip of a 2 inch wafer was about 10 mV. Threshold voltage shift between room temperature and 77 K was less than 30 mV, much smaller than those of MODFETs. The mean value of the threshold voltages observed was 0.25 V. This value is slightly higher than the expected value of 0.1 V. It is attributed to a thin p-type layer formed at the Ge/GaAs interface during activation annealing for implanted layers.

Reduction of AlGaAs gate insulator thickness was studied based on transconductance by MAEZAWA et al. /14/. As expected, transconductance increased to 430 mS/mm on reducing the thickness down to 10 nm. However, further decrease of the thickness down to 5 nm resulted in a decrease in transconductance because of significant increase in the gate leakage current. As the permissible aspect ratio of gate length/thickness is about 4, an extremely short gate length of 0.04 μm can be realized in terms of gate insulator scaling-down.

2.4 Complementary Circuit with MIS-HFETs

Using MISHFETs, it is very easy to integrate n-channel and p-channel FETs on the same wafers. This is because no doping control is necessary for threshold voltage control. Complementary circuits, which included Ge/WSi gate n-channel and WSi gate p-channel MIS-HFETs, were demonstrated by MIZUTANI et al. /15/. Complementary technology is attractive for VLSI applications because of low-power dissipation. The schematic cross-section of these complementary MIS-HFETs is shown in Fig. 2.

The complementary circuit was fabricated as follows. First, an n^+-Ge (200 nm)/undoped GaAs cap (5 nm)/undoped AlGaAs insulator (30 nm)/undoped GaAs (300 nm) heterostructure is grown by MBE. Next, Ge layer is removed except for the regions of n-channel HFETs by RIE, followed by mesa-etching for isolation. After this, WSi gate electrodes are formed by sputtered deposition and patterning by RIE. Source and drain n^+ and p^+ layers are formed by self-aligned ion implantations of Si for n-channel and Be for p-channel HFETs, followed by rapid thermal annealing. After alloyed ohmic contact formation by AuGeNi for n-channel and AuZnNi for p-channel HFETs, interconnection TiAu lines are formed by lift-off. The gate length/width is 1.5 μm/ 20 μm for n-channel and 1.5 μm/40 μm for p-channel HFETs.

The threshold voltages obtained were 0.1 and -0.55 V for n and p-channel devices. The transconductances measured at room temperature were 160 and 30

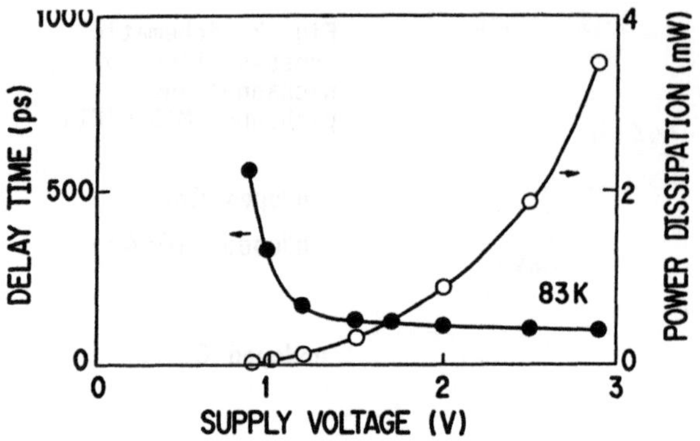

Fig. 3 Switching performance of a complementary MIS-HFET ring oscillator measured at 83 K

mS/mm for n and p-channel HFETs, respectively. A 15 stage ring oscillator was successfully operated with a minimum delay time of 125 ps/gate at 300 K and 100 ps/gate at 83 K. The supply voltage dependence of delay time and power dissipation obtained at 83 K is shown in Fig. 3.

3. HETEROJUNCTION BIPOLAR TRANSISTOR (HBT)

As mentioned in the first section, HBTs have potentially high switching performance which is basically due to high transconductance inherent to bipolar devices and to very high cutoff frequency. Deviation of threshold voltage in HBT is expected to be very small in principle. A standard deviation of threshold voltage smaller than 0.8 mV has been confirmed in large area transistors /16/. This uniformity is of great advantage over GaAs MESFETs. However, there are still many problems in realizing expected characteristics. The problems are mostly associated with material properties of III-V semiconductors, such as short recombination time, high generation-recombination current and poor processing flexibility. Recent progress in HBT fabrication is reviewed next.

3.1 MBE Growth of AlGaAs/GaAs HBTs

An AlGaAs/GaAs HBT wafer is usually prepared by molecular beam epitaxy (MBE). Sufficiently high current gain has been obtained for a precisely controlled and heavily doped base layer with up to about 5 x $10^{19}/cm^3$ hole concentration. In order to achieve a high doping level, relatively low substrate temperatures during the growth($\sim$600 °C) are required for suppressing Be diffusion.

A troublesome process in MBE growth has been the affects of mounting the GaAs substrate on a Mo block with indium solder. Treatment of the wafer after MBE growth and removal from the block generated wafer contamination which is undesirable for the IC process. Consequently, an Indium-free wafer holder has been developed and has already been applied to HBT growth /17,18/. Figure 4 shows the distribution of HBT current gain grown with such a holder, where current gains as high as 80 are obtained with $4x10^{19}/cm^3$ base doping /18/. Very uniform current gain over a 2-inch wafer indicates good substrate temperature uniformity during the growth.

3.2 Current Gains in HBTs

Since an emitter efficiency is very high in HBTs, the current gain is dominated by the base transport factor or g-r current of the emitter/base junction. We must, therefore, give attention to carrier lifetime, which

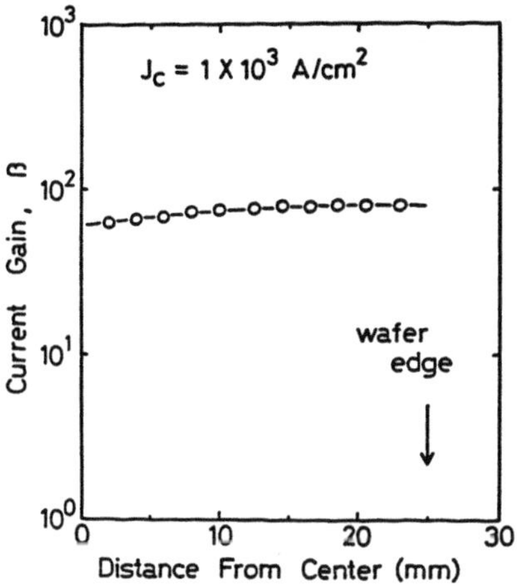

Fig. 4 Current gain distribution of HBTs grown with Indium-free holder

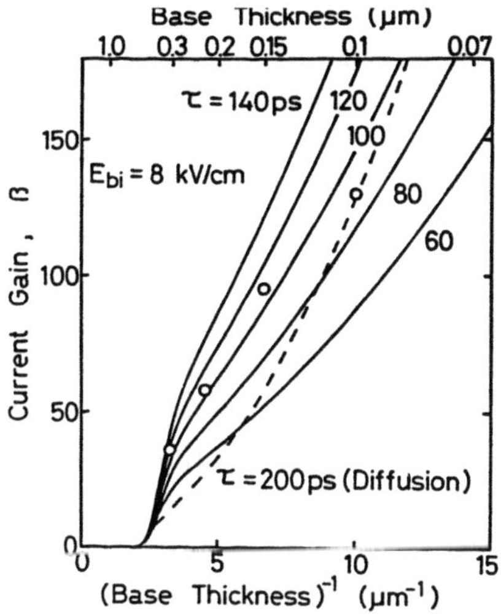

Fig. 5 Current gain vs. base thickness in graded base HBTs: Circles are experimental results, solid lines calculated results with various lifetimes assumed, and broken line a calculated result by diffusion model

depends on doping and crystal quality. Figure 5 shows the dependence of current gain on base thickness from 0.1 to 0.3 μm in graded base HBTs, where the built-in field in the base is kept constant at 8 kV/cm. In spite of Al fractional change from 7 to 20 %, the current gain variation can be explained by a constant electron lifetime, 100 ps, indicating good crystal quality in the AlGaAs base /19/.

It must also be noted that the graded base improves the current gain /19, 20/. A higher base doping or larger base thickness than those for a uniform base are adopted, relaxing the tradeoff between base resistance and current gain.

The graded base structure has successfully alleviated the current gain reduction observed in uniform base HBTs with size scaling down, as shown in Fig. 6 /21/. This is because the vertical built-in field in the graded base suppresses a base current lateral divergence and thereby reduces the excess recombination current at the base periphery.

3.3 Device Fabrication and Performance

Even in HBTs, an appropriate device scale-down is still necessary for high speed operation. It is obvious in the scaling that conventional wet etching has a limitation in defining emitter/base and base/collector areas. Reactive ion etching (RIE) is, therefore, of importance in most process steps. Several self-alignment structures that can minimize parasitic elements have also been proposed /22, 23, 24/. At present, a reduction of collector capacitance is one of concern. Sidewall-Separated Base Electrode (SSBE) structure proposed by NAGATA et al. /22/ is very effective to reduce both collector capacitance and base resistance (Fig. 7). To form an SiO_2 sidewall, highly anisotropic reactive ion-beam etching (RIBE) for GaAs is used.

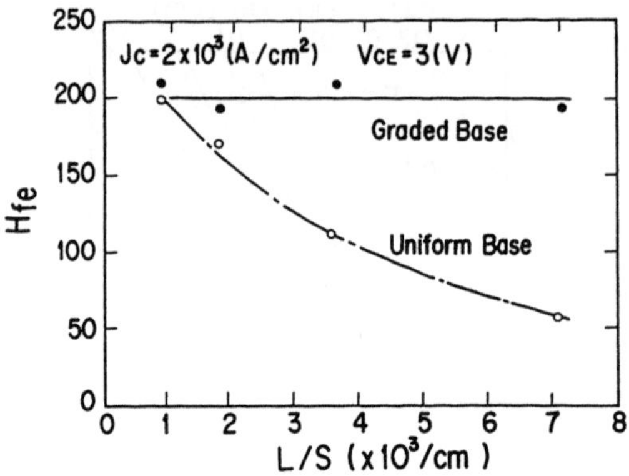

Fig. 6 Comparison of current gain dependences on emitter-base junction size between uniform and graded base HBTs; L/S = junction periphery length over junction area

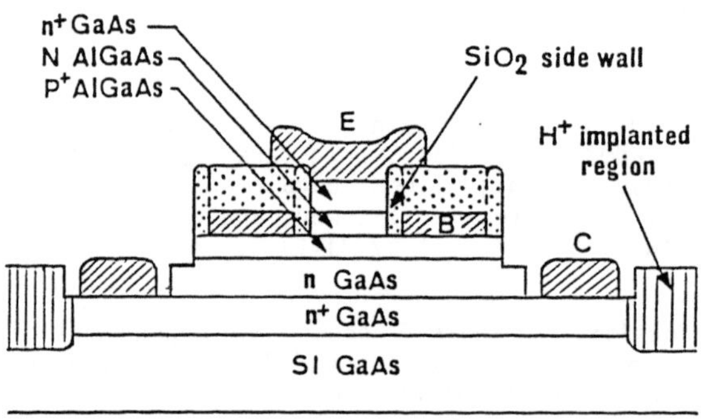

Fig. 7 Schematic cross-section of a sidewall separated base structure (SSBE) HBT

When RIBE is applied, crystal damage on GaAs surface induced by high energy ions must be carefully characterized. The degree of damage has been evaluated by contact resistance through tunneling current in a Schottky diode /25/. The contact resistance increases with ion accelerating voltage, where reactive gas discharge is generated by Electron Cyclotron Resonance reactor. The thickness of the damage layer is below 100 A at an acceleration voltage of 200 V. An anisotropic etching has also been confirmed. These properties of RIBE enable full-dry processing of HBTs.

A propagation delay time of 65 psec in ECL ring oscillator has been reported with devices fabricated by SSBE technique in 3 micron pattern rule. More recently, a delay time of 37 p/gate in CML circuit /24/ and of 22 p/gate in NTL circuit /26/ have been reported. By minimizing parasitics and optimizing a device structure, a propagation delay time below 10 p/gate can be inferred.

4. DOUBLE-HETEROSTRUCTURE LASER DIODE (LD)

Double-heterostructure made laser diodes CW operation at room temperature in 1970. Figure 8 shows "evolution" of the device, where the key concept of the improvement is on the basis of confinement of drive current, electron/hole densities, and a light field in domains of space and energy or wavelength. The material has been extented from GaAlAs/GaAs to InGaAsP/InP and InAlAs/InP, however, the guide line is still true. One fantastic feature of the device is that the lasing light can be easily modulated with

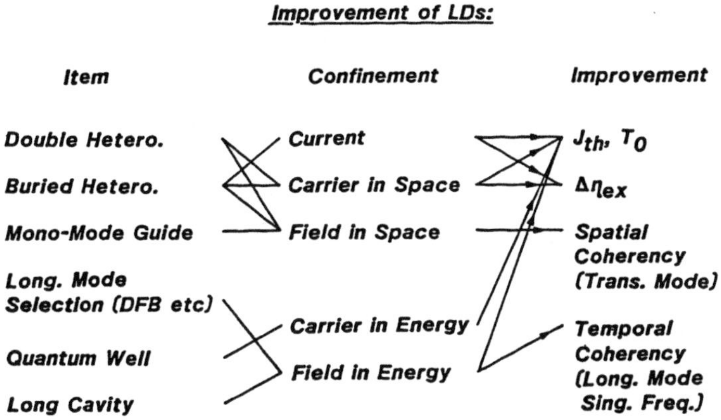

Fig. 8 Evolution of double-heterostructure laser diodes

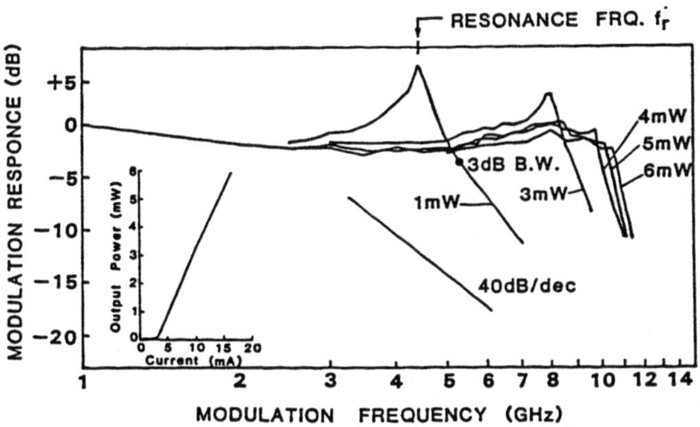

Fig. 9 Typical direct modulation characteristics of a laser diode

the modulated drive current, that is, the direct modulation. Figure 9 indicates the typical direct modulation characteristics /27/, where the modulation response has its peak at a frequency f_r, called resonant frequency. The modulation band width is closely related to f_r, which was expressed by

$$f_r = \frac{1}{2\pi}\sqrt{\frac{A\ p_o}{\tau_p}} \tag{2-1}$$

$$= \frac{1}{2\pi}\sqrt{\frac{A\ n_{th}}{\tau_s}\left(\frac{J_b}{J_{th}} - 1\right)}\ , \tag{2-2}$$

where A, p_o, τ_p, n_{th}, τ_s, J_b, and J_{th} are the differential optical gain constant, the photon density, the photon lifetime (order of ps), the carrier density at the threshold, the carrier lifetime (order of ns), the bias

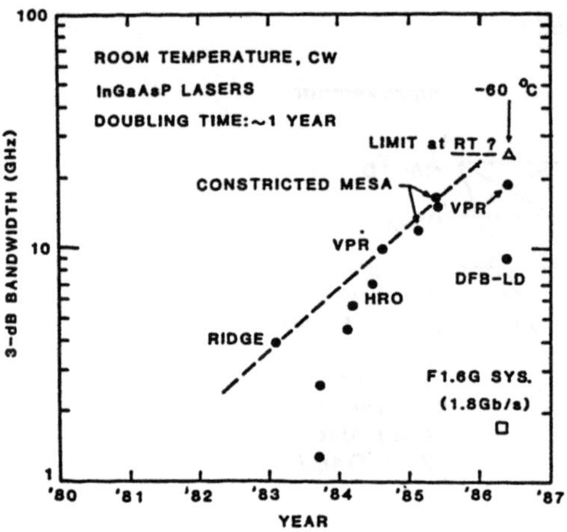

Fig. 10 Historical dependence of 3 dB bandwidths of laser diodes

current, and the threshold current, respectively. Note that the original analysis /28/ covers (2), however, (2-2) can be approximated by $1/\tau_p$ instead of $A\ n_{th}$ at very low temperature where the device operation is like that of four-level lasers (The laser diode could operate only at very low temperature at that time).

Figure 10 shows data of the 3 dB bandwidth of laser diodes with year /29/ and that of Distributed Feedback (DFB) lasers. The improvement has been mainly performed by reducing parasitic capacitance in the device with mesa structures, vapor phase regrown (VPR) technique /30/ and so forth.

The intrinsic bandwidth of the device is governed by (2) and it can be enhanced by controlling the parameters. For example, at low temperatures, A can be enhanced so that the bandwidth increases as shown in Fig. 10. Recently, some nonlinear effect in the device may limit the bandwidth of the device, however, it is under consideration at present /29/.

5. SUMMARY AND CONCLUSIONS

Heterostructure devices are promising for very high-speed integrated circuits, although some technical problems to be overcome remain. Some of them can be applied to ICs for practical use. MIS-HFETs are the most attractive for near future LSI applications because of their controllability in threshold and properties superior to GaAs MESFETs and MODFETs. HBTs are also attractive for future ultra high-speed ICs due to their excellent performance and inherent threshold uniformity. The direct modulation of laser diodes could be used up to several Gb/s in speed. However, an external modulator will be more realistic beyond that speed. High-speed heterostructure electron devices as well as optical devices have an important rule in future advanced information network system.

ACKNOWLEDGEMENTS

The authors are much indebted to K. Kurumada, T. Sugeta, and K. Hirata and their colleagues at NTT Electrical Communications Laboratories for valuable discussions.

REFERENCES

1. J. Drummond, W. T. Masselink, and H. Morkoc: Proc. IEEE 74, 773 (1986)
2. H. Kroemer: Proc. IEEE 70,13 (1982)
3. T. Takada, N. Kato, and M. Ida: IEEE Electron Device Lett. ED-7, 47 (1986)
4. N. J. Shah, S. S. Pei, C. W. Tu, and R. C. Tiberio: IEEE Trans. Electron Devices ED-33, 543 (1986)
5. T. J. Drummond, H. Morkoc, K. Lee, and M. S. Shur: IEEE Electron Device Lett. EDL-3, 338 (1982)
6. J. F. Rochette, P. Delescluse, M. Laviron, D. Delagebeaudeuf, F. Diamand, and J. Chevrier: Int. Symp. GaAs and Related Compounds, Inst. Phys. Conf. Ser. No. 65, 385 (1982)
7. T. J. Drummond, R. Fischer, W. Kopp, H. Morkoc, K. Lee, and M. S. Shur: IEEE Trans. Electron Devices, ED-30, 1806 (1983)
8. A. van der Ziel: Solid-State Electron. 26, 385 (1983)
9. T. J. Drummond, W. Kopp, D. Arnold, R. Fischer, H. Morkoc, L. P. Erickson, and P. W. Palmberg: Electron. Lett. 19, 986 (1983)
10. Y. Katayama, M. Morioka, Y. Sawada, K. Ueyanagi, T. Mishima, Y. Ono, T. Usagawa, and Y. Shiraki: Jpn. J. Appl. Phys. 23, L150 (1984)
11. K. Matsumoto, M. Ogura, T. Wada, N. Hashizume, T. Yao, and Y. Hayashi: Electron. Lett. 20, 462 (1984)
12. P. M. Solomon, C. M. Knoedler, S. L. Wright: IEEE Electron Device Lett. EDL-5, 379 (1984)
13. K. Arai, T. Mizutani, F. Yanagawa: Int. Symp. GaAs and Related Compounds, Inst. Phys. Conf. Ser. No. 79, 631 (1985)
14. K. Maezawa, T. Mizutani, K. Arai, and F. Yanagawa: IEEE Electron Device Lett. EDL-7, 454 (1986)
15. T. Mizutani, S. Fujita, F. Yanagawa: Int. Symp. GaAs and Related Compounds, Inst. Phys. Conf. Ser. No 79, 733 (1985)
16. P. M. Asbeck, D. L. Miller, R. J. Anderson, R. N. Deming, R. T. Chen, C. A. Liechti, and F. H. Eisen: Tech. Dig. 1984 GaAs IC Symp., 100 (1984)
17. H. D. Shih, S .E. Matteson, W. V. McLevige, and H. T. Yuan: J. Vac. Sci. Technol. B3, 793 (1984)
18. H. Ito and T. Ishibashi: Jpn. J. Appl. Phys. 24, 1567 (1985)
19. H. Ito and T. Ishibashi: Inst. Phys. Conf. Ser. No.79, 607 (1985)
20. J. R. Hayes, F. Capasso, A. C. Gossard, R. J. Malik, and W. Wiegmann: Electron. Lett. 19, 410 (1983)
21. O. Nakajima, K. Nagata, H. Itoh, T. Ishibashi, and T. Sugeta: Jpn. J. Appl. Phys. 24, 1368 (1985)
22. K. Nagata, O. Nakajima, Y. Yamauchi, and T. Ishibashi: Int. Symp. GaAs and Related Compounds,Inst. Phys. Conf. Ser. No.79, 589 (1985)
23. M. F. Chang, P. M. Asbeck, D. L. Miller, and K. C. Wang: IEEE Electron Device Lett. EDL-7, 8 (1986)
24. K. Morizuka, K. Tsuda, T. Kobayashi, and M. Azuma: presented at the 44th Device Res. Conf. (1986)
25. K. Nagata, O. Nakajima, and T. Ishibashi: Jpn. J. Appl. Phys. 25, L510 (1986)
26. G. J. Sullivian, P. M. Asbeck, M. F. Chang, D. L. Miller, and K. C. Wang: presented at the 44th Device Res. Conf. (1986)
27 K. Y. Lau and A. Yariv: IEEE J. Quant. Electron. QE-21, 121 (1985)
28 T. Ikegami and Y. Suematu: IEEE J. Quant. Electron. QE-4, 148 (1968)
29 J. E. Bowers and C. A. Burrus: Conf. on Lasers and Electro-Optics, THH1 (1986)
30 R. Olshansky and C. B. Su: Conf. on Optical Fiber Comm., WBB1 (1986)

Hot-Carrier-Excited Two-Dimensional Plasmon in Selectively Doped AlGaAs/GaAs Heterointerface Under High Electric Field Application

T. Kobayashi, N. Okisu, and Y. Sambe

Faculty of Engineering Science, Osaka University, Toyonaka, Osaka 560, Japan

The present paper describes the first observation of the 2D-plasmon in AlGaAs/GaAs heterointerface under the high field application. The FIR emission from the grating-coupled 2D-plasmon grew up to 30[$\mu W/cm^2$] for the applied field of 760[V/cm]. The other related subjects such as the real-space transfer and 2D electron temperature are also demonstrated.

1. Introduction

Along the two-dimensional (2D) electron stream, the 2D-plasmon is excited via the energy relaxation of the hot electrons [1,2]. The plasmon in solids, an elemental excitation, is a self-organized electron motion and serves as a scatterer of the accelerated electrons. According to LUGLI and FERRY's Monte-Carlo simulation, the plasmon in 3D system takes a large part in détermination of the mean-free-path of the injected hot electron [3]. It might be true for the 2D-plasmon in the Modulation-Doping (or Selective doping) system.

Apart from the electron transport, the grating coupled 2D-plasmon at and near the semiconductor surface is a promising candidate for a tunable solid-state FIR emitter. The plasmon emitter serves not only a FIR source not easily obtainable but also a powerful tool to directly measure the excitation of the 2D-plasmon.

Until now, the experiments have been carried out with Si MOS structure, where 2D electrons were induced under the applied gate bias and the plasmon photon came out attenuatively through the thin metal gate, resulting in the FIR emission power as low as 10^{-8}[W/cm^2]. The second point to be noticed is an establishment of higher electron temperature T_e which can grow up the plasmon excitation. Experiments, however, had to be made in the low-field region to avoid the pinch-off effects as long as the MOS structures were used. In the present work, we report the first observation of the plasmon FIR emission and 2D electron temperature under the high electric field application by employing AlGaAs/GaAs selectively doped heterostructures [4,5].

2. Experimental

The FIR emitters were fabricated on MBE grown heterostructure wafers, the details of which are summarized in Tab. I. The donor impurity in n^+-AlGaAs layer was Si. The emitter structure is shown in Fig. 1. The Au-Ge alloy, rather than AuGeNi, provided better ohmic contact to n^+-AlGaAs layer even at the cryogenic ambient temperatures. The metal gratings placed on the top of the epitaxial wafers and between the two electrodes were designed according to the modified dispersion relation [6] given by us so

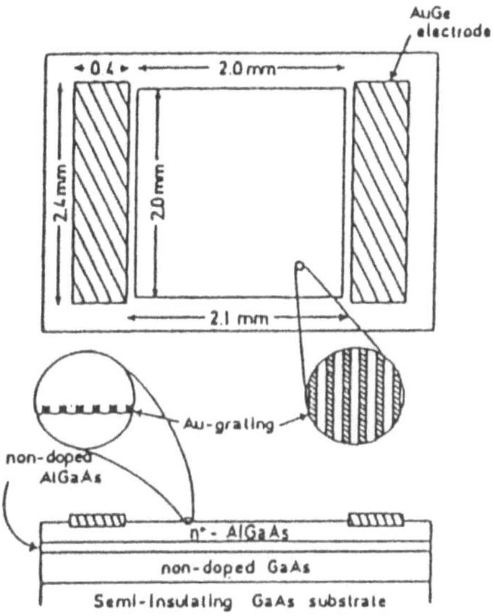

Fig.1 Schematic view of plasmon FIR emitter.

as to match the resonant plasmon FIR frequency with the highly pure ($7x10^{14}$[/cm^3]) GaAs photodetector detection peak (35.5[cm^{-1}]). Prior to the metal-grating formation, thermal FIR measurements were made on all-bare emitters to obtain the electron temperature T_e vs the applied electric field E relations which facilitate our exploring on the plasmon excitation mechanism. After this, Au thin films (1000[Å]) were deposited onto the AlGaAs surface, and the gratings were patterned with the conventional photolithography and wet etching techniques.

Throughout the present work, the emphasis is placed on the 2D plasmon excitation under the high electric field. The hot electron behavior in 3D system has been satisfactorily investigated from various aspects. Whereas, the knowledge on hot 2D electrons is still less. This section, therefore, seeks to demonstrate how and to what extent the 2D electron temperature is increased by the high-field application, and how the 2D electron stream escapes from the real-space transfer effect.

The electron temperatures T_c measured from the thermal FIR power are plotted in Fig. 2 as a function of the input power per an electron $e\mu_0E^2$, where μ_0 is the field-dependent d.c. mobility. The derivation of T_e can be seen in Ref. 7. This figure clearly illustrates a change of the dominant scattering mechanism as the field intensity increases: Up to T_e=100[K], the acoustic phonon scattering determines the 2D electron transport and $T_e^2 \propto e\mu_0E^2$. With further increase in E, the observed T_e falls on the curve calculated on the dominant polar-optical phonon scattering model. As the field intensity increases from 3 to 10 [kV/cm], T_e curve exhibits a steep rise like an occurrence of the polar-runaway and no sign of intervalley-scattering can be found. Anyway, the 2D electron temperature T_e higher than 3000[K] can be established under the high-field application.

A possible mechanism which deteriorates the stabilized plasmon FIR emission is the real-space transfer (RST) of hot electrons. A systematic study has been carried to check how strongly the RST effect depends on the energy-band diagram of the heterostructure. In Fig. 3, the pulsed I-E

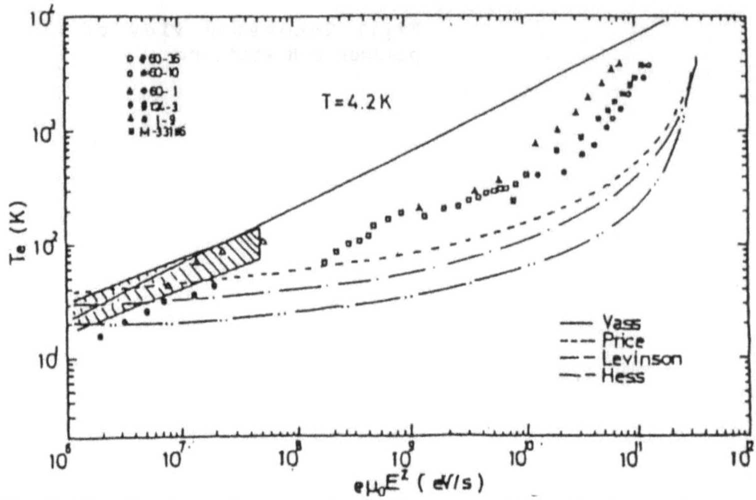

Fig.2 2D electron temperature T_e vs $e\mu_oE^2$. Test samples are with x=0.27 0.3.

Table I. Device parameters.

Sample type	Layer structure							Surface electron density	Mobility
	Non-doped	Buffer layer		Doped layer		Cap layer			
	GaAs	$Al_xGa_{1-x}As$		$Al_xGa_{1-x}As$		$Al_xGa_{1-x}As$			
	(μm)	x,	(Å)	x,	(Å)	x,	(Å)	n_s (cm^{-2})	(cm^2/Vs)
#60	1.4	0.28	80	0.28	1700		0	7.2×10^{11}	58500
#124	1.4	0.28	70	0.28	350	0.28	5000	1.4×10^{12}	8700
#1	1.5	0.27	63	0.27	2400		0	5.1×10^{11}	80000
M-329	1.1	0.29	100	0.29	2000		0	7.3×10^{11}	96500
M-331	1.1	0.30	75	0.30	2000		0	6.8×10^{11}	76100
M-337	1.2	0.53	50	0.53	2000		0	4.7×10^{11}	27100

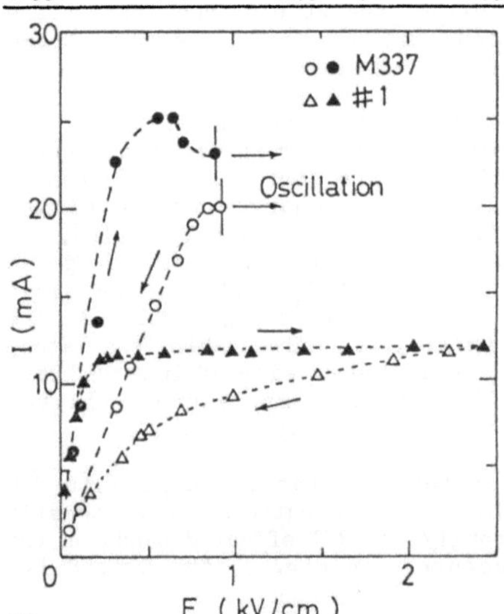

Fig.3 Current-field curve.

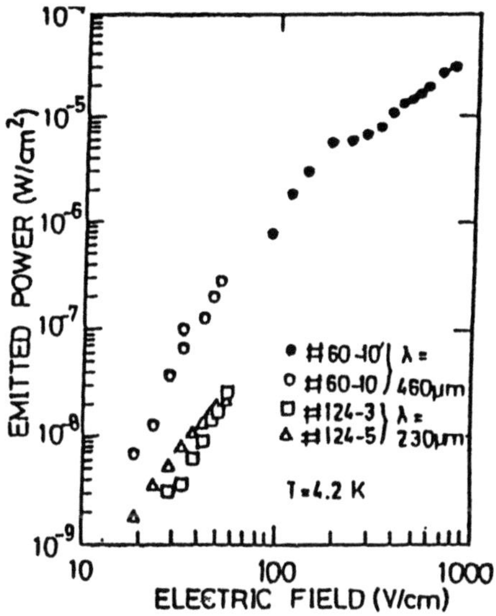

Fig.4 Plasmon FIR power vs E.

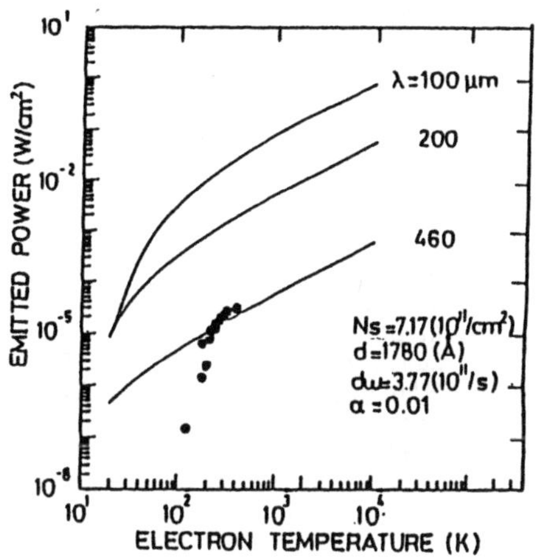

Fig.5 Calculated plasmon FIR power. Modified 2D plasmon dispersion relation was used in the calculation. Experimental results are plotted by solid circles.

curves measured on #337(x=0.53) and #1(x=0.27) are drawn, showing that the sample #337 suffered from more intensive RST effect. No oscillatory behavior was seen in the sample #1 even when the field was increased to 10[kV/cm]. Though data are not given here, the I-V collapse was found to depend on the thickness of n^+-AlGaAs(x=0.27) layer. The result strongly suggests that the adoption of n^+-AlGaAs layer thicker than 1500[Å] is of great favor to stabilize the high-field operation.

The FIR emission in the high-field region is of much interest and the attempt was done in finding the high field 2D plasmon. Figure 4 gives the field dependence of the plasmon FIR emission measured under the applied field ranging from 20 to 760 [V/cm], where the RST effect (reduction of the 2D electron concentration) was almost completely ignored. The highest emission power obtained so far was in excess of 30[μW/cm^2] under the applied field of 760[V/cm] (T_e=400[K]) at the wave length of 460[μm]. This value is more than three orders of magnitude higher than those ever reported and equivalent to emitted photon number density of 7×10^{16} [1/s/cm^2], which is comparable to the conventional LEDs in IR and visible range. Some data of Fig. 4 are reproduced in Fig. 5 with the theoretical curves as a function of the electron temperature T_e, where the grating antenna coupling efficiency is determined to be 0.01 from the curve fitting.

The experimental data agree with the calculations semiquantitatively. The fact that T_e in the heterostructures raised up to 4000[K] when the applied field increased to 3-10[kV/cm] together with calculations given in Fig. 5 strongly suggest that the FIR emission power necessary for the practical use, order of milli-watts, is promised with our radiator and, besides, it will be able to work even at room temperature.

3. Conclusion

The 2D-plasmon at AlGaAs/GaAs heterointerface under the high-field application was investigated from measurements of the plasmon FIR emission. From the structural improvement, the RST effect was avoided to a large extent, leading to the stabilized 2D-plasmon excitation.

The authors would like to thank T. Sakurai and T. Tomita of Sharp Corporation for providing MBE grown epitaxial wafers. They also thank T. Shimada for measurement.

References

1. R. A. Höpfel and E. Gornik, Surf. Sci., **142**, (1984) 412.
2. R. A. Höpfel, E. Vass and E. Gornik, Phys. Rev. Lett., **49**,(1982)1667.
3. P. Lugli and D.K. Ferry, IEEE Electron Device Lett., EDL-6,(1985)25.
4. R. Höpfel, G. Lindemann, E. Gornik, G. Stangle, A. C. Gossard and W. Wiegmann, Surf. Sci.,**113**,(1982) 118.
5. Y. Sambe, N. Okisu, K. Suehiro, T. Kobayashi, K. Fujisawa, T. Tomita and T. Sakurai, Extended Abstracts of the 17th Conference on Solid State Devices and Materials, Tokyo, (1985) 95.
6. N. Okisu, Y. Sambe and T. Kobayashi, Appl. Phys. Lett.,**48**,(1986) 776.
7. R. A. Höpfel, E. Vass and E. Gornik, Solid State Commun.,**49**,(1984)501.

Optical High-Field-Transport Experiments in GaAs Quantum Wells

*J. Shah and R.A. Höpfel**

AT & T Bell Laboratories, Holmdel, NJ 07733, USA

We review three recent experiments in which we have used optical techniques to investigate quantities relevant to high-field transport in semiconductors. We will discuss (1) a measurement of the drift velocity of minority carriers in GaAs quantum wells by a time-of-flight technique, (2) a simultaneous measurement of the luminescence spectra which allows us to determine the carrier distribution function and investigate the energy exchange rate between electrons and holes, and (3) an optical imaging experiment which demonstrates the existence of negative absolute mobility for the minority carriers in modulation-doped quantum wells. The momentum scattering rate between electrons and holes is determined from this experiment.

I. DRIFT VELOCITY BY TIME-OF-FLIGHT TECHNIQUE

Time-of-flight techniques provide a direct means of determining carrier drift velocities in semiconductors [1]. We have determined the drift velocity of minority electrons in p-modulation-doped quantum wells as a function of electric field up to 14 kV/cm [2-3]. The presence of majority holes prevents the measurement problems related to space charge effects. Minority electrons were injected with a 5 psec pulse from a synchronously pumped, cavity-dumped dye laser operating at 6,060Å. High fields were applied parallel to the layers of the quantum wells by ohmic p-type contacts on the multiquantum well sample. Contact separation was typically 50 μm and the laser was focussed to 2 μm width. The sample was mounted on a microwave stripline to permit fast measurements (system response < 50 psec). The photoinjected minority electrons gave essentially a square-topped photocurrent pulse (insert of Figure 1(a)) from which the electron drift velocity was determined directly.

The drift velocity of electrons determined in this manner is plotted in Figure 1(a)). The most interesting feature of the data is that the drift velocity reaches a peak (1.1×10^7 *cm/sec*) and then decreases; i.e., it shows a distinct negative differential mobility. Secondly, the low field drift mobility is approximately 1500 cm^2/Vs and is much smaller than the typical values of >7000 cm^2/Vs observed for electrons in *n*-modulation doped quantum wells. We attribute the lower value in our sample to momentum scattering between photoexcited electrons and the majority holes.

II. SIMULTANEOUS LUMINESCENCE MEASUREMENTS: ELECTRON-HOLE ENERGY EXCHANGE RATE

The observation of negative differential mobility immediately indicates that the electrons are getting sufficiently hot to transfer to either the barrier layers or the higher conduction band valleys. We have made a simultaneous measurement of the luminescence spectra

---

* Permanent address: Institut für Experimentalphysik, Universität Innsbruck, A-6020 Innsbruck, Austria.

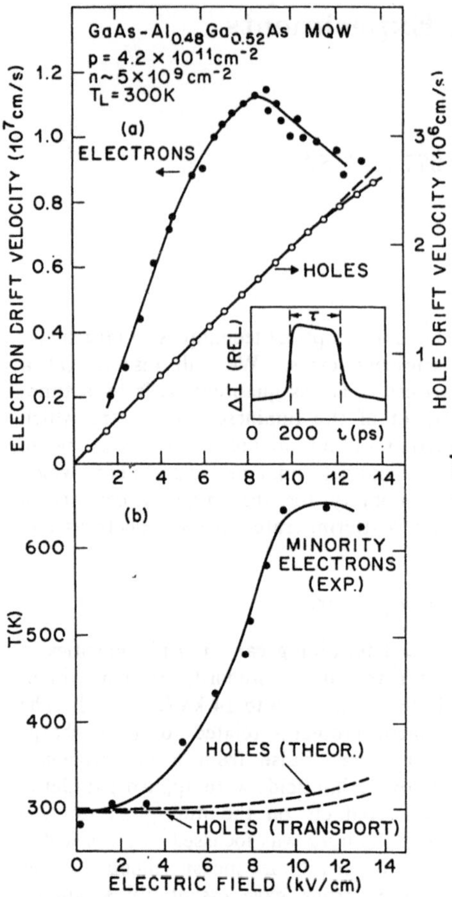

Fig. 1 (a) Electron drift velocity in p-GaAs quantum wells measured by picosecond time-of-flight technique and hole drift velocity from I-V measurements. The inset shows the photocurrent pulse from which drift velocity is determined. (b) Electron temperature as a function of applied electric field deduced from luminescence measurements and hole temperatures as discussed in text.

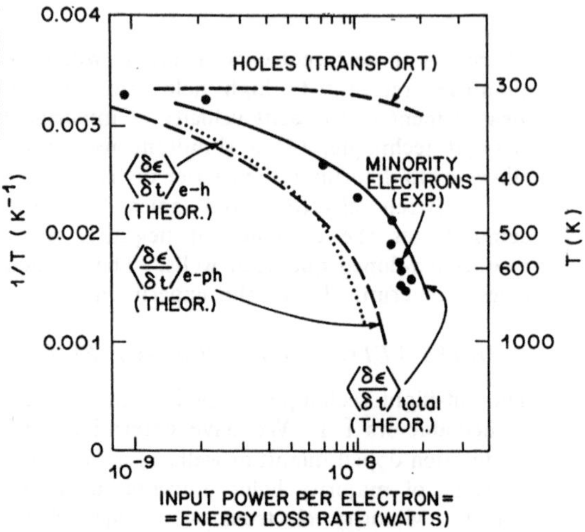

Fig. 2 Inverse of electron temperature vs. electron energy loss rate; circles are experimental and curves are calculated for different scattering mechanisms.

under the influence of applied electric field. The spectra indicate that the distribution function of the hot electrons can be described by a temperature over a range of more than 100 meV near the bandgap. Therefore we discuss the results in terms of a temperature model. We find [3] that the electron temperature increases with electric field (Figure 1b) to a value of about 650K and remains approximately constant in the region of negative differential mobility. We have argued that this temperature is too low for transfer to higher valleys in GaAs and that the observed negative differential mobility arises as a result of real-space transfer to the barrier layers [3].

Figure 1b also shows the hole temperature estimated on the basis of the measured I-V [5] curve as well as from the known [4] hole energy loss rates to the lattice. We see that the holes are at a much lower temperature than the electrons. Since the electron and hole energy loss rates to the lattice are different [4], one expects to find different temperatures for electrons and holes in the absence of electron-hole scattering. Electron-hole scattering will tend to reduce this difference between the two temperatures. When the two temperatures are different, we expect a net flow of energy from electrons to the holes. We plot in Figure 2 the inverse of the electron temperature as a function of the total power loss of electrons (which, in this steady state conditions is equal to the power input from the electric field to the electrons). We compare [5] this with the expected energy loss rate from electrons into the lattice (LO phonons) and find that (Figure 2) the measured energy loss rate is larger than that expected by about a factor of 2. We attribute this difference to energy exchange between electrons and holes. This is supported by a calculation of energy exchange rate between electrons and holes from a simple gas plasma model [5]. While more elaborate calculations more appropriate to the semiconductor case are needed, these measurements provide the first estimate of electron-hole energy exchange rates.

III. NEGATIVE ABSOLUTE MOBILITY OF MINORITY CARRIERS

We describe in this section an unusual manifestation of momentum scattering between electrons and holes in a semiconductor [6]. Electric field was applied parallel to the plane of the layers of a p-modulation doped multiple quantum well sample. Minority electrons were photoexcited with a tightly focussed (diameter 3 micron) picosecond laser (same as in Section I). The sample was mounted in variable temperature cryostat (lowest temperature 15K). Since luminescence occurs only where the minority electrons are present, the motion of the electrons under the applied electric field can be monitored by measuring where the luminescence originates. We have done this by imaging the sample on a plane and scanning a narrow slit across this plane to measure the image of the time-integrated luminescence. Two typical scans are shown as the insert in Figure 3. At 15K, the luminescence shifts towards the negative electrode; i.e., the minority electrons have a negative absolute mobility at low temperatures! The negative drift decreases with increasing temperature and the drift becomes positive above about 90K as shown for the 150K curve in the insert. The negative drift velocity of minority carriers is determined from the luminescence image and the measured luminescence decay time.

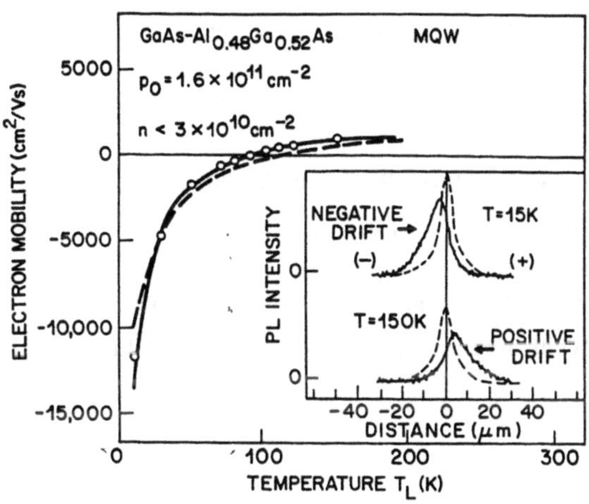

Fig. 3
Minority electron mobility as a function of lattice temperature; the points are experimental and the curve is a comparison with theory. The inset shows the luminescence images, showing negative drift at 15 K and positive drift at 150 K.

The negative drift of electrons is a consequence of the momentum scattering between electrons and holes via Coulomb interaction. The holes in our high mobility samples interact strongly with electrons and drag the electrons with them to the negative electrode. Such an effect was first predicted for the case of *InSb* by McLean and Paige [7] but to our knowledge has not been observed in any semiconductor system until now. We have used a relaxation time approximation to the Boltzmann equation to analyze this case and found that [6] under certain simplifying assumptions, the electron mobility is given by subtracting the hole mobility from the mobility of electrons due only to electron-hole scattering. The latter quantity can be approximated by mobility due to electron-ionized impurity scattering. In Figure 3 we have plotted the measured electron mobility as a function of the lattice temperature and compared it with the difference between electron-ionized impurity mobility (for appropriate impurity density) and the measured hole mobility in our sample. We see that the simple theory gives a good agreement with the observations. From these measurements, we deduce a momentum scattering time of approximately 50 to 100 fsec for electrons in a hole plasma of areal density 1.5×10^{11} per cm^2 per layer.

Recently, we have also observed negative absolute mobility for minority holes in *n*-modulation doped samples. These data will be presented elsewhere. [8]

In summary, time- and spatially resolved luminescence, as well as picosecond photoconductivity, in high electric fields provide excellent tools for studying carrier drift velocities and energy distribution functions in materials and systems relevant for high speed electronics. Using these techniques, we have investigated drift velocities, distribution functions and electron-hole (momentum and energy) scattering rates in GaAs quantum wells.

We acknowledge collaboration with D. Block, T. C. Damen, A. E. DiGiovanni, A. C. Gossard, P. A. Wolff and W. Wiegmann.

REFERENCES

[1] L. Reggiani, in "Physics of Nonlinear Transport in Semiconductors," edited by D. K. Ferry, J. R. Barker and C. Jacobani, Plenum Press; pp. 243-254 (1980).

[2] R. A. Höpfel, J. Shah, A. C. Gossard and W. Wiegmann, Physica 134B, 509-513 (1985).

[3] R. A. Höpfel, J. Shah, D. Block and A. C. Gossard, Appl. Phys. Lett. *48*, 148 (1986).

[4] J. Shah, A. Pinczuk, A. C. Gossard, and W. Wiegmann, Phys. Rev. Lett. *54* 2045-2048 (1985).

[5] R. A. Höpfel, J. Shah and A. C. Gossard, Phys. Rev. Lett. *56*, 765-768 (1986).

[6] R. A. Höpfel, J. Shah, P. A. Wolff and A. C. Gossard, Phys. Rev. Lett. *56*, 2736-2739 (1986).

[7] T. P. McLean, and E. G. S. Paige, J. Phys. Chem. Solids *16*, 220 (1960).

[8] R. A. Höpfel, J. Shah, P. A. Wolff, and A. C. Gossard, Appl. Phys. Lett., to be published (1986).

Optical Time-of-Flight Investigation in Ambipolar Carrier Transport in Specially Designed GaAs/GaAlAs Quantum Well Structures

H. Hillmer[1], *G. Mayer*[1], *A. Forchel*[1], *K.S. Löchner*[2], *and E. Bauser*[2]

[1]IV. Physikalisches Institut, Universität Stuttgart, D-7000 Stuttgart 80, Fed. Rep. of Germany
[2]Max-Planck-Institut für Festkörperforschung, D-7000 Stuttgart 80, Fed. Rep. of Germany

We present a new all-optical time-of-flight method capable of high spatial and temporal resolution. Transport properties are provided from measurements of the flight duration between two quantum wells sandwiching a semiconductor transport layer.

Today the high frequency properties of semiconductor devices are essential for fast computer and fast communication systems. The cutoff frequency of such devices is substantially determined by the transit times and velocities in all involved layers. The most direct approach to these transport properties is obtained by Time-of-Flight (TOF) experiments, in which one derives the carrier velocity for measurements of the flight duration over different known distances. Hence the results critically depend on the accuracy with which the transit time and distance can be measured. Recently, different optical techniques have been developed to increase the time resolution in relation to electrical TOF studies. In particular, carrier-induced reflectivity changes [1] and nonlinear luminescence [2] have been used to probe the carrier movement. The penetration depth of the light and the focus diameter, respectively, limit the spatial resolution of these experiments.

The spatial resolution of our new optical TOF-method is $\leq$ 100 μm. Since, in principle, this method is also capable of high temporal resolution, it seems very promising for transport studies on ultra short distances. Using two quantum wells (QWs) which define the transport distance we get this high spatial resolution. Figure 1 shows the energy gap variation in our LPE grown GaAs/$Ga_{1-x}Al_xAs$ structures. The two QWs (QW1, QW2) have different well widths and therefore different characteristic transition energies, giving good distinction between the optical start (QW1) and stop probe (QW2). The samples are mounted in a temperature controlled cryostat (1.8K-300K).

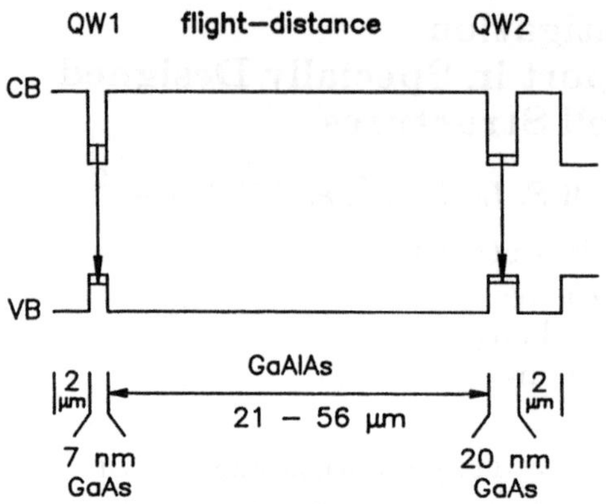

Fig.1

Design of the TOF samples in the GaAs/$Ga_{1-x}Al_xAs$ system. The transport layer is situated between QW1 and QW2

Carriers are generated by short mode-locked laser pulses (Ar, width 150 ps, λ=514 nm, Nd-YAG, 300 ps, λ=532 nm and 1064 nm). The emitted luminescence light is spectrally dispersed and detected by a fast micro-channel-plate photomultiplier (time resolution=50 ps). By short laser pulses a hot ambipolar plasma is generated in the outer GaAlAs layer (absorption layer, in fig. 1 at the left side). Due to nonequilibrium effects (e.g. concentration and temperature gradients) the ambipolar plasma expands and reaches QW1. There about 20% of the carriers are captured [3] and after a fast relaxation [4,5] they populate the lowest subband of the quantum wells. The time-resolved recombination luminescence indicates the arrival of carriers and serves as a start mark. The remaining carrier pairs travel through the GaAlAs-layer (flight distance) and finally reach QW2 where again about 20% of the carriers relax to the deepest bound state (this recombination luminescence is the stop signal).

TOF results of a sample with 21.8 μm flight distance at 10K are shown in fig.2. The time-integrated photoluminescence spectrum (inset) is dominated by the QW1-QW2 emission. The maximum of the QW2 emission is significantly delayed compared with QW1, which nearly follows the laser pulse. An immediate result is the average carrier velocity obtained by the flight distance divided by the time delay between the two emissions (fig. 2: $v=3\times10^6$ cm/s). Another transport property, the diffusivity, can also be obtained directly from the broadening of the profiles during the flight. Using the formula for a gaussian broadening of a δ-profile and the spatial profiles calculated from fig.2 using v, we can estimate D=2300 cm^2/s for the diffusivity which broadens a FWHM from 15 to 45 μm during 0.7 ns.

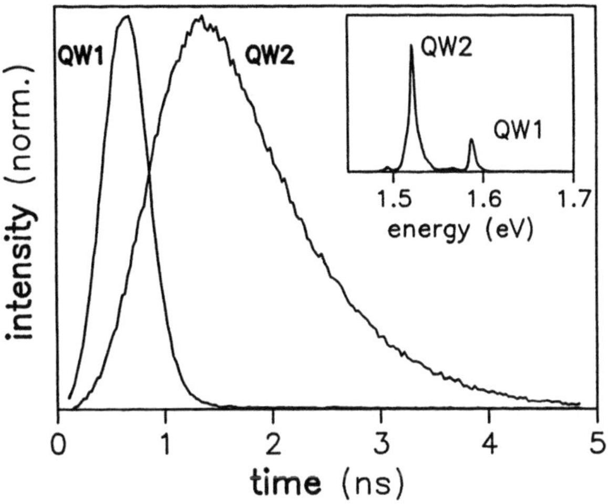

Fig.2
Time-resolved luminescence of the start (QW1) and stop layer (QW2) for a transport distance of 21.8 μm thickness at 10 K. Inset: time integrated luminescence

We have developed a model calculation which accounts for the entire profiles of the time-resolved QW-signals. First we have to make a small correction of the profiles: because the relaxation- and life-times of the QWs will not be infinitely fast, the luminescence profiles (dots in fig.3) will be broadened compared with the concentration profiles (circled) by an effective time constant τ_{eff}. To obtain this τ_{eff} we assume that no transport broadening of the carrier cloud arriving at QW1 has taken place and so the concentration is given mainly by the time dependence of the laser-pulse. So we have to unfold the QW1-profile till it is equivalent to the laser pulse. We have done this by a Fourier transformation which yields a τ_{eff} of 75 ps for the shown example. We obtain the concentration at QW2 unfolding the luminescence profile with the same τ_{eff}. The unfolded luminescence profile is then compared with the results of the equation of continuity.

$$\frac{\partial c}{\partial t} = D \cdot \frac{\partial^2 c}{\partial x^2} - v \cdot \frac{\partial c}{\partial x} \tag{1}$$

The first term on the rhs describes the diffusion of the carriers while the second, which is introduced phenomenologically, accounts for an additional expansion due to nonequilibrium effects. Because the transit time is much shorter than the life-time in the transport-layer, we need no recombination term. The boundary conditions are given by the unfolded QW-profiles.

The full line in fig. 3 displays the best fit with the fit parameters $v=2.1\cdot10^6$cm/s and $D=1810$ cm^2/s. These values are rather

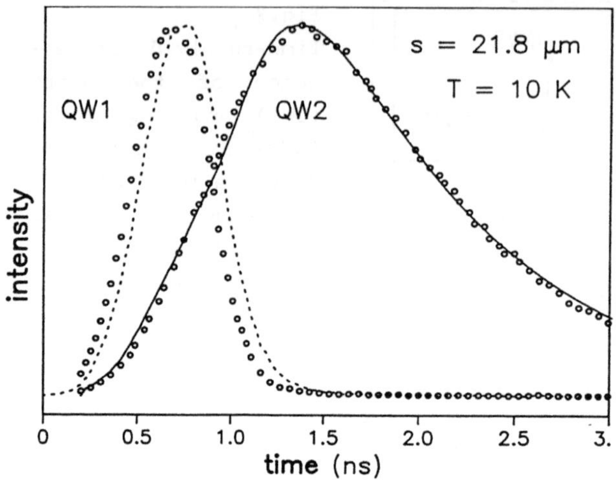

Fig.3
Time-resolved concentration profiles at the QWs (dots).
The dashed curve displays the QW1 emission.
The solid curve represents the best line shape fit of QW2 with the fit parameters $v = 2.1 \cdot 10^6$ cm/s and $D = 1810$ cm^2/s

similar to the above estimates. Using the Einstein relation we calculate from the diffusivity value the mobility $\mu(T)=D(T)\cdot e/kT$ for the ambipolar system: $\mu(10K)=2.1\cdot 10^6$ cm^2/Vs.

Altering the setup slightly, the method demonstrated here for ambipolar transport, can also be applied for unipolar transport. An electric field can be maintained by positioning semitransparent electrodes near and parallel to the QW layers. Illuminating the electrode near QW1, either electrons or holes are accelerated towards QW2 depending on the polarity of the applied bias voltage. A suitable doping inside the QWs or in their vicinity provides the opposite charges, necessary for the radiative recombination of the arriving carrier type.

We thank M. Pilkuhn for stimulating discussions and the Stiftung Volkswagenwerk for financial support.

1. B.F. Levine, W.T. Tsang, C.G. Bethea, F. Capasso: Appl. Phys. Lett. 41 470 (1982)
2. R.A. Höpfel, J. Shah, A.C. Gossard, W. Wiegmann: Proc. 4th Int. Conf. on Hot Electrons in Semic., Innsbruck 1985, Physica 134, Dec. (1985)
3. E.O. Göbel, H. Jung, J. Kuhl, K. Ploog: Phys. Rev. Lett. 51, 1588 (1983)
4. C.V. Shank, R.L. Fork, R. Yen, J. Shah: Sol. State Comm. 47, 981 (1983)
5. A.J. Taylor, D.J. Erskine, C.I. Tang: J. Opt. Soc. Am. B 2, 663 (1985)

An Ultra-Fast Optical Modulator: The Double-Well GaAs/GaAlAs Superlattice (DWSL)

B. Deveaud[1], *A. Chomette*[1], *A. Regreny*[1], *J.L. Oudar*[2], *D. Hulin*[3], *and A. Antonetti*[3]

[1]Centre National d'Etudes des Télécommunications (LAB/ICM), F-22301 Lannion, France
[2]Centre National d'Etudes des Télécommunications (PAB/BAG), F-92220 Bagneux, France
[3]Laboratoire d'Optique Appliquée, ENSTA – Ecole Polytechnique, Centre de l'Yvette, F-91120 Palaiseau, France

ABSTRACT :

A new kind of superlattice system is proposed where wells of two different widths are closely associated through thin barriers.The narrow wells are used as the optically active part of the device where the excitonic absorption can be bleached very rapidly by an external light source. The wider wells trap the excess carriers in a very short time so that the absorption is quickly restored. All optical switches with a time response shorter than 1 ps can be made with such a system.

Multi-quantum-well (MQW) systems have demonstrated their very interesting optical non-linear properties [1,2]. The main characteristic of these systems, as far as optical non-linearities are concerned, is that excitonic resonances persist even at room temperature. This allows one to obtain strong and very fast changes in the optical transmission of the system upon excitation with an external power source, which can be a solid-state laser [1]. One of the possible applications of this property is the passive mode locking of a solidstate laser [3]. An alternative use might be an ultra-fast optical NOR gate [4].

Optical bleaching of the exciton resonance can be obtained in less than 1 ps with an optical pulse creating electron-hole pairs via the screening of the exciton interaction. Alternatively one can use a light giving rise directly to the exciton. Return to the steady state is then obtained after recombination of a sufficient number of electron-hole pairs. This disappearance corresponds to the radiative recombination of the carriers and occurs with a characteristic time of 34 ns at 300 K [1]. This limits the possible uses of the MQWs. Reduction of the lifetime of carriers can be obtained through electron irradiation of the samples [5]. By creating a large number of defects, electron irradiation reduces the lifetime of the carriers. However a large dose also destroys the excitonic resonances which limits the maximum dose to 10^{13} cm^{-2} and the minimum lifetime to 100 ps. A further reduction of the lifetime has to be obtained in order to use fully the possibilities of MQWs.

By studying superlattices (SL) of small periods in which large wells were purposely introduced, we have been able to show that the mean trapping time of an electron hole pair from the superlattice into the large well can be very short [6]. If the barriers are thin enough, this time can be less

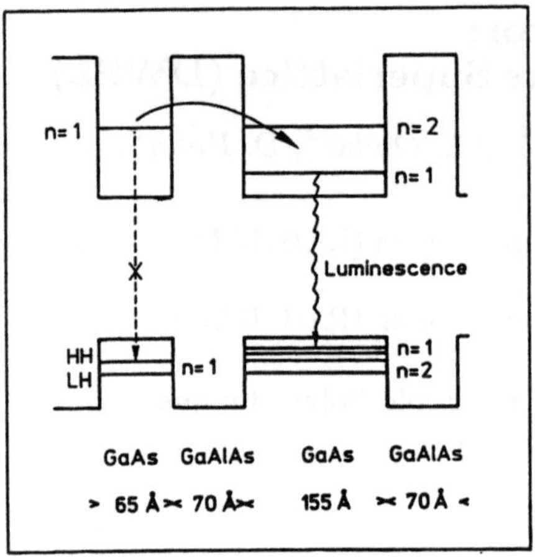

Fig. 1

than 10 ps for one well each 1400 A. We then propose a new kind of superlattice system called a double well superlattice (DWSL), the simplest case of which is shown in Fig. 1. The principle is simply to couple in the same SL two kinds of wells through thin barriers :

- narrow wells will give rise to the excitonic absorption used in the device,
- large wells will be used to trap the excess carriers so as to restore the excitonic resonance in the narrow wells.

The narrower the barriers, the faster the transfert to the large wells but also the larger the power necessary to commute the device. The respective widths of the two kinds of wells must be such that the energy difference between the first quantum level in each well is larger than one LO phonon (36 meV). Different samples of DWSL have been realized by MBE, one typical example being sample N° 422 :

$$L_{Z1} = 65\ \text{Å},\ L_{Z2} = 155\ \text{Å},\ L_b = 70\ \text{Å}$$

Luminescence and luminescence excitation experiments show that the luminescence is dominated by recombination in the large wells, the typical transfer time from narrow wells to large wells is of the order of 0.3 ps (this trapping time is further reduced below 0.1 ps when the barrier width is lowered). Excitonic resonances in both kind of wells persist at room temperature, a critical condition for making a device.

After processing the samples (substrate etching and anti-reflexion coating) absorption experiments can be performed. The spectrum of sample 422 is shown in Fig. 2 : it evidences, at 300 K, the excitonic resonances in the large wells (1.44 eV) and in the narrow wells (1.50 eV).

Due to the very fast transfer through 70 Å thick barriers, one can imagine that a better device quality may be obtained by using 4 or 5 narrow wells for one large. The trapping time would be a few picoseconds but the excitonic absorption in the narrow well would be much larger than the background absorption of the large wells.

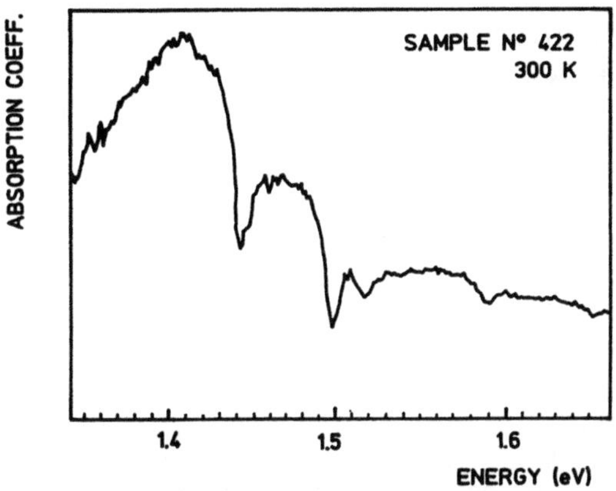

Fig. 2 : Absorption spectrum at 300 K of sample N° 422 showing excitonic resonances in the large wells (1.44 eV) and in the narrow wells (1.50 eV).

In conclusion, a new kind of superlattice system named DWSL has been proposed to realize optical switches. Typical times of less than 1 ps can be obtained with such structures.

[1] D.S. Chemla, D.A.B. Miller : J. Opt. Soc. Am. B 2, 1155 (1985)

[2] H.M. Gibbs, S.L. McCall, T.N.C. Venkatessen, A.C. Gossard, A. Passner, W. Wiegmann, : Appl. Phys. Lett. 35, 451 (1979)

[3] Y. Silberberg, P.W. Smith, J. Eilenberger, D.A.B. Miller, A.C. Gossard, W. Wiegmann, : Opt. Lett. 46, 70 (1984)

[4] A. Migus, A. Antonetti, D. Hulin, A. Mysyrowicz, H.M. Gibbs, N. Peghambarian, J.L. Lewel : Appl. Phys. Lett. 46, 70 (1985)

[5] Y. Silberger, P.N. Smith, D.A.B. Miller, B. Tell, A.C. Gossard, W. Wiegmann, Appl. Phys. Lett. 46, 701 (1985)

[6] A. Chomette, B. Deveaud, J.Y. Emery, A. Regreny, B. Lambert Solid State Commun. 54, 75 (1985)

High-Velocity Vertical Transport in Graded Gap GaAs/GaAlAs Superlattices

B. Lambert[1], *B. Deveaud*[1], *A. Chomette*[1], *A. Regreny*[1], *and R. Romestain*[2]

[1]Centre National d'Etudes des Télécommunications, F-22301 Lannion, France
[1]Université de Grenoble, Laboratoire de Spectrométrie Physique, BP 68, F-38402 St. Martin d'Heres, France

Abstract : We have experimentally estimated the carrier mobility (vertical transport) in a compositionally graded superlattice layer grown by molecular beam epitaxy; the experimental method is an "all-optical" one. In low period superlattice the mobility compares favourably with results obtained with GaAlAs graded alloys.

The use of graded gap alloys is a very promising technique for future devices /1/. In the present paper, we propose a new way of realizing graded gap structures and show that it might provide better characteristics than the originally proposed graded gap alloy. We have realized very short period superlattices with a grading of the alloy composition (i.e. of the barrier height), leading to a shift of the conduction miniband edge. The vertical transport in such structures (transport along the growth axis) is studied by an "all-optical" approach. The principle of the experiment is to perform luminescence on samples where a well larger than the other ones has been introduced purposely about one micron away from the surface.

The samples are grown in a home modified MBE 500 Riber system. The growth process and the photoluminescence (PL) experimental conditions have already been described /2/.

For the present study three series of samples have been grown. First we have grown superlattices (SL) with equal barrier (L_B) and well (L_Z) widths (30Å and 20Å) the Al concentration (x) in the barrier being kept constant at a value of about 30 %. Starting from the surface a 9000Å thick SL is followed by an enlarged well (width : L_Z + 5 monolayers) then by 1000Å of SL.

Secondly we have grown graded gap superlattices (G.G.S.L.) ; the value of L_Z and L_B are 30Å, 20Å and 10Å. From the surface to the substrate we have a 2000Å wide high gap SL followed by nine decreasing gap 800Å wide SLs and finally a 2000A low gap SL in which an enlarged well (E.W.) is included ; in these samples the Al concentration x is ranging from 35 % to 20 %. These structures are schematically shown in figure 1.b.

Finally, in order to compare the properties of SLs and equivalent GaAlAs we have grown two GaAlAs samples : one sample with x = 14 % and another one with x ranging from 18 % to 10% . These GaAlAs samples have the same geometrical parameters as the SLs described above (in particular they have an E.W.). Our GaAlAs samples have the following electrical characteristics : $N_A - N_D = 7\text{-}1 \times 10^{14}\,cm^{-3}$ and $\mu = 300 cm^2 V^{-1} s^{-1}$ at room temperature. The properties of all these samples are summarized in Table 1.

A typical luminescence spectrum is displayed in figure 2, the PL spectrum of a 20Å - 20Å G.G.S.L.. The PL experiments are performed using the 4979Å line of an Ar laser : thus photo-carriers are created in the first 1000A of the SL near the surface. We observe ten lines labelled 1 to 10 corresponding to

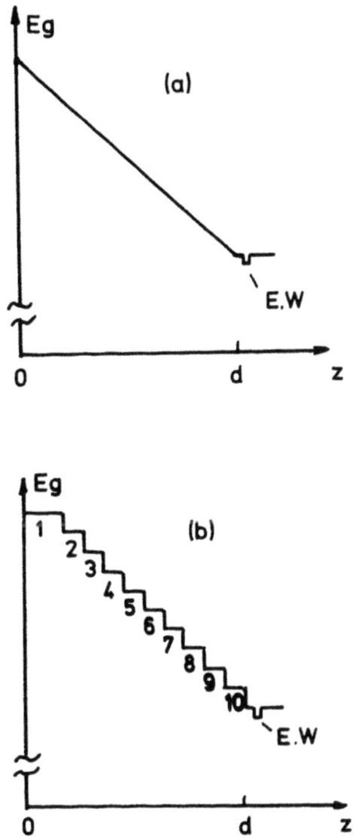

Figure 1 : Dependence of the forbidden band width Eg on the coordinate z : (a) in the case used in the model. (b) grown MBE samples

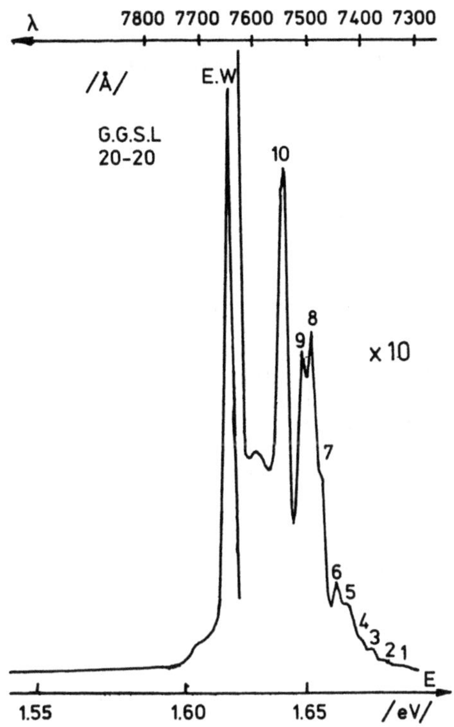

Figure 2 : Luminescence spectrum of a graded-gap superlattice : E.W. denotes the luminescence arising from the enlarged well and the labels 1-10 correspond to the different SLs.

Table 1

L_Z-L_B (Å)	R drift (T= 2K)	L (cm)	l (cm)	ΔE_g(meV)	F (V/cm)	$\mu_{2K}(cm^2V^{-1}s^{-1})$
30-30	0.15	3.5 10^{-5}	4.10^{-5}	37	220	364
20-20	3.0	8.5 10^{-5}	7.10^{-4}	46	275	5090
10-10	27			148	890	
GaAlAs	4	4.10^{-5}	5.2 10^{-4}	130	780	1 330

each distinct SL and at a lower energy an intense line corresponding to the E.W. ; the relative intensities of the lines indicate the importance of the carrier transfer. This transfer can be characterized by the ratio R between the integrated line intensity and the sum of the 10 integrated SL lines intensities. The value of R increases by about two orders of magnitude from the 30-30 sample to the 10-10 one (see the column R drift in table 1) ; this corresponds to a dramatic increase of the transfer efficiency when the SL period decreases. The R drift value for the graded gap GaAlAs is about equal to the value of the 20-20 G.G.S.L.

In order to explain the shape of the ten PL lines (as displayed in figure 2) and the different R drift values, we have used a simple model which has been built for variable-gap semiconductors /3/ ; it corresponds to the case of a forbidden gap the width of which is varying linearly with the coordinate z in the range $0 < z < d$ (see figure 1.a). This calculation is valid in the case of weak continuous illumination : in fact we have checked that the R value is constant independently of the density Φ_i of the exciting (incident) light. So we have calculated the distribution n(z) of the photo-carrier density(see the continuous line in the figure 3.)

In order to fit with experimental results two parameters must be determined: the diffusion length L and the drift length l (we have neglected the surface recombination). The procedure is the following : i) using the R diffusion values as deduced from PL experiments performed on the first series of SL samples (without graded gap) we determine the diffusion length L (see table 1). ii) using the R drift value as deduced from the G.G.S.L. photoluminescence spectra, we determine the drift length l (see table 1). iii) finally we compare the n(z) density profile with the 10 SLs line intensities. The agreement (see figure 3) is quite satisfactory bearing in mind the simplicity of the model. In particular, we are able to explain the difference between the profiles of the PL lines in the 30-30 and the 20-20 G.G.S.L.. However, three important assumptions have been made in this calculation: i) the carrier life time τ is the same in the SLs and in the E.W. ii) every carrier which is located at $z > d$ is collected by the E.W. iii) the two structures as depicted in figure 1 have the same properties ; in our view this is true because we have shown /4/ that in short period SLs the coherence length is greater than 1500Å (each SL step is 800 Å wide).

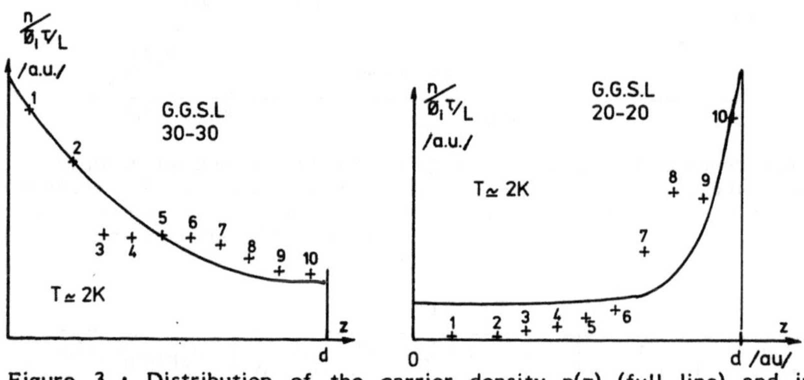

Figure 3 : Distribution of the carrier density n(z) (full line) and intensity of the different (1-10) PL lines related to the SL (crosses).

With the l value it is easy to calculate the drift speed $v = l/\tau$ if the lifetime τ is known. We have recently studied by picosecond PL the decay times of SLs at liquid helium temperature /5/ : in this temperature range, the PL recombination being mainly excitonic, the τ values are much lower than at higher temperature. The complete study is now in progress and will be published elsewhere ; preliminary results indicate that $\tau \simeq 500$ ps for the samples studied at low temperature ($T < 20$ K). Now it is possible to access the mobility $\mu = v/F$ with an estimation of F where F is the quasi-field in the G.G.S.L. ; the gap energy variation ΔEg (table 1) is obtained from the PL lines 1-10 and F is deduced with an estimation of $\Delta E_C = 0.6 \Delta Eg$ and a value $d = 1 \mu m$. The μ values are given in the last column of the table 1. We see that the 20-20 SL mobility is higher than the mobility of our GaAlAs reference sample ; even for the 10-10 SL preliminary results (see for example the value Rdrift = 27) are quite promising.

We have measured the R(T) variations at higher temperatures ; our range is limited to T <20 K because at higher temperatures the PL intensity dependence versus Φ_i is no more linear and, as a consequence, the ratio R increases with Φ_i. When T increases up to 20 K we observe an increase of the Rdrift value of the GGSL even when the GaAlAs Rdrift one is quite constant ; as a matter of fact we calculate a mobility μ = 13 100 $cm^2V^{-1}s^{-1}$ at T = 10 K for the 20-20 GGSL. The R(T) dependance in a SL has already been reported and is interpreted as due to more and more carrier de-trapping from the tail states when T increases /6/ ; these tail states are due to a slight disorder existing in a SL.

All these experimental results indicate a very high velocity vertical transport in our good quality superlattices. They favourably compare with results obtained with GaAlAs graded alloys /7/.

REFERENCES

1. F. Capasso : Physica 129 B, 92 (1985)
2. B. Deveaud, J.Y. Emery, A. Chomette, B. Lambert and M. Baudet Appl. Phys. Lett. 45, 1078 (1984)
3. A.S. Volkov and G.V. Tsarenkov : Sov. Phys. Semicond. 11, 1004(1977)
4. A. Chomette, B. Deveaud, J.Y. Emery, A. Regreny and B. Lambert Solid State Commun. 54, 75 (1985)
5. B. Deveaud, A. Chomette, B. Lambert, A. Regreny, R. Romestain and P. Edel : Solid State Commun. 57, 885 (1986)
6. A. Chomette, B. Deveaud, A. Regreny and G. Bastard, to be published
7. B.F. Levine, C.G. Bethea, W.T. Tsang, F. Capasso, K.K. Thornber, R.C. Fulton and D.A. Kleimann : Appl. Phys. Lett. 42, 769 (1983).

Modelling of Mobility Degradation in Submicron MOSFETs After Electrical Stressing

B. Cabon-Till and G. Ghibaudo

Laboratoire de Physique des Composants à Semiconducteurs, UA-CNRS 840, ENSERG, 23 Av. des Martyrs, F-38031 Grenoble, France

1. INTRODUCTION

In submicron MOSFETs, aging effects increase because of the strong electrical field during the device operation. Hence, the strengthening of Channel Hot Electron injection (C.H.E) is well known to give rise to a higher degradation of transconductance and threshold voltages /1/. This type of degradation occurs for $V_G > V_D$ and for normal values of V_D corresponding to the MOSFET operation biases.

2. EXPERIMENTAL RESULTS

From experimental $I_D(V_G)$ measurements in ohmic regime is derived the conductance $G_d = I_D/V_D$, the normalized transconductance $g_G = \delta G_D/\delta V_G$ and the field effect mobility μ_{FE} defined by :

$$\mu_{FE} = \frac{L}{W\,C_{ox}}\, g_G \qquad (1)$$

where L and W are the channel length and width, and C_{ox} is the gate oxide capacitance.

The inflexion point of the $I_D(V_G)$ characteristics gives the most important parameters for aging investigations : V_{Text}, the extrapolated threshold voltage, V_{Gm}, the mobility threshold voltage corresponding to the maximum value of the field effect mobility, μ_m /2/.

Experimental results relative to field effect mobility μ_{FE} and threshold voltage degradation obtained on short-channel MOSFETs with two different lengths 0.4 and 1.4 μm are reported herein. Fig. 1 shows that the mobility degradation occurs mainly at the maximum mobility and not in the strong inversion region, so the maximum mobility μ_m and the corresponding gate voltage V_{Gm} are more sensitive to the aging process than other usual parameters such as the extrapolated threshold voltage V_{Text} /3/.

Figure 1 shows that the most important degradation of mobility threshold voltage occurs at highest stress biases V_G.

A worthy analysis of the degradation mechanism has in turn to be conducted by plotting the correlation curves between the relative mobility change $\Delta\mu_m/\mu_m$ and the mobility threshold voltage shift ΔV_{Gm}. In this case, as time interferes as a parameter, these correlation diagrams become independent of the degradation kinetics.

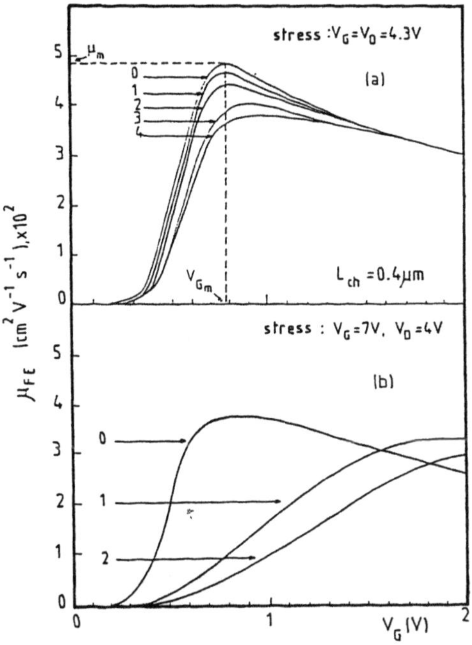

Fig. 1 : Field effect mobility μ_{FE} vs gate voltage V_G for several stress durations (0, 1=2hours, 2=10hours, 3=50hours, 4=70hours)

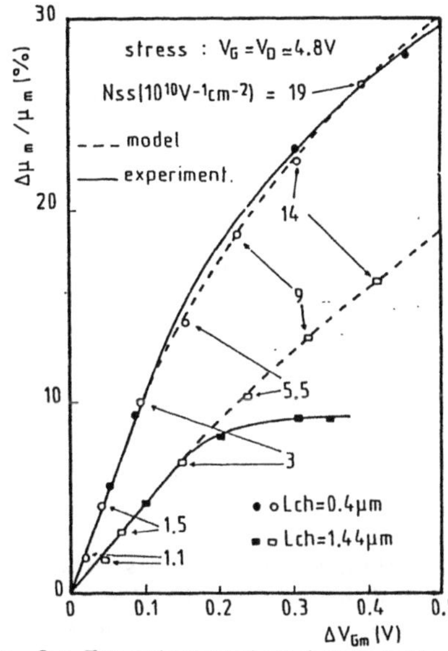

Fig. 2 : Experimental and theoretical correlation plots of relative mobility degradation $\Delta\mu_m/\mu_m$ against mobility threshold voltage shift ΔV_{Gm} (N_{ss} theoretical parameter)

Figure 2 gives an example of such a correlation diagram for two MOSFETs of different channel length stressed under the same bias condition. Note that a saturation in the mobility degradation is observed for the longer channel. Moreover, for short stress times, the slope of the linear part of the curve is strongly dependent on the channel length through the series resistance R_{SD} via the equation :

$$\mu_m = \mu_m^\circ/(1 - 2 R_{SD} G_d^\circ) \qquad (2)$$

where G_d° is the intrinsic channel conductance and μ_m° is the intrinsic maximum mobility /4/.

3. DISCUSSION

As mobility degradation is mainly observed in weak and moderate inversion regimes, we suggest an interpretation of mobility degradation based on an inversion charge dependent scattering process, such as that due to potential fluctuations arising from charge trapping and surface state generation /5/.

It can be shown from Kubo-Greenwood transport calculations, that the effective mobility fits an empirical law that accounts for mobility reduction with increasing inversion charge /5/ :

$$\mu_{eff} = \frac{\mu_0}{1 + Q_i/Q_c} \qquad (3)$$

where μ_0 is a constant mobility, Q_i is the inversion charge and Q_c is a critical charge at which μ_{eff} begins to decrease because of degeneracy.

Because carrier localisation occurs at the band edge due to potential fluctuations, μ_0 becomes /5/ :

$$\mu_0 = \mu_{00} \exp(- \frac{q\,\sigma_s}{kT}) \qquad (4)$$

where μ_{00} is the mobility value in absence of potential fluctuations and σ_s is the standard deviation of the surface potential related to the standard deviation of the charge distribution σ_q by a capacitance relationship :

$$\sigma_s = \frac{\sigma_q}{C_{ox} + C_D + C_{ss} + C_i} \qquad (5)$$

where C_D, C_{ss}, C_i are respectively the depletion, surface state and inversion layer capacitances.

With usual laws describing Q_i, C_i, C_D and V_G with surface potential, Rel. 3 which is valid from weak inversion to strong inversion, allows a description of μ_{FE} over the whole range of V_G.

In this context, the degradation is attributed to an increase of the potential fluctuation rate. As a result, the correlation between the maximum mobility degradation $\Delta\mu_m/\mu_m$ and the corresponding mobility threshold voltage shift ΔV_{Gm} can be properly described by a model in which the rate of potential fluctuations arising from the spatial fluctuations of the interface charge distribution is monitored by the density of generated surface states N_{ss} /3/. For a Poisson charge distribution, the variance of the charge distribution σ_q is quadratically dependent on the density of states N_{ss} /3/:

$$\sigma_q = [\alpha N_{ss}]^{1/2} \qquad (6)$$

where α is a fitting parameter (in our case α is about $2x10^{10} q^2 eV/cm^2$).

Figure 2 indicates that good agreement between theory and experiment can be achieved except when a mobility degradation saturation occurs.

Furthermore, it is worth noting that, in such a model, only one parameter, namely the density of surface state N_{ss}, can drive both the degradation of mobility and mobility threshold voltage and also the extrapolated threshold voltage shift ΔV_{Text} related to ΔV_{Gm} and $\Delta\mu_m$ by /3,5/ :

$$\Delta V_{Text} = \Delta V_{Gm} + G_{dm}/g_m(\Delta\mu_m/\mu_m) \qquad (7)$$

where G_{dm} = conductance and $g_m = dG_{dm}/dV_G$ at $V_G = V_{Gm}$.

This last expression is useful to interpret the two main degradation regimes that are experimentally observed for V_{Text} : i) a dominant mobility degradation resulting in a very weak V_{Text} shift and ii) a dominant surface state generation process giving rise to a prevailing positive shift of V_{Gm} and V_{Text} /3/.

In addition, in contrast to standard results, we have observed that the final mobility degradation is always stronger for higher gate stress biases. This can be explained by a strong effect of the saturation velocity phenomena in the whole range of gate biases for shorter devices. This results in a weaker dependence of substrate current with gate voltage and in a stronger impact ionization rate in the whole range $V_G > V_T$. This is why Channel Hot Electron injection into gate oxide is finally enhanced at higher gate voltage.

4. CONCLUSION

In conclusion it is worth emphasizing that the mobility maximum μ_m and the corresponding gate voltage V_{Gm} are suitable parameters for monitoring aging effects. The use of degradation correlation diagrams $\Delta\mu_m/\mu_m$ vs ΔV_{Gm} is also very fruitful for a proper interpretation of the degradation processes.

REFERENCES

1- E. Takeda, A. Shimizu, T. Hagiwagara : IEEE-EDL4, 9 (1983) p.329.
2. Chu Hao, B. Cabon-Till, S. Cristoloveanu, G. Ghibaudo : SSE, 23 (1985) p. 1025.
3. B. Cabon-Till, G. Ghibaudo : Rev. Phys. Appl., 21 (1986) p.305.
4. B. Cabon-Till, G. Ghibaudo, S. Cristoloveanu : Electron. Lett., 21 (1985) p. 457.
5. G. Ghibaudo : Phys. Stat. Sol. (a), 95 (1986) p. 323.

Negative Differential Mobility and Drift Velocity Overshoot in a Single Quantum Well of AlGaAs/GaAs/AlGaAs Heterostructure

C. Hamaguchi, K. Miyatsuji, and H. Tanimoto

Department of Electronics, Osaka University, Suita City, Osaka 565, Japan

Monte Carlo simulation is carried out to investigate high-field transport properties of two-dimensional electrons in a single quantum well of AlGaAs/GaAs/AlGaAs with 500A GaAs-well and $N_s=1X10^{12}cm^{-2}$. Negative differential mobility and drift velocity overshoot are found.

1. INTRODUCTION

High-field transport of two-dimensional (2D) hot electrons is very important in heterostructure devices in which very high-speed operation and new effects are expected. For example, high electron mobility transistor (HEMT)/1/ and real-space transfer devices/2/ have been proposed as future electron devices. However, their high-field characteristics are known only a little from the points of both experimental/3/ and theoretical aspects/4,5/. This is mainly due to the facts that electrons populate several subbands, resulting in a complicated situation to analyze the transport properties, and that treatment of the scattering processes of the 2D electrons is not well established. In this paper we report for the first time that the electrons transferred into the L-valleys are two-dimensional and that the electron transfer from the Γ-valley subbands to the L-valley subbands results in negative differential mobility.

2. ENERGY BAND STRUCTURE

Energy band structures of GaAs and $Al_xGa_{1-x}As$ (x=0.3) can be calculated by using the empirical pseudopotential method or k.p perturbation method. In the present work we used the empirical pseudopotential method/6,7/. We combined the calculated values with the experimental values/8/ and we chose ΔE(Γ)=0.30eV, ΔE(L)=0.20eV and ΔE(X)=0.15eV. Quantization of electrons in X-points appears only when the conduction band discontinuity exceeds 0.1eV for 500A well and thus we choose ΔE(X)=0.15eV for the X-points. The value of 0.15eV at the X-points is a little bit larger but the following calculations are not influenced so much even if we choose different values. We calculate the subband energies, wave functions and the population of the electrons, at the Γ-point, using self-consistent method/9/. These calculations determine the energy band bending at the Γ-point and thus at the L- and X-points. Using the effective masses 0.222m in the L- valleys and 0.58m in the X-valleys we obtained the subband structures in these valleys. In the present analysis we assumed that the width of the GaAs-well is 500A and the electron sheet density is $1X10^{12}$ cm^{-2}. In addition we assumed that the subband structures are not affected by the repopulation of the electrons. This simplification is supported by Monte Carlo simulation of 2D electrons in Si MOS inversion layers, where the repopulation of electrons was taken into account.

3. SCATTERING PROBABILITIES OF 2D ELECTRONS

The wave function and energy of 2D electron in i-valley are given by

$$\Psi_i = F_{in}(z)\cdot\exp(i\vec{k}\cdot\vec{r}) \ , \quad E_{in}(\vec{k}) = \frac{\hbar^2 k^2}{2m_i^*} + E_{in} \ , \tag{1}$$

where $\vec{k}$ is the wavevector in the plane perpendicular to the z direction, and $F_{in}(z)$ and E_{in} are given by solving Schrödinger and Poisson equations self-consistently. Scattering probability of an electron in i-valley with phonon emission or absorption from a state $\vec{k}_1$ and energy E_1 in the subband m to a state $\vec{k}_2$ and E_2 in the subband n is given by PRICE/10/ as

$$W^{\pm}_{i,m\to n}(\vec{k}_1,\vec{k}_2) = \frac{2\pi}{\hbar}\,|M^{\pm}_{i,mn}(\vec{q})|^2\cdot\delta[E_{in}(\vec{k}_2) - E_{im}(\vec{k}_1) \pm \hbar\omega_q] \ , \tag{2}$$

where the matrix element $M^{\pm}_{i,mn}(\vec{q})$ is

$$M^{\pm}_{i,mn}(\vec{q}) = \left(\frac{n_q + \frac{1}{2} \pm \frac{1}{2}}{V}\right)^{\frac{1}{2}}\cdot\delta_{\vec{k}_1,\vec{k}_2\pm\vec{Q}}\cdot J^{\pm}_{i,mn}(q_z)\cdot C_q \ . \tag{3}$$

Here phonon wave vector is $\vec{q} = (\vec{Q},q_z)$, n_q is the phonon occupation number and

$$J^{\pm}_{i,mn}(q_z) = \int F_{in}(z)F_{im}(z)\exp(\pm iq_z z)dz. \tag{4}$$

In the case of acoustic phonon deformation potential scattering we have

$$C_q^{\,2} = \hbar D^2 q^2/2\rho\omega_q \ , \tag{5}$$

where D is the deformation potential and ρ is the density of the material. For polar optical phonon scattering

$$C_q^{\,2} = (e^2\hbar\omega_{LO}/2\varepsilon_0 q^2)(1/\kappa_\infty - 1/\kappa_0) \ , \tag{6}$$

where $\varepsilon_0\kappa_0$ and $\varepsilon_0\kappa_\infty$ are the static and optical dielectric constants, respectively. For intervalley phonon scattering we have

$$C_q^{\,2} = \hbar D_{ij}^2/2\rho\omega_{ij} \ , \tag{7}$$

where $\hbar\omega_{ij}$ the intervalley phonon energy. At low electric fields impurity scattering is very important. Here we used the results of PRICE/11/ for the impurity scattering screened by 2D electrons.

4. MONTE CARLO SIMULATION OF 2D ELECTRONS IN A SINGLE QUANTUM WELL

Using the results of the self-consistent calculations we obtained the scattering probabilities in a single quantum well. The parameters used in the present simulations are those reported by LITTLEJOHN et al./12/ because electrons are mainly confined in the GaAs-well and thus their transport is determined by the parameters in GaAs. Donors of $1\mathrm{X}10^{18}$ cm^{-3} are assumed to distribute uniformly in the barrier regions of AlGaAs one side with a spacer region of 60A and the other side without a spacer region. Monte Carlo simulation of 2D electron was reported by TERASHIMA et al./13/ for MOS inversion layers and by MIYATSUJI et al./14/ for single quantum well. Simulations are carried out to obtain the steady state velocity-field curves and the drift velocity overshoot in uniform electric fields at 77K.

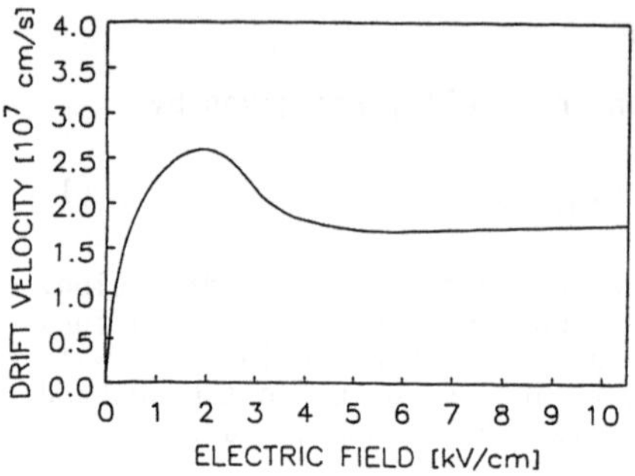

Fig.1 Steady-state drift velocity vs. electric field in SQW with 500A-well and $N_s=1X10^{12}$ cm^{-2} at 77K.

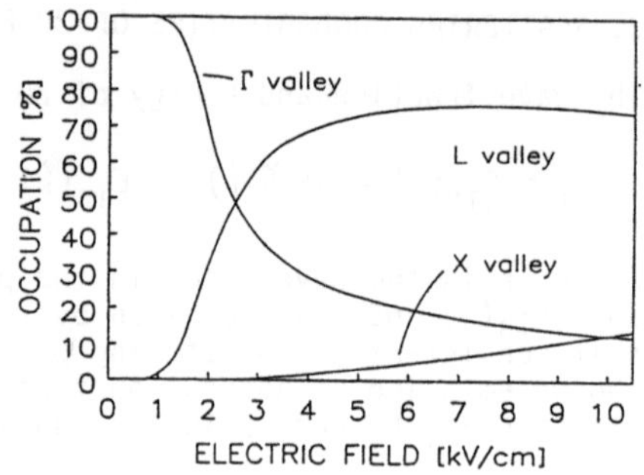

Fig.2 Occupation of electron as a function of electric field in Γ-, L- and X-valleys at 77K.

Figure 1 represents the calculated drift velocity as a function of the electric field at 77K, where we find that the maximum drift velocity is about $2.6X10^7$ cm/s at 2 kV/cm and negative differential mobility appears beyond this field. It is very interesting that the ohmic mobility is quite high and reaches $8x10^4$ cm^2/Vs at 77K which is very close to the value measured in a sample with the same structure described here. This means that the parameters of the Γ-valleys used in the present work are precise. In Fig. 2 we present the occupation of the electrons in the three valleys. Here we find that electrons in the Γ-valley subbands decrease with increasing the electric field and the electron population in the L-valley subbands increases at electric field beyond 2kV/cm. The population in the X-valley is not appreciable and thus we can neglect intervalley transfer to the X-valleys in a simplified case. This indicates also that the energy band ordering between GaAs and AlGaAs at the X-point is not so important in AlGaAs/GaAs heterostructures.

Figure 3 represents time evolution of the drift velocity in uniform electric field 1, 5 and 10 kV/cm at 77K, where the average drift velocity is shown. The maximum overshoot $9.5X10^7$cm/s appears at about 0.7ps which is larger than the values obtained in a bulk GaAs, $6X10^7$cm/s at 77K. In order to clarify the valley transfer of the electrons we plotted the drift

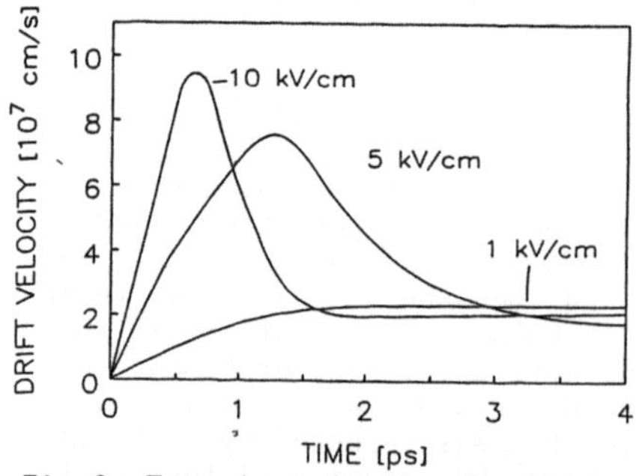

Fig.3 Transient drift velocities in electric field of 1, 5 and 10 kV/cm at 77K.

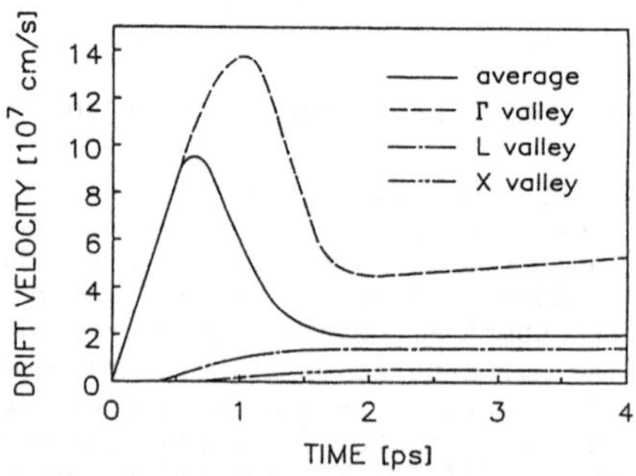

Fig.4 Transient drift velocities in Γ-, L- and X-valleys at 10kV/cm and at 77K.

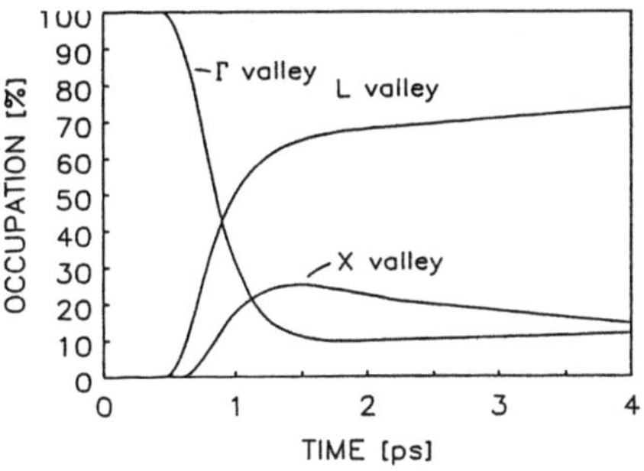

Fig.5 Occupation of electrons as a function of time in Γ-, L- and X-valleys at 10kV/cm and at 77K.

velocities as a function of time in Γ-, L- and X-valleys in Fig. 4 at 10kV/cm, where we find the maximum drift velocity in the Γ-valley is about 14X10^7 cm/s which is much larger than the maximum reported in bulk GaAs. Such a high value seems to reflect that the scattering probabilities of 2D electrons are quite different from those of 3D electrons. In Fig. 5 occupation of the electrons in the three valleys is plotted as a function of time for the case of 10kV/cm. Note that increase in the upper valley population corresponds to the point where the drift velocity maximum occurs.

ACKNOWLEDGEMENT

We are thankful to Dr. S. Hiyamizu and S. Sasa for useful discussions on the high-field transport of 2D electrons in a quantum well and also for providing Hall mobility data in a quantum well prior to publication.

References

1. T.Miura,S.Hiyamizu,T.Fujii and K.Nanbu:Jpn.J.Appl.Phys. 19 (1980) L225.
2. K. Hess, H. Morkoc, H. Shichijo and B. G. Streetman: Appl. Phys. Lett. 35 (1979) 469.
3. M. Inoue, S. Hiyamizu, M. Inayama and Y. Inuishi: Jpn. J. Appl. Phys. 22, Suppl. 22-1 (1983) 357.
4. M. Tomizawa, K. Yokoyama and A. Yoshii: IEEE Electron Device Lett. EDL-5 (1984) 464.
5. K. Yokoyama and K. Hess: The 2nd Int. Conf. on Modulated Semiconductor Structures (Kyoto, 1985) 533.
6. M. L. Cohen and T. K. Bergstresser: Phys. Rev. 141 (1966) 789.
7. H. Hazama, T. Sugimasa, T. Imachi and C. Hamaguchi: J. Phys. Soc. Jpn. 55 (1986) 1282.
8. H. J. Lee, L. Y. Juravel, J. C. Woolley and A. J. SpringThorpe: Phys. Rev. B21 (1980) 659.
9. K. Miyatsuji, H. Hihara and C. Hamaguchi: Superlattices and Microstructures 1 (1985) 43.
10. P. J. Price: Ann. Phys. (N.Y.) 133 (1981) 217.
11. P. J. Price: J. Vac. Sci. Technol. 19 (1981) 599.
12. M. A. Littlejohn, J. R. Hauser and T. H. Glisson: J. Appl. Phys. 48, (1977) 4587.
13. K. Terashima, C. Hamaguchi and K. Taniguchi: Superlattices and Microstructures 1 (1985) 15.
14. K. Miyatsuji, H. Tanimoto and C. Hamaguchi: 1986 Int. Conf. on Solid State Devices and Materials (Tokyo, 1986).

Monte Carlo Study of Hot Electron Transport in GaAs-AlGaAs Quantum Wells

S.M. Goodnick [1] *and P. Lugli* [1]

[1] Physik Department E 16, Technische Universität München, D-8046 Garching, Fed. Rep. of Germany

[2] Dipartimento di Fisica, Universita di Modena, Via Campi 213/a, I-41100 Modena, Italy

The rapid advancement of epitaxial growth technology for semiconductor compounds has resulted in a host of new device possibilities in which spatial dimensions are reduced to the atomic scale. Spatial confinement of the carriers results in quantization of the electronic motion, and thus the non-equilibrium behavior of such systems under external excitation may differ considerably from the bulk semiconductor situation. In the present work, we report on the results of a Monte Carlo simulation of hot carriers confined to a GaAs-AlGaAs quantum well. Here we have included the effects of degeneracy, electron-electron interaction, and non-equilibrium polar optical phonons (POP) in order to study the low-temperature transient and steady state response of spatially quantized electrons driven by high electric fields and laser excitation.

As a model system we have considered a single square well representing the conduction band discontinuities of an AlGaAs-GaAs-AlGaAs well. The subband energy eigenvalues are determined by the solutions of the one-dimensional effective mass equation for this potential. Results using self-consistent wave functions will be discussed elsewhere /1/. We simulate the quasi-free electronic motion parallel to the well using an ensemble Monte Carlo calculation including both intra- and intersubband scattering within the central and satellite L valleys (considered parabolic for simplicity). The material parameters for GaAs were the same as those used in bulk Monte Carlo simulations /2/.

The dominant energy-loss mechanism in GaAs is due to POP scattering, at least for hot carriers. Here we neglect confinement effects in the phonon spectra and treat the phonon modes as bulk-like interacting with quantized quasi-two-dimensional electrons via the Fröhlich Hamiltonian. The scattering rates resulting from this coupling are calculated fully numerically over the electronic envelope functions without analytic approximation. Without the presence of upper satellite valleys in the simulation, non-stationary behavior is found in the transient response of the system with POP scattering alone at sufficiently high electronic fields (greater than 5kV/cm at 300K), indicating a polar runaway similar to the bulk. With upper valleys present, inter-valley transfer allows the electrons to cool and thus prevents runaway. For the quantum well, we calculate the inter-valley scattering rate from the optical deformation potential between quantized states in both central and satellite valleys. The calculated transient velocity versus time at 300K is not significantly different from the results for bulk GaAs. This is also found to be true for the velocity-field characteristics and arises from the fact that in the multi-subband system, the total POP scattering rate as a function of energy follows the bulk value fairly closely.

In order to simulate low temperatures and high electron concentrations, it is necessary to consider the effects of degeneracy on the electron scattering rates. Degeneracy in the quasi-2D system is included through a generalization of the bulk method described in /3/. A grid on the two-dimensional k-space is assigned to each subband, and the electron distribution function is defined over this grid. In an ensemble Monte Carlo simulation, the distribution is simply given by the occupation of each cell, normalized to account for the nominal concentration modeled in the calculation. As the simulation evolves, the distribution function is continuously updated and at each scattering event, a self-scattering rejection technique is used to account for the occupancy of the final state.

Electron-electron scattering is included into the 2D Monte Carlo simulation through use of the rejection method given by Brunetti et al /4/. This technique was generalized to the multi-subband quantum well including both inter- and intrasubband e-e scattering in which the various electrons in the ensemble are allowed to interact via a screened Coulomb interaction. In this initial work, we used a single $\vec{q}$ independent screening constant given by the 2D Fermi-Thomas function, which results in an almost negligible contribution due to inter-subband e-e scattering. In contrast, the intra-subband scattering rate for electron-electron interaction (in which neither electron changes subband from initial to final state) is found to be at least an order of magnitude greater than the POP scattering rate.

At low temperatures, one has to consider the effect of impurity scattering as well. We approximate the effect of this scattering mechanism by a sheet of charge located in the AlGaAs layer, representing the ionized donors in a modulation doped structure. The interaction is treated in the usual way /5/ by the screened Coulomb potential with the impurity concentration taken to be approximately equal to the electron concentration.

The results of this model for several applied electric fields at 5K are shown in Fig. 1. Here we show the steady state distribution function corresponding to 100, 500 and 750V/cm applied along the channel of a 250 Å well ($n=4.0 \times 10^{11}/cm^2$) which corresponds to one of the samples studied by Shah et al. /6/ using steady state photoluminescence. Several features are of importance. The DFs are clearly Fermi-Dirac with a high-energy tail charac-

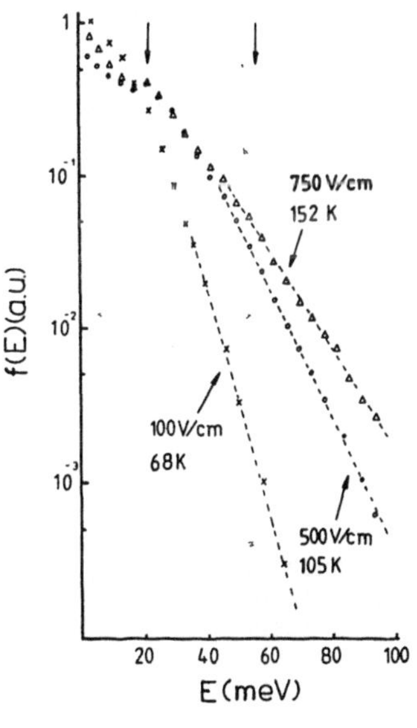

FIG.1. Steady state electron distribution function for 3 different electric fields. The electron temperature calculated from the slope of the high-energy tail is also shown.

terized by an electron temperature given by the slope of the curve. The arrows indicate the positions of the first and second excited subbands in which structure is observed in the DF, similar to photoluminescence results. The absence of structure at the phonon energy (35meV) is an indication of the effectiveness of e-e scattering. Without this interaction in the simulation, a depletion of carriers above the phonon emission threshold is found. For 750V/cm, the average time for e-e interaction is found to be about 0.06ps, compared with an average time for phonon emission of 3.1ps. The electron temperature of 151K calculated here for E=750V/cm is in good agreement with the value of 165K measured experimentally. We also calculate the energy loss rate per carrier as a function of the electron temperature from the net emission of POP phonons during the simulation. For T_e=151K, the power loss rate per carrier is calculated to be $1.8x10^{-9}$W, close to the experimental results. At lower electric fields, the agreement is not as good, possibly due to the effect of physical phenomena that been neglected (i.e. dissipation due to acoustic phonons, hot phonons, etc.)

The electron mobility may be calculated from the ratio of the drift velocity and the electric field. At low fields and temperatures, the mobility in the quantum well is controlled by remote impurities in the AlGaAs. For an impurity concentration of $5x10^{11}/cm^2$, the estimated mobility was determined to be $53000cm^2$/V-s at E=28V/cm. At higher fields, the electron distribution heats up, and POP emission becomes the dominant process, causing the mobility to decrease to a value of 21000 at 750V/cm.

The relaxation towards equilibrium of photoexcited electrons in a quantum well has also been studied. Due to the strong phonon emission of the high energy carriers injected into the sample, the phonon population is driven from its equilibrium value which in turn affects the electron energy loss rate. In the present work, we have included hot phonon effects (see/7/ for more details) by keeping a detailed balance of phonon emission and absorption events during the simulation. The carrier scattering rates are updated continuously with the evolving phonon distribution while the excess phonons in each mode are allowed to decay phenomenologically.

We have considered a system in which a background density of $2.5x10^{11}/cm^2$ exists and to which an injected density of $5x10^{11}/cm^2$ is created over 0.6ps. Fig. 2 shows the details of the initial electron relaxtion in a 150A well with four subbands considered in the and L valley. The energy distribution function is plotted at different times during and after laser excitation. The con-

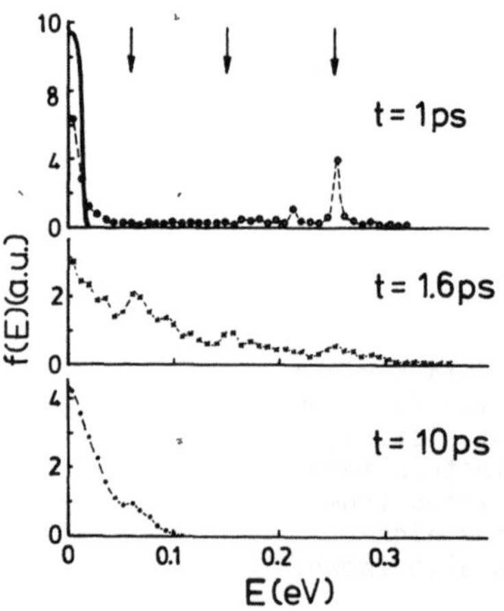

FIG.2. Time dependence of the electron distribution during and after the laser pulse.

tinuous curve in the top figure represents the initial cold background electrons. At the maximum of the laser pulse (t=1ps in this plot), a significant exchange of energy has already occurred between the photoexcited electrons (injected 0.25eV above the lowest subband) and the cold background. The secondary peak below the injection level represents POP emission by the photoexcited electrons. Shortly after the end of the pulse (t=1.6ps), strong intercarrier scattering creates a broad distribution where the subband minima (indicated by the arrows) clearly appear. Within each subband, the distribution function starts to exhibit a Fermi-like appearance, which is fully established at longer times as shown in the bottom curve.

The time evolution of the average electron energy is shown in Fig. 3. As indicated in the figure, the presence of POP hot phonons (x,o) clearly reduces the electron relaxation rate due to reabsorption at later times of the phonons emitted during the initial excitation. The result without hot phonons is also shown for comparison (□). We have performed the calculation for a 250 Å well (including 6 subbands) and as shown in Fig. 3 (X), no real dependence on well width was found, contrary to the experimental evidence of Ryan et al. /8/, although this could be due to the different background density and higher injection densities used in their measurements. The continuous curves in Fig. 3 refer to the total electron energy (kinetic plus potential) calculated as ensemble averages during the simulation. For degenerate systems, this quantity can vary considerably from the electron temperature, which can be rigorously defined only in the presence of a Fermi distribution. The electron temperature calculated from the average energy, assuming the DF is Fermi-like, (dashed curve) corresponds to the slope of the tail of the DF.

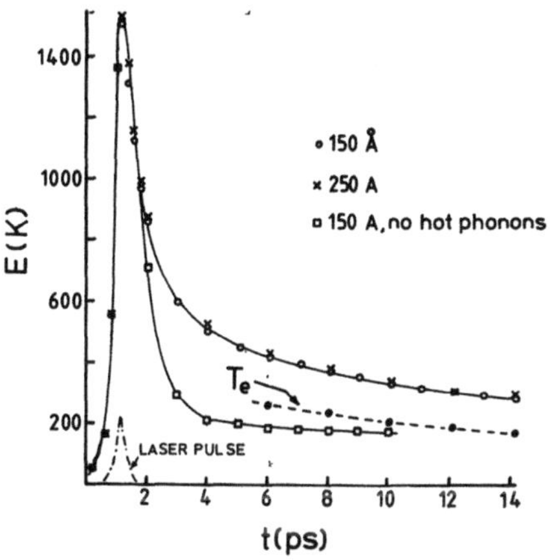

FIG.3. Average electron energy as a function of time during and after the laser pulse. The electron temperature is shown by the dashed line.

The authors would like to acknowledge the support of ERO and AvH foundation.

1. P. Lugli, S.M. Goodnick, and F. Koch, to be published in Superlattices and Microstructures
2. P. Lugli and D.K. Ferry, Physica 134b, 364 (1985)
3. P. Lugli and D.K. Ferry, IEEE Trans. Electron Dev., ED-32, 2431 (1985)
4. R. Brunetti and C. Jacoboni, Physica 134b, 369 (1985)
5. T. Ando, A.B. Fowler and F. Stern, Rev. Mod. Phys. 54, 437 (1982)
6. J. Shah, A. Pinczuck, H.L. Störmer, A.C. Gossard, and W. Wiegmann, Appl. Phys. Lett. 44, 322 (1984)
7. S.M. Goodnick and P. Lugli, to be published in "Proceedings of 18th Int. Conf. on the Phys. of Sem.", Stockholm 1986
8. J.F. Ryan, R.A. Taylor, A.J. Turberfield, F.M. Worlock, Surf. Sci., 170, 511 (1986)

Part III

High-Speed Electronic Devices

High-Speed Bulk Unipolar Structures in Silicon

J.M. Shannon

Philips Research Laboratories, Redhill, Surrey, RH1 5HA, UK

Some years ago, hot electron transistors with high current gain and good power gain were made in silicon using ion implantation doping [1]. These monolithic hot electron transistors were based on the fact that the bulk unipolar diode, in particular the camel diode, is an efficient collector of hot electrons. In this case the hot electrons were generated using a reverse biased, barrier raised Schottky diode. The F_T of these devices was ≈3 GHz and in good agreement with that calculated, the major factor determining F_T being the charging time of the tunnel emitter. To improve the frequency response, a low capacitance unipolar diode emitter is required and the barrier heights of the unipolar diode have to be controlled very precisely.

The high frequency response of camel diodes has been verified by using them as mm-wave mixers. Gallium arsenide devices have been operated in the mm-wave region with performances comparable with the best Schottky diode devices [2]. A problem with bulk unipolar diode structures arises because in some circumstances holes are able to accumulate in the barrier region to such an extent that the barrier height is modulated. Under these conditions, the frequency response is very poor since it is controlled by holes moving in or out of the junction. The generation of holes is a complex process and can arise from tunnelling processes within these structures, ionisation by the hot electrons or simple generation processes.

Clearly the design of the bulk unipolar diode has a major influence on the high frequency performance of device structures. In general, the thinner the barrier reigon the more ideal is the diode behaviour. In principle, technologies such as MBE, ion implantation and solid phase epitaxy [3] may be combined together to make diodes having a quality approaching that of a Schottky diode, but this has yet to be done. Significant advances have been made, however, in the fabrication of high quality bulk unipolar camel diodes.

Although useful current gains have been reported in monolithic hot electron transistor structures it would be comforting to know that useful performance can be obtained in structures containing camel emitters. Recently, measurements have been made on GaAs structures containing camel emitters and collectors [4]. High current gain has been obtained over useful base widths and furthermore this has been achieved at current densities $\sim 10^3\,\mathrm{A\,cm^{-2}}$ and temperatures ~450 K.

References

1. J.M. Shannon, J.A.G. Slatter: Jpn. J. Appl. Phys. **22**, Sup. 22–1, 259 (1983)
2. S.J. Battersby, J.J. Harris: To be published in IEEE Trans.-ED
3. J.M. Shannon, M.J. King, B.J. Goldsmith, J.B. Clegg: Proc. 1st Int. Symposium Silicon MBE, Electrochem. Soc., Vol. 85–7, p. 242 (1985)
4. J.M. Woodcock, J.J. Marris, J.M. Shannon: Physica **134**B, 111 (1985)

Silicon Bulk Barrier Diodes Fabricated by LPVPE

H. Beneking, L. Vescan, A. Gruhle, J.M. Cloos, and M. Marso

Institute of Semiconductor Electronics, Aachen Technical University, Sommerfeldstraße, D-5100 Aachen, Fed. Rep. of Germany

Bulk unipolar (camel) diodes have been realized in Si using Low - Pressure Vapour Phase Epitaxy (LPVPE) of $SiCl_2H_2/H_2$. Ideality factors of 1.08 - 1.2 and barrier heights 0.46 eV to 0.81 eV have been obtained by varying the thickness and doping of the p^+ region.

1. Introduction

There are several types of homojunction bulk unipolar diodes. Two of them are based on the existence of a depleted region, sandwiched between a highly and a lightly doped region leading to the formation of an assymmetric potential barrier (reach-through pnp or npn structures). The structure proposed and realized by SHANNON /1/, the camel diode, has for a given potential barrier an intermediate p-type region which is highly doped so that its thickness is of the order of the mean free path of electrons. The bulk barrier diode realized by MADER /2/ has a lower doped intermediate region with a thickness much higher than the mean free path of the carriers. This structure was succesfully applied for high-gain photodetectors. However, if high-frequency operation with minimum storage effects is wanted a camel diode should be used.

Up to now the technology used to realize in silicon these structures was ion implantation combined with preamorphization and rapid thermal annealing/1,2,3,4/. This paper reports the use for the first time of gas phase epitaxy for fabricating Si hot electron devices. Multilayer structures have been grown with in situ n- and p-type doping using LPVPE.

2. Experimental

The epitaxial growth was performed in a cold wall radiantly heated reactor using $SiCl_2H_2/H_2$, PH_3 and B_2H_6. The growth conditions were: pressure: 2 mbar and temperature 750°C - 900°C. Under certain conditions the epitaxial growth is selective, and the growth rate can be varied in the range of 5 - 30nm/min /5/. The substrates used were 3mΩcm n-type (100) Si wafers , thermally oxidized and patterned with a mask for diodes. After chemical cleaning, the wafers were loaded into the reactor, prebaked at 930°C and then layer sequences of the type $n^-p^+n^+$ were epitaxially deposited. Silicon grows selectively inside the oxide windows, the oxide walls providing a good device isolation. Two types of diode structures were investigated: mesa isolated respectively selectively grown structures. For the top contact Ti/Pt/Au and for the backside AuSb were evaporated. The I-V curves were recorded before and after annealing the contacts at 200°C for 1 min.

3. Results

LPVPE allows epitaxial growth down to 750°C. The epilayers are mirror smooth and the residual disorder is only 3.5% as determined by RBS and

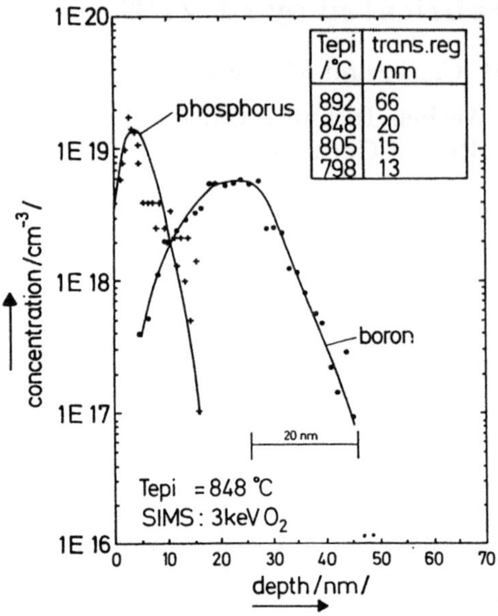

Fig.1 SIMS profiles of P and B in an epilayer grown at 848°C. The insert shows the width of the transition region in the B profile in dependence of the deposition temperature

channeling measurements/5/. In order to realize camel diodes the p-type doping was investigated in respect to abrupt profiles in the temperature range 800° - 900°C. Figure 1 shows doping profiles of P and B of a multilayer grown at 848°C. It results that the transition region over two order of magnitude in the B profile extends over 20nm. The transition region becomes steeper by lowering the epitaxial temperature being for 798°C only 13 nm. For comparison, very low energy ion implantation of B into preamorphized samples leads to a transition region in the B profile broader than 20nm /3/.

For camel diodes three layers ($n^{-}p^{+}n^{+}$) were epitaxially grown at 823°C on 3mΩcm n^{+}-substrate. The thickness of the lightly doped n-region was varied between 0.5 - 0.9μm, the doping was N_{D-}: 10^{16} - 10^{17} cm^{-3}. The thickness of the p^{+} region was varied between 7 and 25 nm, while the highly doped region was always 10nm thick with $N_{D+} \sim 5\times10^{19} cm^{-3}$. I-V characteristics of selectively grown diodes were similar to mesa isolated. However, some selectively grown diodes showed leakage currents, very probably due to facets formation at the vertical interface Si/SiO_2. The results given below were carefully proved to be only minimal affected by leakage currents. Figure 2 shows I-V curves of mesa-etched camel diodes. The currents scale with the diode area which proves the bulk nature of the diode current. Selectively grown camel diodes with different barrier heights are shown in fig. 3. The corresponding ideality factors and the barrier heights are given in table 1.

By solving the Poisson equation for uniform doping of the n^{+}, p^{+} and n^{-} regions and assuming thermionic and diffusion-limited current, the potential and electric field profiles and the I-V curves were calculated for given concentrations and thicknesses of the n^{+}, p^{+} and n^{-} regions. With this simplified model the experimental I-V curves were fitted with calculated ones. The results of this fit are included in table 1. The experimental ideality factors and barrier heights fit quite well with those calculated. One can therefore conclude that there is only little contribution from generation-recombination centres to the current.

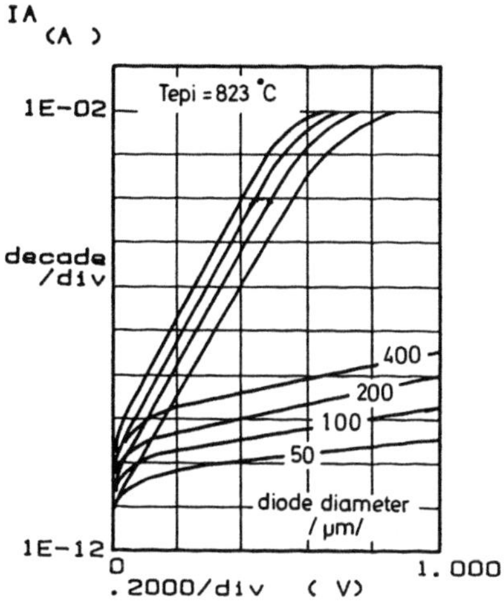

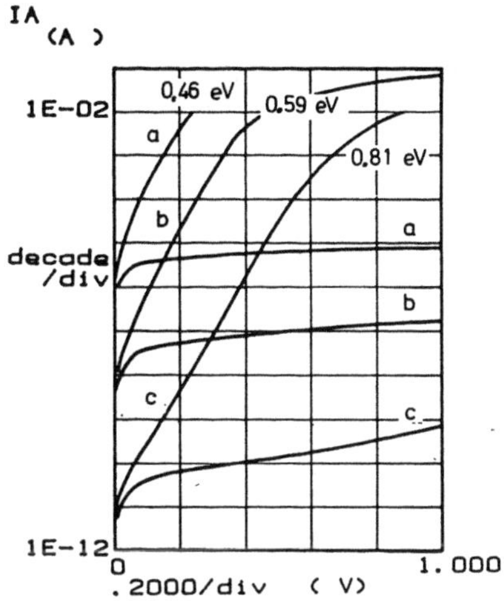

Fig.2 Forward and reverse I-V characteristics for mesa-isolated camel diodes grown at 823°C. Parameter: diode diameter

Fig.3 Forward and reverse characteristics of selectively grown camel diodes; diode diameter was 50μm

Table 1. CAMEL DIODE DATA

Curve	measured				calculated				
	n	Φ_B/eV/	d/μm/	t/nm/	n	Φ_B/eV/	N_A/cm^{-3}/	N_D/cm^{-3}/	t/nm/
a	1.08	0.46	0.58	7	1.15	0.47	8.2x10^{18}	1x10^{16}	7
b	1.16	0.59	0.65	10	1.10	0.58	5.0x10^{18}	1x10^{16}	11
c	1.20	0.81	0.84	20	1.13	0.77	5.0x10^{18}	5x10^{16}	14

t: thickness of the p^+ region , d- thickness of the lightly doped region.

In order to evaluate the influence of diffusion broadening on the barrier height, the Poisson equation was solved numerically using finite differences for arbitrary doping profiles. As the p^+ region is the critical one in the camel diode in the following a nonuniform profile was assumed only for this region. The following assumption was made: the total surface density of acceptors is constant i.e. $\int N(x)dx=$ const. For the volume acceptor density N(x) an analytical form was chosen which looks like a diffusion broadening: $N(x) = N_A \mathrm{x} 0.5 \mathrm{x} (1-2\mathrm{erf}(x-t'/\Delta))$. Figure 4 shows that the effect of acceptor-profile broadening is an increase of the barrier height, an increase of the "kT layer" and a shift of the potential maximum to lower acceptor concentration. Therefore, the corresponding mean free path of electrons increases, also remaining, at least for these concrete camel structures, higher than the "kT layer". Concluding, diffusion broadening will not lead in this case to a transition from thermionic to diffusion-limited current, but minority carrier storage effects are to be expected.

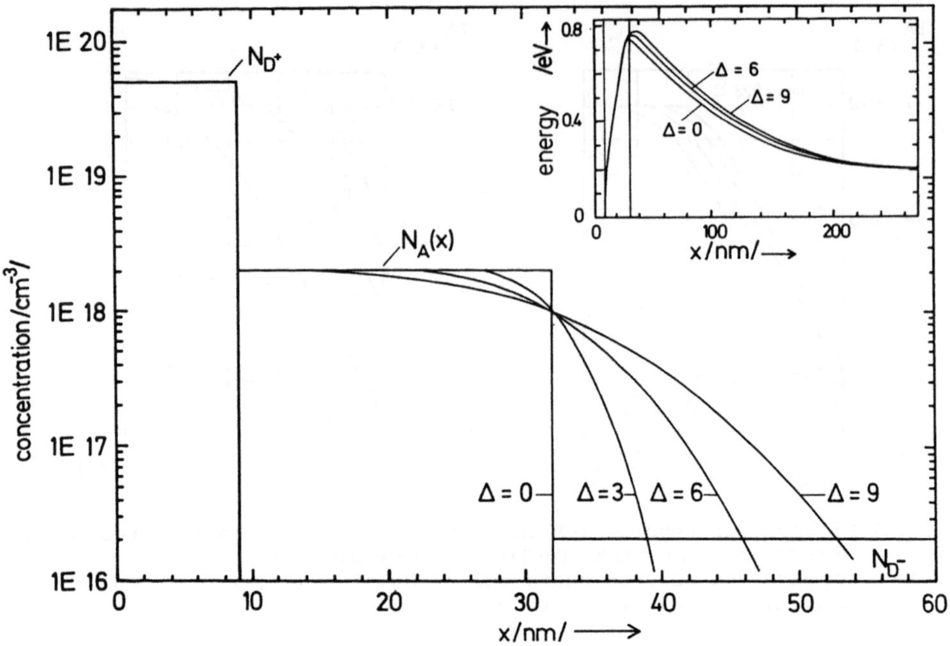

Fig.4 Acceptor profiles N(x) used to evaluate the influence of diffusion broadening on the potential profile; the parameter Δs in nm

In summary, the present investigation demonstrates that: 1) the LPVPE is suitable for the fabrication of Si multilayers, allowing for instance B profiles as steep as 13 nm, 2) further work is necessary for the elimination of facets formation and 3) camel diodes fabricated by LPVPE have ideality factors down to 1.1 and low leakage and low generation and recombination currents which is very promising for ballistic devices.

The authors wish to thank BMFT for the financial support and H.Ottenheim and U.Breuer for the material characterization.

/1/ J.M.Shannon: Appl.Phys.Lett., 35, 63 (1979)

/2/ H.Mader: IEEE Trans.El.Dev., ED-29, 1766 (1982)

/3/ J.M.Shannon and J.B.Clegg: Vacuum, 34, 193 (1984)

/4/ L.Vescan, J.Splettstößer and H.Beneking, Electr.Lett., 22, 493 (1986)

/5/ L.Vescan, H. Beneking and O.Meyer, MRS Spring Meeting, Palo Alto, Cal. April 1986

Impact Ionization Breakdown of GaAs Current Limiters

K. Lehovec

University of Southern California, Departments of Electrical Engineering and Materials Science, University Park, MC-0483,
Los Angeles, CA 90089, USA

Breakdown in GaAs saturated drift velocity resistors is attributed to incipient electron-hole pair generation by electron impact. Transit of the holes leads to a non-uniform field distribution and a breakdown I-V which has been analyzed quantitatively.

1. Introduction

GaAs current limiters consist of an n-doped GaAs channel on semi-insulating substrate with two ohmic contacts [1]. Product of dopant concentration and channel length is chosen to be less than $10^{12}cm^{-2}$ to prevent Gunn domain formation. The current saturates with increasing voltage when electron velocity reaches its steady state saturation limit. These devices are of interest as active load components in inverters and as current overload limiters in gate circuits.

Recently breakdown phenomena have been observed [2] in such current limiters (Fig. 1). The average field at breakdown, about 4×10^4 V/cm, is an order of magnitude lower than the field for avalanche breakdown in GaAs IMPATT diodes and P-I-N detectors. It also occurs at voltages much lower than the 4.8 eV of energy conversion into electron-hole pairs in single event upsets.

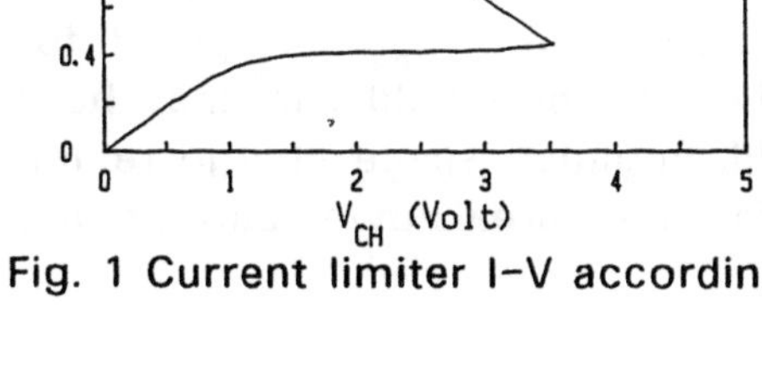

Fig. 1 Current limiter I-V according to [2]

2. General Theory

The breakdown phenomenon is ascribed to incipient impact ionization by hot electrons. However unlike the case of the avalanche diode, the probability αL for an electron to generate an electron-hole pair over the length L of the sample is quite low so that no avalanche occurs. Thus the hole content is $I_n \alpha L\, t_p/q$ where t_p is the lifetime of an impact generated hole. The electron content in the electrically neutral sample is then $W\, D + I_n \alpha L\, t_p/q$ where W is the width of the sample and D is the donor implant dose per unit area. Dividing the electron content by the transit time of electrons t_n and equating it to I_n/q gives

$$I_n = \frac{Wq\, D/t_n}{1 - \alpha L\, t_p/t_n}. \tag{1}$$

Breakdown instability occurs for

$$L\alpha\left(E_{BR}\right) = t_n/t_p. \tag{2}$$

Thus the instability arises from conductivity modulation by electron impact generated holes.

The lifetime t_p of impact generated holes can be the transit time through the sample; a prolonged lifetime by trapping before the hole exits at the cathode; or a recombination lifetime. Noise measurements suggest slow moving charges [3] which we identify with the motion of repeatedly trapped holes. The breakdown phenomenon here described does not occur in transistors since impact generated holes are removed through the gate depletion layer, i.e. their lifetime in the channel of the transistor is negligible. However some affect on the transistor characteristics is possible due to holes injected and trapped in the semi-insulating substrate.

3. Non-uniform Field Distribution After Breakdown

At the onset of breakdown in Fig. 1 there exists a large excess of average field $-V_{BR}/L$ over the minimum field $-V_{SAT}/L$ required for electron drift velocity saturation. Thus the bulk field can snap back providing the voltage for a high field, negative space charge region at the anode where the holes for the breakdown current are generated.

A quantitative analysis of the non-uniform after-breakdown field distribution for the case that holes are removed by transit time sweep-out at the cathode is based on (i) Poisson's Eq.; (ii) the steady state current density being the sum of drift currents by electrons and holes; and (iii) the generation rate of holes by electron impact

$$di_p/dx = i_n \, \alpha(E) = i_n A e^{+BE}. \tag{3}$$

Boundary conditions are V = 0 and p = 0 at x = 0; $V_B^{II} = 0$, $E = E_B$ at $x = x_B$; and $V = V_A < 0$ at x = L. Expressing the electron and hole concentrations in terms of V^{II} and $I^* = i/Nq\,v_n = I/I_{SAT}$, using (i) and (ii), and inserting them into (3) gives a differential equation for $V^{II}(E)$ with the approximate solution

$$E = E_B - \frac{2}{B} \ln \cos a(x - x_B), \tag{4}$$

where $a^2 = \alpha_B^2 P(I^* + v_p/v_n)/2$; $V_0^{II} = (Nq/\varepsilon)(I^* - 1) = (2a/B)\,\mathrm{tg}\,(ax_B)$ and $P = B(Nq/\varepsilon)/2\alpha_B$. The resulting breakdown characteristic

$$V_A = -E_B L + \frac{1}{B\alpha_B/P\,(I^* + v_p/v_n)^{1/2}} \int_0^{ax_B} \ln \cos^2 u\,du \tag{5}$$

is shown in Fig. 2 for various values of P assuming that E_B is independent of current.

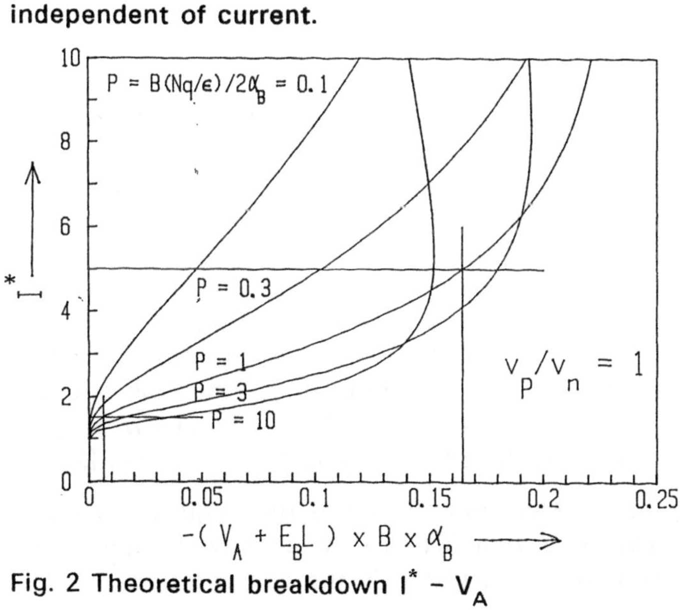

Fig. 2 Theoretical breakdown $I^* - V_A$

4. Comparison with Experiment

The breakdown field $E_{BR} = -V_{BR}/L$ derived from the observed breakdown voltage $V_{BR} = -3.7$ V of Fig. 1 provides α_{BR} by (2) with $t_n/t_p = 2v_p/v_n = 2$. The threshold voltage $V_T = 3.0$ V gives the bulk field $E_B = -V_T/L$. Matching the average slope $S = (5 - 1.5)/0.55 = 6.4\ V^{-1}$ of the measured breakdown I-V to the theoretical curves of Fig. 2 gives $B\,\alpha_B$ as function of P. From $\alpha_{BR}/\alpha_B = \exp(\Gamma\,\alpha_{BR}/\alpha_B)$ with $\Gamma = B\,\alpha_B(E_{BR} - E_B)/\alpha_{BR}$ we obtain α_B, then B and then N by using P.

Pertinent values are listed in Table 1, for $L = 10^{-4}$cm.

Table 1

	P = 0.3	P = 1
α_B	$1.86 \times 10^4 cm^{-1}$	$1.78 \times 10^4 cm^{-1}$
B	1.02×10^{-5} cm/V	1.69×10^{-5} cm/V
N	$0.68 \times 10^{16} cm^{-3}$	$1.31 \times 10^{16} cm^{-3}$
A	$1.37 \times 10^4 cm^{-1}$	$1.07 \times 10^4 cm^{-1}$

5. Conclusions

Sample length and derived dopant concentration agree with nominal design objectives. Breakdown voltage is expected to increase in proportion to sample length. However, external influences such as a single event upset may trigger the breakdown mode at any lower voltage exceeding V_T. The origin or significance of the current-independent bulk field $E_B = 3 \times 10^4$ V/cm might be sought in the requirement for holes to reach drift velocity saturation, or else to "preheat" the electron to generate a hole by collision in the space charge layer. The impact generated holes leave the high bulk field region at the cathode and may assume there thermal energies before recombination. This explains the observed narrow band to band emission of recombination radiation from these devices [2].

I thank R. Zuleeg of the McDonnell Douglas Astronautics Company who prepared the devices, J. Roach and H. Wieder who supplied the measurements and H. Bao who assisted in the computations.

1. K. Lehovec, R. Zuleeg: IEEE Trans. ED-27, 1074 (1980)
2. J. Roach, H. Wieder, R. Zuleeg: Workshop on Microwave Materials and Devices, San Francisco, California, January 10–13, 1986
3. A. Peczalski, A. Vander Ziel, R. Zuleeg: Solid-State Electron. 26, 861 (1983)

The New High Speed Devices: The Barrier Transistor and the TEG-Base Transistor

C.Y. Chang, W.C. Liu, and Y.H. Wang

Semiconductor and System Laboratories, National Cheng Kung University, P.O. Box 7-290 Tainan, Taiwan, Republic of China

Two new high speed transistors, namely the V-groove barrier transistor (VBT) and the V-groove Injection -Induced transistor (VI^2T) are presented. The former device (VBT) possesses MBE grown $n^+i\delta p^+in^+$ structure, which is a majority carrier (hot) transport device with a modulation of the barrier directly by the V_{EB} bias. The latter one (VI^2T) possesses the same transport mechanism, except the structure is different by using a two dimensional electron (or hole) gas as the base, therefore the structure is n^+GaAs (emitter)/graded n^-AlGaAs/i GaAs (TEG base)/n^-AlGaAs/n^+GaAs. These devices are very promising due to many significant advantages over some other state-of-the-art devices.

1. INTRODUCTION

The V-groove technique applied to the ultra thin base transistor provides an easy way to realize the barrier transistor /1/ (VBT) and the TEG-base transistor /2,3/.

The cross-section of the grooved barrier transistor is $n^+i\delta p^+in^+$ in which δp^+ is the very thin base (100Å~500Å). The δp^+ base contact pad was formed by depositing Al on the etched V-groove along the emitter periphery. The same technique was applied to the TEG-base transistor to make a contact to the base. The cross-section of the device is n^+/i-$Al_xGa_{1-x}As$($x=0\rightarrow0.4$)/ i-GaAs quantum well base /i-$Al_{0.4}Ga_{0.6}As$/n^+GaAs.

A two-dimensional electron gas is formed in the quantum well base. The former device possesses the very simple structure with fairly good operational performances (f_T~30GHZ). The latter device possesses very good high frequency high power capability ($f_T\geq$200GHZ) with very low noise. In following section, we will talk about the VBT first followed by TEG-base transistor and finally a summary will be given.

2. V-GROOVE BARRIER TRANSISTOR

In a barrier transistor, the potential barrier is produced by a heavily doped, very thin layer [designated as $\delta(p^+)$ or $\delta(n^+)$] lying between two lightly doped regions. In this paper, we present the three-terminal n^+-i-$\delta(p^+)$-i-n^+ V-groove barrier transistor (VBT) which has been successfully fabricated by molecular beam epitaxy (MBE). In order to make contact to the $\delta(p^+)$ base, the V-groove was etched along the emitter periphery. Aluminum was evaporation deposited to the V-groove and ohmic contact was obtained to the $\delta(p^+)$ thin base layer. Fig.1 shows the structure and the barrier formed.

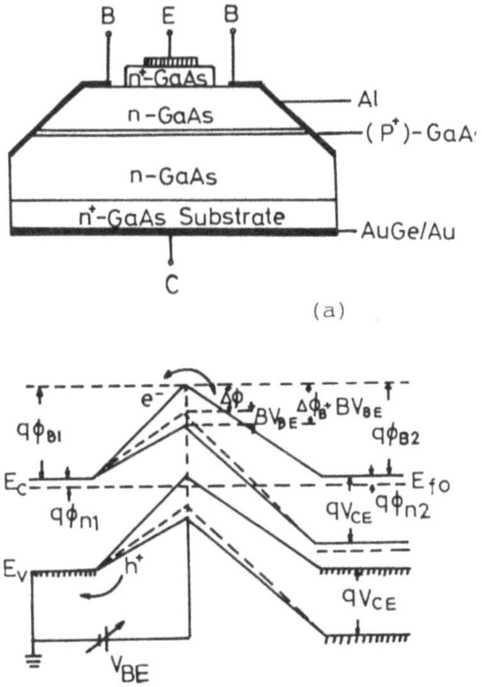

Fig.1 (a) The cross-section of the V-groove barrier transistor.
(b) The potential barrier of the VBT under V_{CE} and V_{BE} bias conditions.

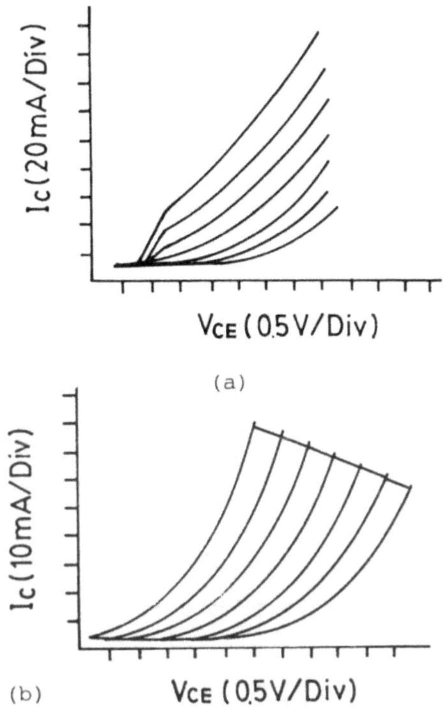

Fig.2 The I_{CE}-V_{CE} characteristics with V_{BE} as parameters (from right to left 0.5V per step from zero), Vert: 10mA/step.

The demonstrated devices have many advantages: (1) ease of fabrication; (2) higher frequency (f_T >30GHz) operation; (3) high-power handling capability; (4) high reliability due to no interface problem. (e.g. AlGaAs/GaAs, MIS etc.) The reason for (2) is owing to the almost negligible transit time but is related to the delay time of $r_{bb'}$, C_b, and $(gm/c)^{-1}$ which are almost independent of area, thus high-power operation is feasible. Furthermore, VBT possesses low-noise capability due to carriers transporting by thermionic emission and being controlled by the base-emitter bias. Fig.2 shows the I/V characteristics.

3. INJECTED AND INDUCED BASE TRANSISTOR

A novel hot-electron transistor, named injected and induced base transistor (I^2BT), has been fabricated successfully by MBE technique /4/. Fig.3 and Fig.4 show the structure and the barrier formed between emitter and collector. This device preserves the main attractive feature (high speed) of a metal base transistor, while avoiding its inherent drawback of a low common base current gain α. The key new idea of I^2BT is that of an induced and injected rather than doped-base, which can be fabricated as thin as 100Å without a loss in its sheet conductance. The undoped barrier structure

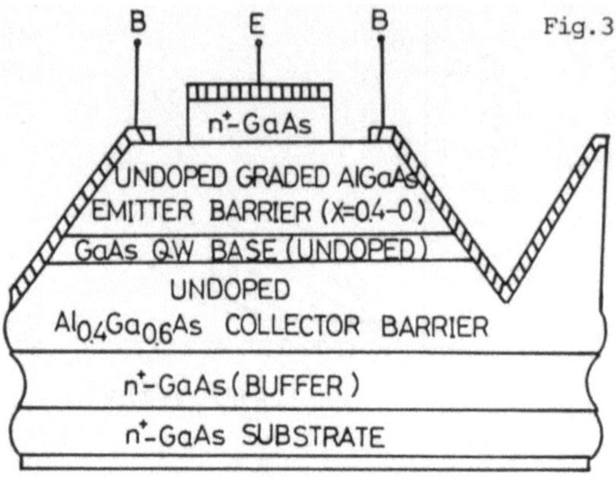

Fig.3 The cross section of the induced transistor.

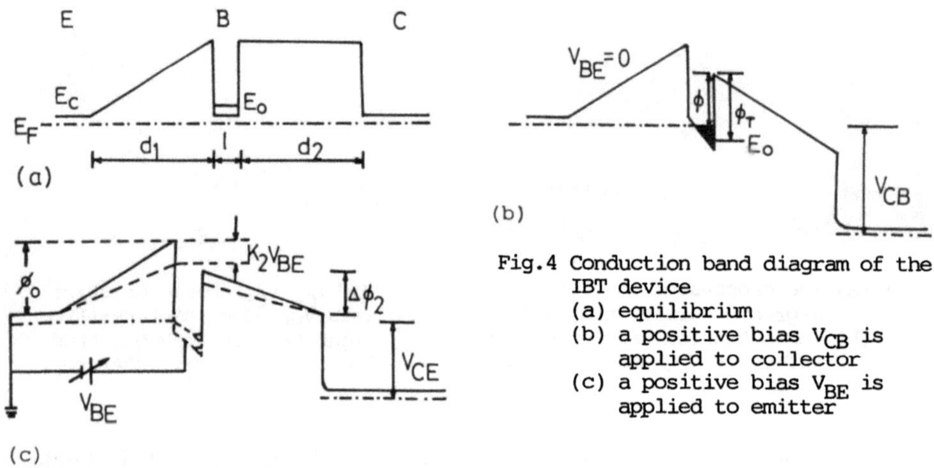

Fig.4 Conduction band diagram of the IBT device
(a) equilibrium
(b) a positive bias V_{CB} is applied to collector
(c) a positive bias V_{BE} is applied to emitter

is an advantage of the I^2BT, one should minimize the variations in potential barrier height owing to dopant fluctuations. The V-groove etched down to the base layer to make base contact provides the easy fabrication of I^2BT. Fig.5 shows the I/V characteristics. The common base current gain has been measured to be 0.96 under a collector bias of 2.5V and emitter current of 3mA. Fig.6 shows the base sheet resistance changes with biases.

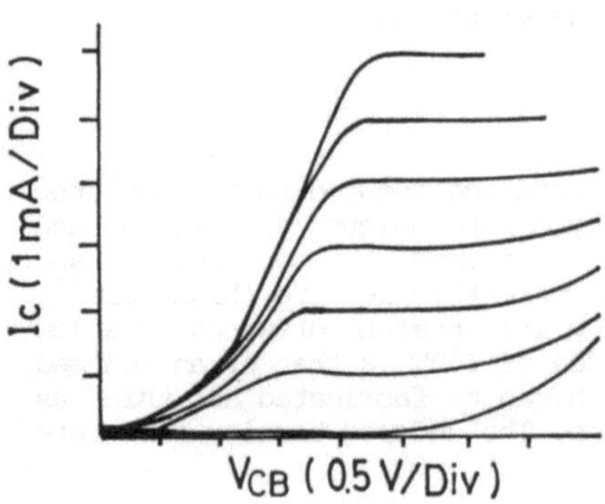

ig.5 I_C-V_{CE} characteristics of IBT with V_{BE} as parameters (from right to left 0.2V per step from zero)

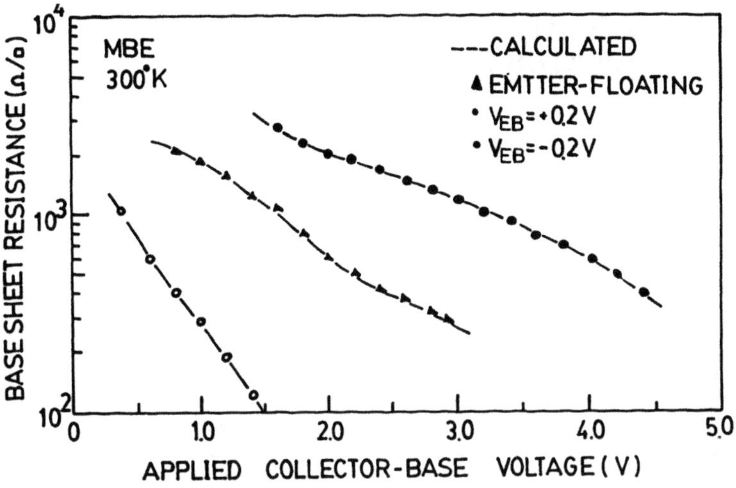

Fig.6 The relationship between the sheet base resistance and the applied collector-base voltage.

4. SUMMARY

Two kinds of new high speed transistors, namely the V-groove barrier transistor and the V-groove quantum well base transistor are presented. Both of them adopted the simple fabrication method of V-groove. However, the former possesses high speed (30GHz), low noise while the latter (I^2BT) bears the ever higher speed (~200GHz) and high gain ($\alpha \sim 0.96$) characteristics.

REFERENCES

1. C.Y. Chang, Y.H. Wang, W.C. Liu and S.A. Liao: IEEE Trans. Electron Device Lett. Vol-EDL-6 (1985) p.123.
2. S.Luryi: IEEE Trans. Electron Device Lett., Vol.EDL-6 (1985) p.178.
3. The idea of using two-dimensional carrier gas as the base has also been proposed by Chang, see, for example, C.Y. Chang, Y.H. Wang, W.C. Liu, S.A. Liao and K.Y. Cheng: Appl. Phys. Lett. Vol.46 (1985) p.1084.
4. C.Y. Chang, W.C. Liu, M.S. Jame, Y.H. Wang, S. Luryi and S.M. Sze: IEEE Electron Device Letters, EDL-7, No.9 (1986).

Monte Carlo Investigation of the High Electron Mobility Transistor*

U. Ravaioli and D.K. Ferry

Center for Solid State Electronics Research, Arizona State University, Tempe, AZ 85287, USA

The High Electron Mobility Transistor (HEMT) exhibits a conduction channel at the AlGaAs/GaAs interface, where the electrons are distributed in quantized subbands and constitute a quasi-two-dimensional electron gas (Q2DEG). We present, in this work, an investigation of submicron HEMT structures, based on an Ensemble Monte Carlo (EMC) model, in which the Q2DEG in the conduction channel is modeled using a two subband triangular well approximation. This simple approach allows an easy determination of the levels and of the wavefunctions in the subbands, which are necessary to evaluate the 2-D scattering rates as a function of the electron positions in the channel. This model is the development of the one presented in [1]. In particular, scattering of hot electrons, between 2-D confined states in the channel and 3-D states, has been investigated in detail, including intervalley scattering to the L-valley. In this way, the mechanisms responsible for real-space transfer of hot electrons from the conduction channel into the AlGaAs layer, as well as the confinement properties of the channel, are properly taken into account.

The first two subbands are considered for the Q2DEG conduction, assuming acoustic and polar optical phonon scatterings , while impurity and remote impurity scatterings are neglected in the channel, since the GaAs is not appreciably doped and a buffer layer is present in the AlGaAs at the interface. The 3-D electrons are treated using a full three-valley EMC model which includes all the major scattering mechanisms. The 2-D scattering rates are essentially dependent on the local subband levels, determined by the local carrier concentration in the channel. In order to account for the non-uniformity of the channel, we fix a reference surface-charge density which is not likely to be exceeded, and obtain the total scattering rate Γ_{max}. We use this to calculate the self-scattering rate and to normalize all scattering probabilities. At the end of each free flight the charge density n_s around the electron is evaluated to select the scattering rates for the random choice of the scattering event.

Poisson's equation is solved at intervals of 10^{-14}s during the simulation time. A finite difference three-level multigrid algorithm is used for the solution [2,3]. This method consists of a series of partial iterative relaxations on grids with increasing mesh size. The intermediate solution is passed between levels using interpolation, and the procedure is stopped when convergence is achieved on the finest grid. The multigrid methods usually converge much faster than other iterative schemes, because long spatial wavelength error components of the solution are efficiently eliminated in the beginning by the iterations on the coarser grids. The algorithm is also ideal for an extension to local refinement schemes, since finer grid levels can be introduced in any

*This work has been supported by the U.S. Army Research Office.

region of the device, and this can be applied to the narrow conduction channel of the HEMT.

The device structure shown in Fig. 1 has been used for the simulation. The distance between source and drain is 0.7μm and the gate length is 0.3μm. The thickness of the AlGaAs buffer layer is 60nm, of which 10nm constitute the undoped buffer layer. The number of electrons simulated is chosen so that 20,000 electrons would be present in the volume considered, without depletion effects. The donor density in the AlGaAs is 2×10^{17} cm^{-3}. The doping level in real devices is usually higher, but this would require the introduction of degeneracy effects in the model, with a considerable increase of computational costs. The energy discontinuity at the hetero-interface is assumed to be 0.25eV.

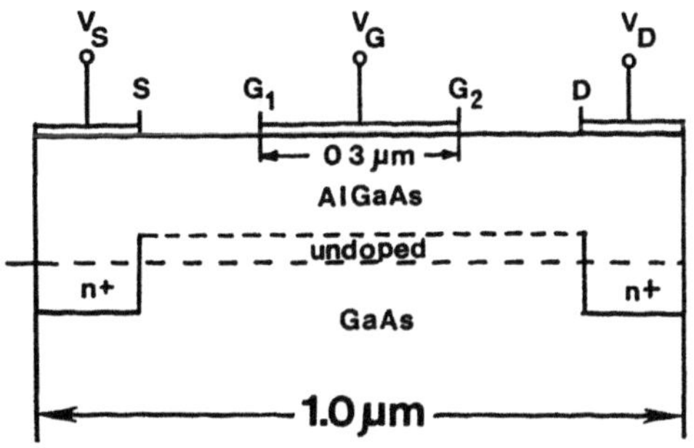

Fig. 1 - Structure of the HEMT used in the simulations.

Figure 2 shows the drain current I_{ds} as a function of the drain to source voltage V_{ds}, for various gate bias conditions. Φ_B indicates the Schottky barrier height at the gate contact. The values are scaled for a transistor with lateral width of 100μm. The curve corresponding to flat-band condition on the gate ($V_g-|\Phi_B|=0.V$) is also included for reference.

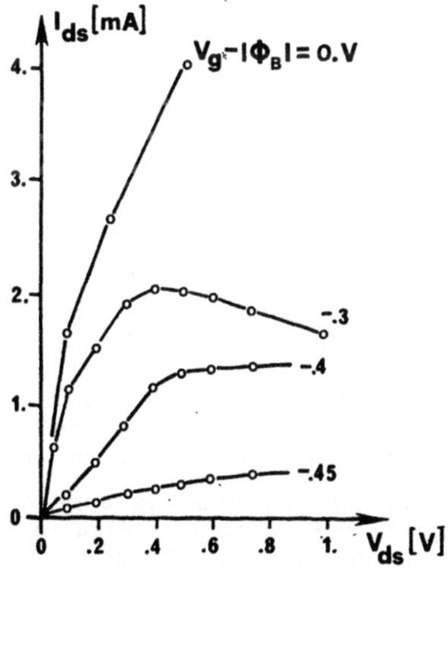

Fig. 2 - $I_{ds}(V_{ds})$ for various gate bias conditions.

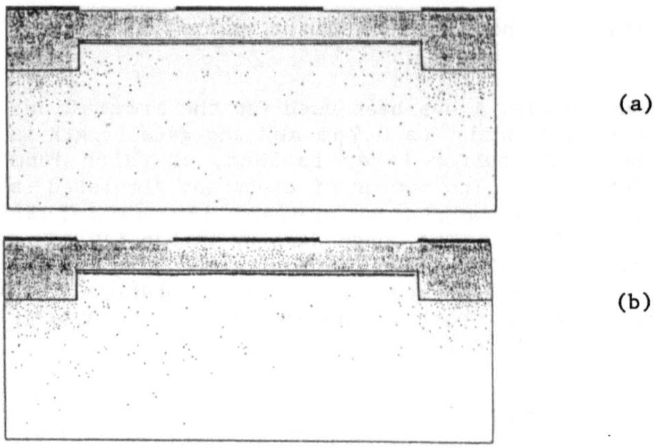

Fig. 3 - Carrier distribution inside the device for V_{ds}=0.2V (a), and V_{ds}=0.5V (b), with $V_g-|\Phi_B|$=-0.3V.

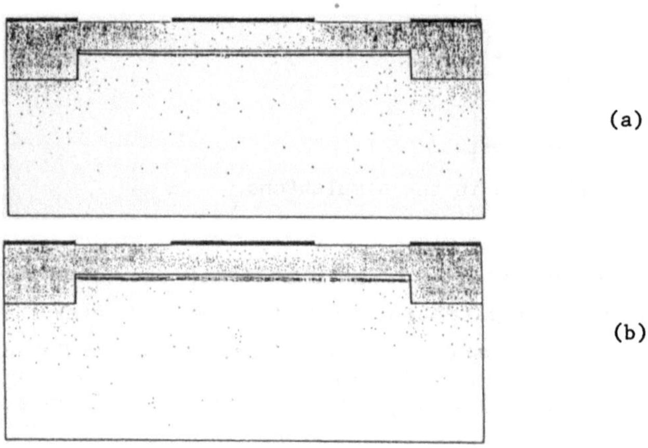

Fig. 4 - Carrier distribution inside the device for V_{ds}=0.2V (a), and V_{ds}=0.5V (b), with $V_g-|\Phi_B|$=-0.4V.

In this case the current can flow freely through the AlGaAs layer. Conduction through the AlGaAs plays an important role in the case when $V_g-|\Phi_B|$=-0.3V . At higher drain voltages, a parallel channel in the AlGaAs is created, with a reduction in the drain current, thus suggesting the presence of a negative differential mobility regime. The AlGaAs has in fact much lower mobility, due to the high impurity scattering and to the higher effective mass. The satellite valleys are also much closer in energy to the Γ valley than in GaAs, and intervalley scattering easily occurs. For $V_g-|\Phi_B|$=-0.4V, the potential barrier between source and drain, due to the Schottky gate, is considerably larger, and the current through the AlGaAs layer at high drain voltage is only minimal. Figures 3 and 4 show the carrier distribution inside the device for the two gate bias conditions discussed above, with V_{ds}=0.2 and 0.5V, and from these is possible to appreciate the amount of carrier flow through the AlGaAs layer.

The model allows us to keep track of the particles crossing the heterojunction, therefore an analysis of real-space transfer of carriers from the channel into the AlGaAs layer is possible. At lower drain voltages, transfer occurs mainly after 2D-3D scattering within the Γ-valley, while, for increasing drain voltages, transfer through intervalley scattering becomes more and more predominant. For the choice of geometry and doping made, real-space transfer is not strong enough to cause effects of negative differential mobility. This is more likely to occur in devices with larger spacing between gate and drain, and with higher channel concentration, which tends to shift the quantized energy levels closer to the edge of the triangular well. In the simulations presented here, the ratio between the number of electrons which leave the channel through real-space transfer and the net number of electrons injected into the n+ regions of the contacts, is on the order of 10-15% for the higher drain voltages considered.

References

[1] U. Ravaioli and D.K. Ferry, IEEE Trans. on Electron Dev., ED-33, 677 (1986).

Excess Gate Current Due to Hot Electrons in GaAs-Gate FETs

D.J. Frank, P.M. Solomon, D.C. La Tulipe, Jr., H. Barattе, C.M. Knoedler, and S.L. Wright

IBM Thomas J. Watson Research Center, P.O. Box 218, Yorktown Heights, NY 10598, USA

We report on the first observations of hot electron effects in the gate current of GaAs-gate FET's. We have made and tested a variety of these FET's using MBE-grown material with 60nm and 35nm thick $Al_{.4}Ga_{.6}As$ gate insulator layers. In measurements at 300K and 77K these devices show drain-voltage-dependent gate current substantially exceeding that which would be expected on the basis of simple vertical transport measurements. We attribute this current to the real-space transfer of hot electrons from the channel of the device into the (Al,Ga)As/GaAs barrier, from which they are collected into the gate.

The GaAs-gate FET is a new variety of heterojunction transistor which has been described in several recent articles [1,2]. It consists of an n^+GaAs gate, an (Al,Ga)As layer which acts as a gate insulator, and undoped GaAs for the channel. It offers several improvements over the conventional MOdulation-Doped FET (MODFET), including improved threshold control and the absence of DX-center related problems. We report new, detailed measurements of the characteristics of this device which show that it exhibits a large increase in gate current due to hot electron effects under certain conditions. A somewhat similar result has quite recently been reported for Heterostructure Insulated Gate FET's (HIGFET's) [3]. The details of these effects should prove quite valuable in the study of hot electron dynamics, which become increasingly more important to understand as dimensions continue to be scaled down.

We have investigated these effects in a variety of GaAs-gate FET's made on MBE material with either 60nm $Al_{.4}Ga_{.6}As$ or 35nm $Al_{.5}Ga_{.5}As$ insulator layers. These FET's were fabricated in the same manner described previously [1], using a self-aligned process in which the gate metallurgy serves as part of the mask for the ion-implanted source and drain. The resulting devices had gate lengths ranging from 0.7μm to 20μm. Figure 1 shows a schematic cross section of the device.

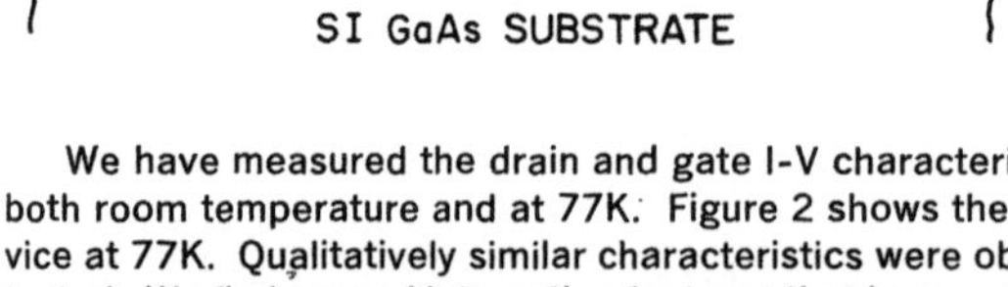

Figure 1. Schematic cross section of a GaAs-gate FET.

We have measured the drain and gate I-V characteristics of many of these GaAs-gate FET's at both room temperature and at 77K. Figure 2 shows the drain and gate I-V curves for a typical device at 77K. Qualitatively similar characteristics were obtained for nearly all of the working devices tested. We find several interesting features that have not previously been discussed in the literature in connection with these devices. There is a negative differential resistance (NDR) region in the drain current I_D vs drain voltage V_D characteristics at high gate voltage, as shown in Fig. 2(a). There is also slight NDR in the gate current I_G vs gate voltage V_G curves at moderate gate voltage and low

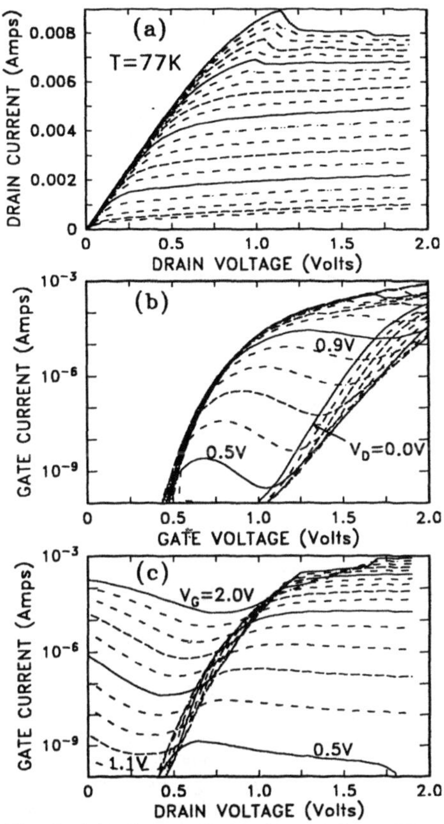

Fig. 2. I-V characteristics for a $1.5\mu m \times 25\mu m$ FET at 77K. (a) shows I_D vs V_D for $V_G = 0$ to 2.0V in steps of 0.1V. (b) shows I_G vs V_G for $V_D = 0$ to 1.9V in steps of 0.1V. (c) shows I_G vs V_D for $V_G = 0.5$ to 2.0V in steps of 0.1V (the lowest curve on the left is $V_G = 1.1V$ and on the right is $V_G = 0.5V$).

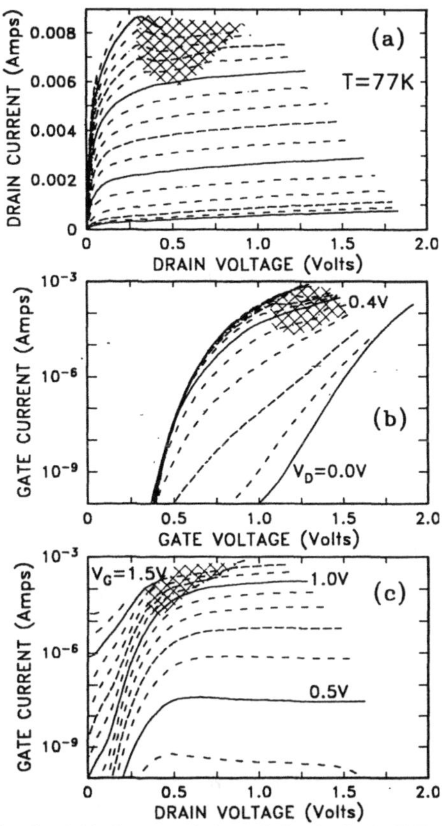

Fig. 3. I-V characteristics of the intrinsic FET deduced from Fig. 2, calculated using $R_S = 45.8\Omega$, $R_D = 48.8\Omega$, and a parasitic gate resistance of approximately 440Ω. (a) shows I_D vs V_D for $V_G = 0$ to 1.8V in steps of 0.1V. (b) shows I_G vs V_G for $V_D = 0$ to 1.6V in steps of 0.1V. (c) shows I_G vs V_D for $V_G = 0.4$ to 1.8V in steps of 0.1V.

drain voltage, as in Fig. 2(b). Finally, the gate current is generally higher than would be expected in the absence of hot electrons and depends strongly on V_D, as can be seen from Fig. 2(b) and (c). It decreases at low V_D primarily due to the voltage drop across the parasitic source resistance, which decreases the channel-to-gate voltage. It then increases sharply because of hot electron injection from the channel into the gate. This hot electron injection current finally saturates at larger values of drain bias, while still increasing with gate bias.

At room temperature the characteristics of these devices are similar. The drain I-V curves are qualitatively the same as Fig. 2(a), except that the magnitudes of the drain currents are reduced due to the lower transconductance. The gate leakage current continues to show the increase with V_D, but the thermal leakage is much higher and overwhelms the NDR effect seen in Fig. 2(b).

As a function of gate length, we find that at 77K the hot electron current is roughly constant for gate lengths of 0.7, 1.3 and 1.7μm, decreases somewhat for 4μm, and is still measurable in 20μm devices. At 300K, the current falls off more rapidly as a function of gate length. This is probably because more voltage drops along the channel of the device due to the lower room temperature mobility.

Unlike the HIGFET, where real-space transfer may directly cause the NDR [3], we find that in our GaAs-gate FET's the NDR features can be accounted for by the parasitic resistances at the three contacts. The NDR on the drain curves occurs when the rapidly rising I_G in this regime, passing through the gate contact resistance, causes the internal gate voltage to fall fast enough that the drain current actually decreases. This NDR causes circuit dependent RF oscillations which distort the average DC currents and produce the sharp slope discontinuities visible in the data. We have observed these oscillations to be in the 100kHz to 100MHz range, depending on circuit conditions. There should also be some decrease in I_D due to the transfer of current to the gate, but this effect seems to be masked by the mechanism above. The NDR in the gate I-V curves is due to the parasitic source and drain resistances. As the gate voltage increases, I_D increases, causing the internal source-to-drain voltage to decrease. If it decreases rapidly enough, the hot electron injection current can actually decrease, even though V_G is increasing. These negative resistances are not surprising when one notes that the hot electron injection effect is inherently regenerative, since the current flows in opposition to the drain bias. In general, when a regenerative effect is combined with passive external circuit elements it can lead to NDR characteristics.

To study the curves quantitatively and to uncover the hot electron injection characteristics of the intrinsic device, we used an HP4145A parameter analyzer to take detailed I-V curves (1599 points each for I_D and I_G) for several FET's at both 77K and 300K. These were then transferred to a mainframe computer which was used to subtract the parasitic resistance voltages. Interpolating among the resulting data yielded I-V curves that correspond to the intrinsic device.

The sum of the source and drain resistances, R_S+R_D, was determined from an extrapolation based on the initial slope of the drain I-V curves as V_G became large. $R_S - R_D$ was deduced by comparing the transconductance at moderate voltages with the transconductance at the same voltages when the source and drain were reversed, assuming the intrinsic FET to be symmetric. For the FET in Fig. 2 we obtained $R_S = 45.8\Omega$ and $R_D = 48.8\Omega$. When these two resistances are taken into account, the negative differential resistance in the gate I-V curves (Fig. 2(b)) is removed. The parasitic gate resistance I-V curve was estimated by assuming that the source transconductance, $g_S = \partial I_S / \partial V_G$, which typically peaks around $V_G = 0.6V$, ought to remain constant for higher values of V_G (due to velocity saturation), and that the extent to which it does not is solely caused by voltage developed on the parasitic gate resistance. This approximation neglects a slight decrease in intrinsic transconductance due to gate depletion with increasing V_G. The parasitic gate resistance estimated in this fashion for the FET in Fig. 2 is 440Ω for I_G below $500\mu A$ and decreases somewhat for higher currents. This is in reasonable agreement with estimates based on the resistivity of the n^+GaAs gate and on measurements of the contact resistance of the gate metallurgy.

Figure 3 shows the intrinsic FET characteristics deduced for the device in Fig. 2 by removing the parasitic resistances. The only remaining NDR is in the cross hatched areas which indicate regions of distortion due to the RF oscillations which accompanied the original measurements. All other NDR effects are indeed removed by this process. Furthermore, it appears likely that if the distortions could be removed there would not be any NDR in the cross hatched areas either. This is not absolutely certain, however, since in other experiments we have observed NDR in the I_D of asymmetric FET's even in the absence of hot electron gate current. Having removed all of the parasitic effects, Figs. 3(b) and (c) now show very clearly the intrinsic excess gate current, which we attribute to hot electron injection.

As Fig. 3(c) shows, the hot electron injection current rises very rapidly for intrinsic V_D between 0.2 and 0.5V and tends to saturate for V_D above 0.5V. While the details of these characteristics are difficult to account for because of the complex nature of the two-dimensional FET geometry, the rapid rise in I_G can be explained in general terms as follows. As the source-drain field increases, it heats more and more of the electrons in the GaAs channel to energies exceeding the barrier height posed by the (Al,Ga)As insulator. Scattering then causes real-space transfer [4] of these electrons into the conduction band of the (Al,Ga)As. Since, unlike MODFET's, there is no barrier to prevent these electrons from flowing into the gate, they do so and are collected as leakage current. Since this mechanism requires a 'cold' electron at approximately source potential to be able to acquire enough energy to pass over the (Al,Ga)As barrier into the gate, it should be activated when both the source-drain voltage and the gate voltage exceed the barrier height ($\sim$ 0.3eV), as is indeed the case. We note that this effect is similar to those seen in NERFET's [5], which is to be expected, since the structures are similar. The reason the effects are smaller in GaAs-gate FET's is that they only collect those hot electrons that are directly under the gate, while in the NERFET, the collector electrode extends under the entire device and hot electrons are collected everywhere.

We believe the saturation of the electron injection at higher V_D values is due to the field distribution in the FET when drain voltages are high enough to create a stationary Gunn-domain. We estimate that domain formation should occur when V_D is about 0.5V (for $\sim 1\mu m$ gate length), which roughly agrees with the voltage at which saturation occurs in our data. Furthermore, two dimensional device simulations using a drift/diffusion model [6] show that for $V_D > 0.5V$ the fields are such that only a limited number of hot electrons can both transfer into the (Al,Ga)As and be collected by the gate. As V_D increases, more electrons can real space transfer into the (Al,Ga)As, but the fields in the (Al,Ga)As direct most of them back into the channel, leaving I_G approximately constant. A different way of stating this is that most of the hot electron gate current comes from the leading edge of the domain, which is shielded from further increases in V_D by the domain itself. Other physical processes which may be involved in this behavior include the transfer of hot electrons to upper valleys and reduced wide angle scattering of hot electrons at high source-drain fields.

This hot electron gate current should not adversely effect the usefulness of the GaAs-gate FET in logic or memory applications. We have not seen semi-permanent threshold voltage shifts like those in MOSFET's, which leads us to conclude that electrons do not become trapped in the (Al,Ga)As. Furthermore, this excess gate current is only large when both V_G and V_D are large. In logic circuits the only time they are both large is during the switch-on transient of a device. Gate current during the transient can slow the switching slightly, but for reasonable logic levels even the maximum possible gate currents are insignificant. For example, if the logic level were 1.0V, the largest possible gate current for the device in Fig. 2 would only be about $10\mu A$. Circuit simulations show that for supply voltages below 1.5V, the increase in delay is less than 1%.

In summary, we have displayed several interesting features in the I-V characteristics of GaAs-gate FET's. These include NDR in the drain and gate I-V curves and substantial excess gate current which is dependent on source-drain voltage. We have shown that the NDR results from the device currents coupled with the parasitic contact resistances of the device. We have carried out a computer-aided analysis of the data to remove the parasitic effects and to display the intrinsic device characteristics. These characteristics show an excess gate current which rises very rapidly for V_D and V_G of order the GaAs/(Al,Ga)As barrier height and saturates for somewhat higher drain voltages. We believe these effects to be caused by real-space transfer of hot electrons from the channel of the device into the (Al,Ga)As, from whence they are collected into the gate.

1. P. M. Solomon, C. M. Knoedler and S. L. Wright: IEEE Elec. Dev. Lett., EDL-5, 379 (1984)
2. K. Matsumoto, M. Ogura T. Wada, N. Hashizume, T. Yao and Y. Hayashi: Electron. Lett., 20, 462 (1984)
3. M. S. Shur, D. K. Arch, R. R. Daniels and J. K. Abrokwah: IEEE Elec. Dev. Lett., EDL-7, 78 (1986)
4. K. Hess, H. Morkoc, H. Shichijo and B. G. Streetman: Appl. Phys. Lett., 35, 469 (1979)
5. A. Kastalsky, S. Luryi, A. C. Gossard and R. Hendel: IEEE Elec. Dev. Lett., EDL-5, 57 (1984)
6. J. Y.-F. Tang: IEEE Trans. Elec. Dev., ED-32, 1817 (1985)

Potential Barriers in Doped GaAs by OM-VPE

P. Roentgen, G. Fernholz, and H. Beneking

Institute of Semiconductor Electronics, Aachen Technical University, Sommerfeldstraße, D-5100 Aachen, Fed. Rep. of Germany

Controlled growth of ultra thin doped GaAs layers during atmospheric pressure OM-VPE is investigated in detail. It is demonstrated that by choice of layer parameters potential barriers can be precisely adjusted in unipolar semiconductor allowing the realization of novel high speed electronic devices. Growth of planar doped GaAs:Se makes it possible to create potentials in which a 2-dim. electron gas is confined. Thin p-doped layers allow the realization of GaAs Schottky-barrier-enhanced- and camel-diodes. The results promise optimum conditions for the realization of high speed GaAs hot electron devices.

1. Introduction

The advanced technologies for vertical structurization offer new possibilities for physics and high speed electronic devices. Due to the possibility of selecting nearly any desired dopant, OM-VPE is an interesting candidate for growth of thin doped multilayers. The aim of our investigations is to study the applicability of the OM-VPE process to adjust precisely potential barriers by the growth of thin doped GaAs layers. These potential barriers make it possible to create new electronic devices, where carriers are confined or the current flow of carriers can be controlled /1,2,5/.

2. Experimental

Growth experiments are performed at atmospheric pressure in a lamp heated OM-VPE reactor described earlier /3/. TMG and AsH_3 are used as sources for GaAs in a H_2 carrier gas flow of $4000cm^3$/min corresponding to a flow velocity of 10cm/s. Typical growth rates of 0.5µm/h are adjusted using a solid TMG source at V/III-ratios of 10. 1000ppm H_2Se in H_2 is used as n-doping gas whereas due to the non-present memory effect, M_2Cp_2Mg is used as p-dopant source /3/.

3. Growth and characterization of planar doped GaAs:Se

Planar doped GaAs:Se is inserted between undoped GaAs layers without interrupting growth. During Se incorporation the TMG flow is reduced and if the growth rate scales down with the flow of TMG, a growth rate of 50pm/s results. Under these conditions, incorporation of Se mainly depends on growth time and vapor pressure of Se over GaAs, resulting in a time dependence as obtained in a conventional adsorption process /4/. This incorporation mechanism makes it possible to adjust sheet carrier concentrations up to $5x10^{13}cm^{-2}$. This is about a factor of 10 higher than achievable by Si doping and due to the non-amphotheric behavior the Hall mobilities are significantly higher than those of comparable Si-doped samples /4/.

The planar doping permits the creation of V-shaped potential wells as shown in Fig. 1 in isotype GaAs where the conduction band is pulled several 100meV below the Fermi level depending on the sheet donor concentration /5/. Hence the effective Γ-L separation is significantly reduced. The I-V-characteristics of 100μm FET channels show clearly negative differential resistance and reproduce the v(E)-distribution. The 2-dim. nature of electrons confined in several subbands of the potentials wells is verified by magnetotransport measurements and the subband energies are determined by tunnelling spectroscopy /5/. Fig. 2 shows Shubnikov-de Haas oscillations of electrons in 2 populated subbands for B⊥N measured on a sample with $N_\square = 1.3 \times 10^{12} cm^{-2}$. A comparison of the results with self-consistent calculations shows no measurable broadening of the atomic sheet Se-donor distribution and suggests ultra sharp Se-transitions /5/.

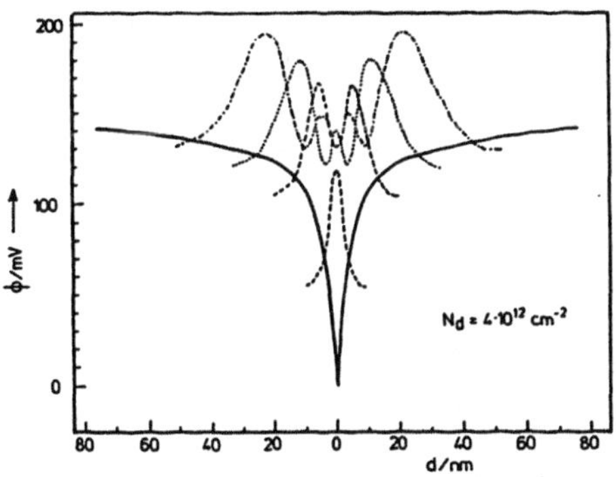

Fig. 1: Calculated potential and wavefunctions for planar n-doped GaAs with $N_\square = 4 \times 10^{12} cm^{-2}$

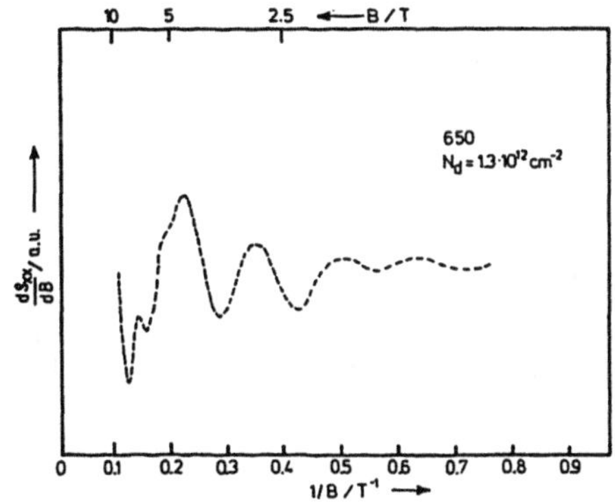

Fig. 2: Shubnikov-de Haas oscillations for B⊥N measured on a planar doped GaAs:Se sample with $N_\square = 1.3 \times 10^{12} cm^{-2}$

4. Growth of thin p-doped GaAs layers for majority carrier devices

Although sharp Zn doping profiles can be realized by OM-VPE, strong adsorption and memory effects limit the use of this dopant /3/. Therefore Mg is choosen as alternative p-dopant. Whereas growth of p-doped layers with thicknesses > 100 nm causes no problems and memory effects are not observed in the case of Mg, reproducible growth of thinner layers becomes difficult. Our investigations suggest that adsorption of dopant as molecules on reactor walls play an important role /3/. One way to overcome this problem is to adjust very low growth rates, typically about 0.5μm/h. This in combination with a solid TMG source makes it possible to grow thin Mg doped layers with precise control over thickness and acceptor concentration. This is verified by the characterization of doping superlattices by SIMS and PL /3/. Diffusion of Mg at 600°C during the growth time is smaller than the depth resolution of SIMS. Fig. 3 gives an example of the Mg distribution profile as measured by SIMS on a hot electron camel transistor test structure.

To verify the accuracy of adjusted layer parameters for electronic devices Ti-n-GaAs barrier-enhanced Schottky diodes /6/ are realized. Here the Schottky barrier height can be modified by inserting a fully depleted thin p-layer between metal and n-GaAs. Fig. 4 shows the j-V-characteristics of such diodes consisting of an 0.5μm $n = 10^{16} cm^{-3}$ epilayer on n^+-substrate followed by p-layers with thicknesses of 0, 15, 25 and 35 nm with acceptor

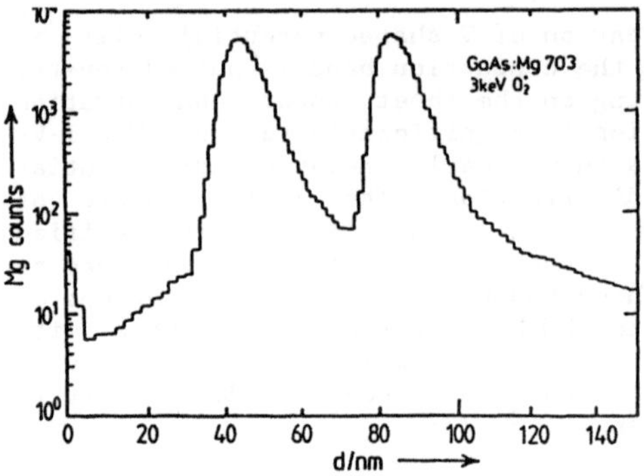

Fig. 3: Mg depth profile of a camel transistor structure as measured by SIMS using 3keV O_2^+ ions

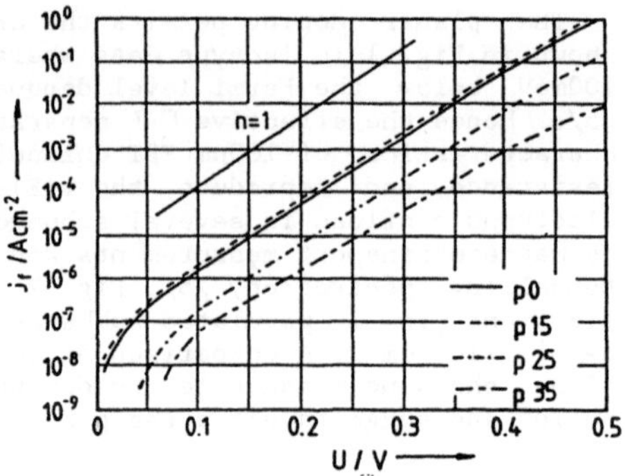

Fig. 4: Dependence of forward current density for a Ti-n-GaAs Schottky diode and barrier-enhanced-Schottky diodes with depleted p-surface layers of 15, 25 and 35nm thickness and $p=5x10^{17}cm^{-3}$

concentrations of $5x10^{17}cm^{-3}$. With increasing thicknesses of the depleted p-layers barrier heights between 0.83V for the conventional Schottky diodes and 1.1V for the diodes with 35nm p-layer are determined by I-V, I-V(T) and C-V measurements in agreement with theoretical calculations.

Although for high frequency devices in GaAs an enhancement of Schottky barrier height seems to be not useful, these structures promise the realization of Schottky contacts on future high speed materials like n-GaInAs. However these devices will be furtheron mainly determined by interface states between metal and semiconductor. Therefore 0.5µm $n=10^{15}cm^{-3}$ - p - 0.1µm $n=10^{19}cm^{-3}$ layer sequences are grown on n^+-substrate by varying the thicknesses of the depleted p-layers between 15 and 40nm and the acceptor concentrations between $5x10^{17}$ and $4x10^{18}cm^{-3}$. Such structures are known as camel diodes /1/ where the barrier for the majority carriers is inside the semiconductor and free of surface states. The properties of the adjustable barriers mainly depend on the parameters of the p-layers as shown in Fig. 5. Also indicated are the measured barrier heights for preselected layer parameters. There still seems to be a systematical error between experimental and theoretical values but the principal dependence shows good agreement. This demonstrates clearly that such thin layers can be reproducibly grown by OM-VPE with precise control over doping concentration resulting in adjustable barrier heights between 0.3 and 1.1V.

The forward I-V characteristics of a camel diode with $p=2x10^{18}cm^{-3}$ and 30 nm p-layer thickness is shown in Fig. 6. The best diodes show comparable properties to those published in /7/ and grown by MBE.

In conclusion, the high potential of OM-VPE for growth of ultra thin doped layers due to the free choice of any desired dopant is demonstrated. Optimization of growth parameters allows the precise creation of nearly arbitrarily shaped potentials for high speed electronic devices. Combining two camel barriers should make possible high speed GaAs hot electron transistors /7/. Concerning the growth of the needed layer sequences there are optimum conditions and highly planar doped GaAs:Se promises ultra low base resistances.

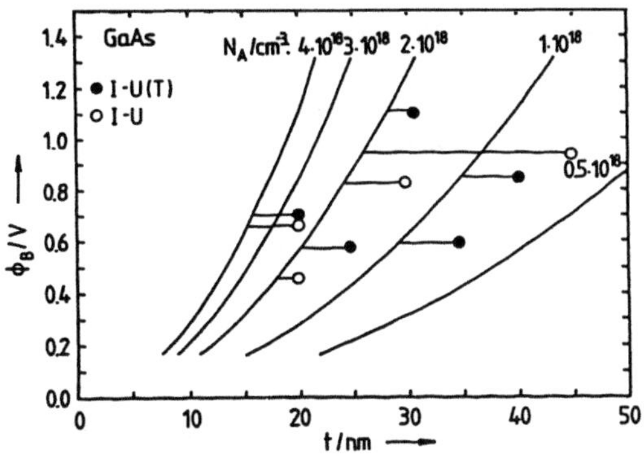

Fig. 5: Calculated dependence of barrier heights on p - layer thicknesses and acceptor concentration for GaAs camel diodes and measured barrier heights for preselected layer parameters

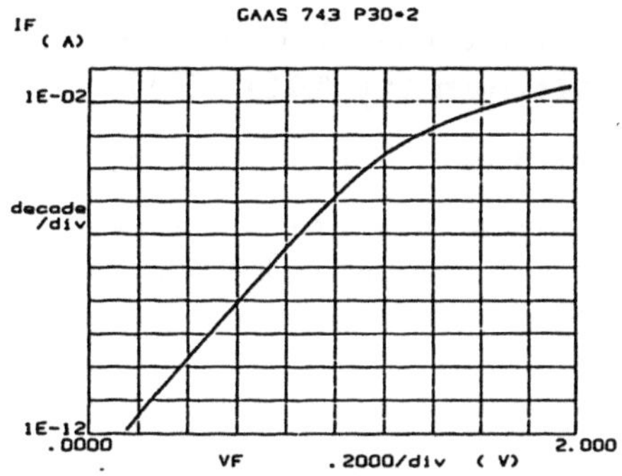

Fig. 6: Forward I-V-characteristics of a GaAs camel diode with $p=2x10^{18}cm^{-3}$ and 30 nm thickness resulting in a barrier height of 1.1V

The authors are indebted to the Basislabor for SIMS measurements and the group of Prof. F. Koch of TU München which performed magnetotransport and tunnelling measurements. This work has been supported by Stiftung Volkswagenwerk.

References

/1/ J.M. Shannon, Appl. Phys. Lett., 35, (1979), 63
/2/ R.J. Malik, K. Board, L.F. Eastman, C.E.C. Wood, T.R. AuCoin and R.L. Ross, Inst. Phys. Conf. Ser., 56, (1981), 697
/3/ P. Roentgen and H. Beneking, Inst. Phys. Conf. Ser., 79, (1986), 145
/4/ P. Roentgen and H. Beneking, to be published by TMS in Proc. "Semiconductor-Based Heterostructures: Interfacial Structure and Stability", Murray Hill, USA, May 1986
/5/ F. Koch, A. Zrenner and M. Zachau, in: "Two-Dimensional Systems: Physics and New Devices", Eds.: G. Bauer, F. Kuchar and H. Heinrich, Springer, Berlin 1986, 175
/6/ J.M. Shannon, Sol. St. Electron., 19, (1976), 537
/7/ J.M. Woodcock, J.J. Harris and J.M. Shannon, Physica B+C, 134, (1985), 111

Low Noise High Electron Mobility Transistors Grown By MOVPE

M. Wolny, P. Chambery, A. Briere, and J.-P. Andre

Laboratoires d'Electronique et de Physique Appliquée*,
3, Av. Descartes, F-94451 Limeil-Brevannes Cedex, France

Low noise HEMTs have been fabricated from MOVPE wafers. A 1.2 dB noise figure with 11 dB associated gain has been measured at 12 GHz. These results demonstrate MOVPE capabilities for growing HEMT structures.

1. INTRODUCTION

Since the successful fabrication of GaAlAs/GaAs heterostructures, HEMTs have attracted a great deal of interest for their high-speed logic capabilities and their excellent low noise performances at high frequencies. These devices are based on the properties of modulation doped GaAlAs (N)/GaAs heterojunctions : because of the higher electron affinity of GaAs, free electrons in GaAlAs layer are transferred to the GaAs layer where they form a quasi 2 DEG and exhibit high mobilities because of the spatial electron-ionized impurities separation.

A lot of attention has been focused on fabricating low noise HEMTs for their superiority on GaAs MESFETs : 0.5 µm long gate HEMTs exhibit performances identical to those of 0.25 µm long gate MESFETs. MBE is up to now intensively used to grow HEMT structures because of its capability to grow extremely abrupt junctions. However, surface quality and high throughput still remain considerable drawbacks for this growth method with the aim of a future industrial development.

One year ago, we obtained preliminary results on HEMTs grown by MOVPE /1/. Improvements in the device performances have been recently performed and a 1.2 dB noise figure with 11 dB associated gain has been measured on 0.55 µm long gate devices at 12 GHz.

2. DEVICE FABRICATION

The epi-growth has been performed in a vertical reactor and involves the thermal decomposition of TMG, Ga $(CH_3)_3$, and TMA, Al $(CH_3)_3$ and of arsine at 720°C. The growth rate is about 2 Å/s. The heterostructure (fig. 1) consists of a GaAlAs buffer layer, an undoped GaAs layer (N type, N = 5.10^{14} cm^{-3}, e = 7000 Å), a GaAlAs spacer layer (20 Å), an N-type Si doped $Ga_{0.72}Al_{0.28}As$ layer (N = 1.10^{18} cm^{-3}, e = 375 Å) and finally an N-type GaAs cap layer (N = 1.10^{18} cm^{-3}, e = 250 Å). Hall measurements have been performed by the Van der Pauw method on structures with 70 Å spacers and without GaAs cap layers : the sheet carrier concentration and the Hall mobility are respectively $8.3.10^{11}$ cm^{-2} and 7200 cm^2 V^{-1}s^{-1} at 300 K, $6.5.10^{11}$ cm^{-2} and 120 000 cm^2 V^{-1} s^{-1} at 77 K.

*"LEP - Laboratoires d'Electronique et de Physique Appliquée, A member of the Philips Research Organization"

The process begins with a boron implantation for device isolation. Then AuGe/Ni ohmic contacts are evaporated and alloyed. The contact resistance is less than 0.2 Ω mm, the sheet resistance being 800 Ω. The 0.55 μm long gate (200 μm width) is patterned by standard photolithography in a 2.7 μm source drain spacing. After a wet etching of the cap layer, a 2300 Å TiPtAu gate is evaporated and lifted off. Finally a 5000 Å TiAu overlayer is deposited.

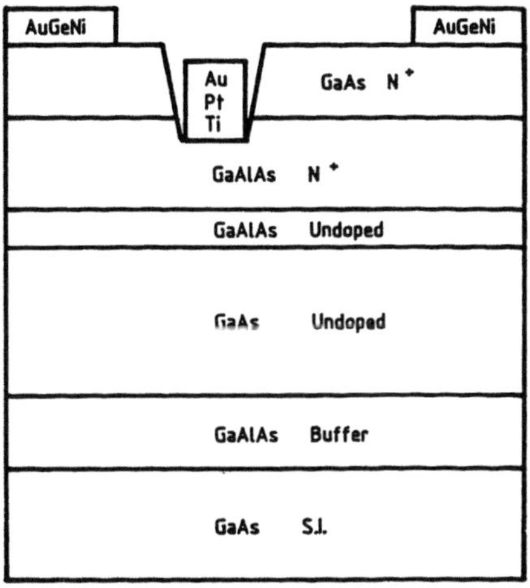

Figure 1 : Schematic figure of the HEMT structure

3. RESULTS AND DISCUSSION

Maximum transconductances as high as 345 mS/mm have been measured on N-off devices at 300 K (480 mS/mm at 77 K). On N-on devices, the maximum transconductances are typically 260 mS/mm at a 110 mA/mm drain current density. After mounting on standard chip carriers, microwave noise figure and associated gain have been measured at 12 GHz. The minimum noise figure is 1.2 dB with 11 dB associated gain at a 12 mA drain current. Figure 2 shows the drain current dependence of NF and G_{ass}. Equivalent circuit element values (fig. 3) have been determined from S-parameter measurements from 2 to 15 GHz. The deduced cut-off frequency is 38 GHz.

The best results ever reported at 12 GHz are summarized in table I. Three groups (LEP, Sony, Toshiba) have now demonstrated the MOVPE capabilities for growing HEMT structures. Table 1 shows that MBE and MOVPE performances are quite comparable taking into account the differences in gate length and thickness (only 2300 Å in our case).

Table I : Low noise HEMT performances at 12 GHz

	MBE (NEC) /3/	MBE (GE) /4/	MBE (Fujitsu) /5/	MBE (Toshiba) /2/	MOVPE (Toshiba) /6/		MOVPE (Sony) /7/	MOVPE (LEP)
Lg (μm)	0.5	0.25	0.4	0.25	0.25	0.5	< 0.5	0.55
SD (μm)	1.5	3			3	3	3	2.7
NF_{min} (dB)	0.95	0.9	1.08	0.66	0.75	0.97	0.87	1.2
G_{ass} (dB)	10.3	12	12.7	11.8	11.1	10.8	12.5	11

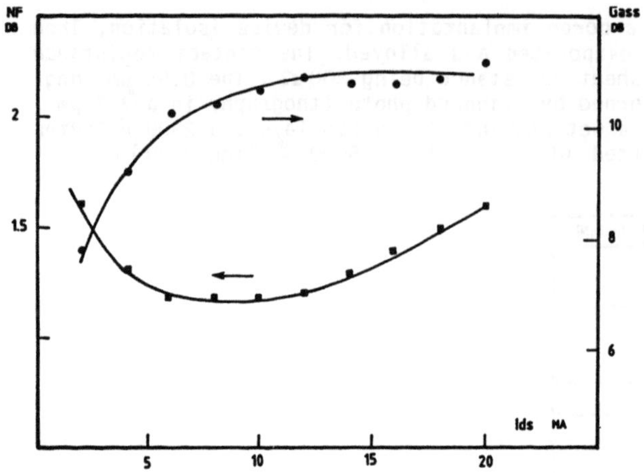

Figure 2 : Optimized noise figure and gain versus drain current (V_{ds} = 3 V)

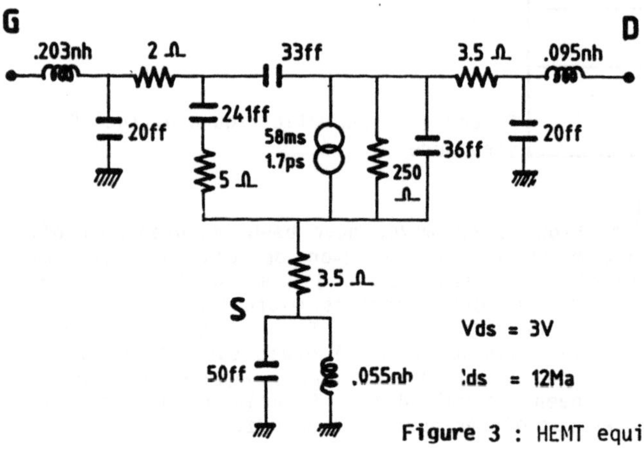

Figure 3 : HEMT equivalent circuit

4. CONCLUSION

A 1.2 dB noise figure with 11 dB associated gain has been measured at 12 GHz on HEMTs grown by MOVPE. Further improvements are expected by optimization of the device structure and of the process. However, these results along with those previously reported /6/, /7/, clearly show that MOVPE can supply as high quality HEMT structures as MBE, while keeping its traditional advantages (surface quality, high throughput...).

REFERENCES

1. J.P. André, M. Wolny, M. Rocchi, Int. Symp. on GaAs and Rel. Comp., Karuizawa (J), (1985), p. 379
2. K. Kamei and al., ibid, p. 541
3. H. Hida and al. IEEE Trans. on Electr. Dev., ED-33, 5, (1986)
4. U.K. Mishra and al., IEEE Electr. Dev. Lett., EDL-6, 3, (1985)
5. T. Mimura and al., Fujitsu Sci. Tech. J., 21, 3, (1985)
6. H. Kawasaki and al., Trans. of the IECE of Japan, vol. E 69, 4, (1986)
7. K. Tanaka and al., Elect. Lett., vol. 22, 9, (1986)

Emitter-Coupled Logic Ring Oscillators Implemented with GaAs/GaAlAs Single and Double Heterojunction Bipolar Transistors: A Comparison

C. Dubon-Chevallier, P. Desrousseaux, A.M. Duchenois, C. Besombes, J. Dangla, C. Bacot, and D. Ankri

Centre National d'Etudes des Télécommuncations, Laboratoire de Bagneux, 196 Av. Henri Ravera, F-92220 Bagneux, France

The influence of different parameters (structure, technological process, circuit design) on the propagation delay time is discussed, taking into account the results of the simulations. A multilayer has been processed and tested leading to a propagation delay time of 240 ps.

I - INTRODUCTION

The use of GaAs/GaAlAs heterojunction bipolar transistors (HBT's) in integrated circuits presents a number of advantages such as high threshold uniformity, drive capabilities and packing densities /1/. In order to investigate the performances of circuits implemented with HBT's, ring oscillators using the Emitter Coupled Logic have been fabricated. We will describe the simulations which have been carried out to investigate the influence of different parameters on the propagation delay time. The investigated parameters were the fan-in (1,2 or 5), the HBT's structure (DHBT's or SHBT's) and a technological parameter (ρ_B), the geometrical parameters being unchanged. The transistors had emitter contact dimensions of 3 μm x 8 μm.

II - FABRICATION TECHNOLOGY

Molecular beam epitaxy (MBE) and Metal Organic Chemical Vapor Deposition (MOCVD) were used to grow the multilayers. The classical structure is presented in Fig. 1. The layers were grown on a semi-insulating GaAs substrate. The base region was 1500 Å thick and there was a grading of the Al composition at the E-B interface over 500 Å.

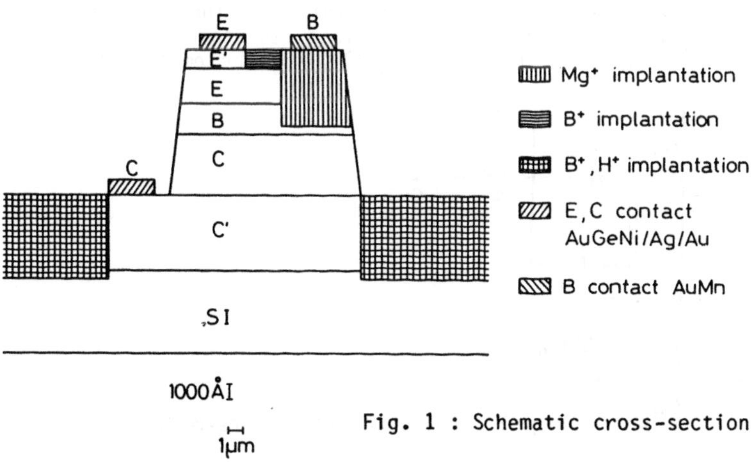

Fig. 1 : Schematic cross-section

Ring oscillators were processed with an 11 masks technology. After the registration marks patterning, a Mg implantation was performed to contact the base, followed by a rapid thermal annealing to activate the dopant. Then a chemical etch was achieved in order to contact the collector buffer layer. A boron implantation was performed to destroy the lateral diode created by the p type implantation with the n^+ layer. Then a high energy B^+, H^+ implantation was used to achieve the transistors electrical isolation. During this very step, the circuit resistances were realized in the collector buffer layer. AuGeNi/Ag/Au was deposited and annealed to contact the emitter and the collector, while AuMn was used to contact the base. The interconnections were obtained with evaporated TiAu films and the isolation between the interconnections was realized with Plasma-Enhanced Chemical Vapor Deposition (PECVD) silicon nitride. The final structure is presented in fig. 1. 2" wafers were processed with optical contact lithography. The emitter contact size is 3 μm x 8 μm, and the emitter crystal is 8 μm x 8 μm.

This work was partly supported by Esprit project 971.

Table 1 : Epitaxial structure of HBT's

Layer	Thickness/μm/	type	doping/cm^{-3}/	Al composition
5	0.1	n^+	4 10^{18}	0
4	0.35	N	3 10^{17}	0.30
3	0.15	p^+	5 10^{18}	0
2	0.4	n	5 10^{16}	0
1	0.5	n^+	2 10^{18}	0
substrate		S.I.		

III - CIRCUIT SIMULATION

Simulations /2/ were carried on with the Astec 3 program using the modified Ebers-Moll model presented in fig. 2. Some d.c. parameters used for the simulations were based on measurements of devices characteristics, others like the capacitances were calculated taking into account the areas and the layers parameters. The parameters are shown in Table 2. The ring oscillator consisted of a chain of ECL inverters where the NOR output level is adapted by an emitter follower stage. A 5 stages oscillator was considered for the simulations. It has been designed for an optimum operation at 1 mA, the values of the resistances being 400 Ω for the load resistance, 1.95 kΩ for the differential pair emitter resistance and 390 Ω for the emitter follower

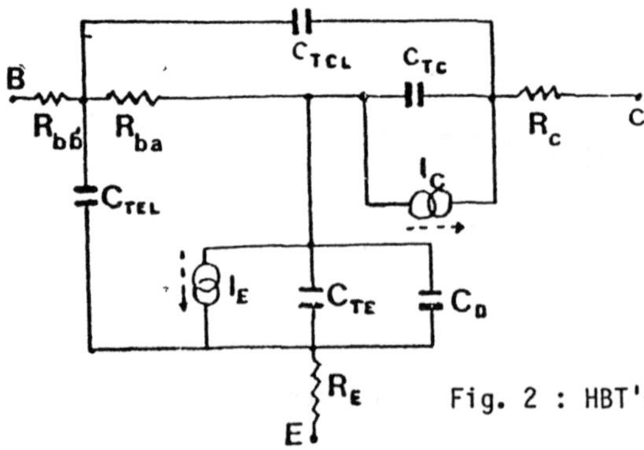

Fig. 2 : HBT's modified Ebers-Moll model

R_{GA}/Ω/	6.5
R_{BL}/Ω/	160
R_E/Ω/	6.5
R_C/Ω/	4.5
C_{TE_A}/fF/	5
C_{TE_L}/fF/	1
C_{TL_A}/fF/	6
C_{TC_L}/fF/	4
I_E/mA/	1.5
I_C/mA/	1.4

Table 2 :
Model parameters of 3 μm x 8 μm HBT's

stage resistance. The influence of different parameters on the propagation delay time has been investigated. These parameters were related to the transistors structure (SHBT's or DHBT's), the technological process (influence of the base contact resistance) or the circuit design (fan-in of 1 or 5).

The Double Heterojunction Bipolar Transistor (DHBT) is a very promising component because of the emitter collector interchangeability which permits to improve the ECL circuit design. But it decreases the performances when implemented in the ring oscillator circuit. The results of the simulations are given in table 3.

Table 3 : Comparison of the performances of oscillators implemented with HBT's or DHBT's

Al composition	t_{pd}/ps/SHBT's	t_{pd}/ps/DHBT's
0.25	25	30

Among the technological parameters, the base contact resistivity seems to act considerably on the propagation delay time. The results of the simulation are given in table 4. The decrease of the propagation delay time with the base contact resistivity is more convincing when the contact resistance is not negligible versus the base layer resistance (ω_B = 2000 Å).

Table 4 : Influence of the base contact resistivity

ρ_B/Ωcm^2/	10^{-5}	10^{-6}	10^{-7}
t_{pd}/ps/(ω_B.1000 Å)	25	20	20
t_{pd}/ps/(ω_B.2000 Å)	40	30	25

The influence of the fan-in condition has also been investigated. The results are presented in table 5 in the case of SHBT's with different Al compositions. There is no variation of the propagation delay time when the fan-in being 1 or 2, the fan-out is changed from 1 to 5.

Table 5 : Influence of the fan-in condition

Al composition	fan in = 1	fan in = 5
0.25	40	70
0.35	38	60

IV - CIRCUIT REALIZATION

A multilayer whose characteristics were presented in table 1 has been processed. A SEM micrograph of a 5 stages oscillator is presented in fig. 3. An 11 stages circuit has been tested and exhibited a propagation delay time of 240 ps (fig. 4). This value, higher than the expected one, could be due to the integrated resistors which were twice as resistive as expected, or to the optimistic values implemented in the Astec 3 program.

Fig. 3 : Processed ring oscillator

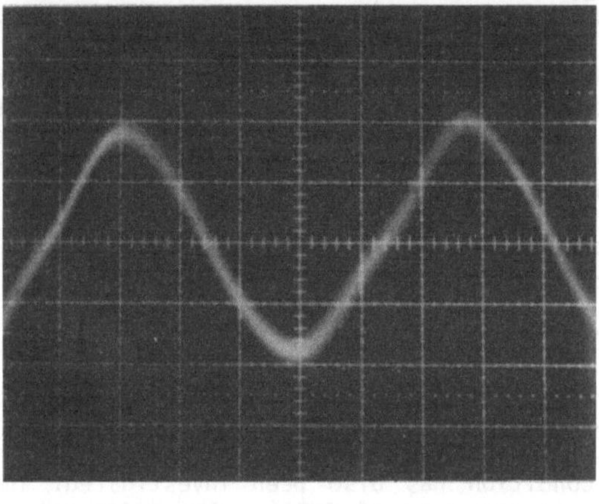

Fig. 4 : Output waveform

V - CONCLUSION

Simulations have been carried out to determine the influence of different parameters on the propagation delay time. The best result, 20 ps, was obtained in the case of SHBT's with a very low contact resistivity (10^{-7} Ωcm^2) when the fan-in was equal to 1. Layers have been processed and the circuit performance has been measured on an 11 stages circuit leading to a value

(240 ps) higher than the expected one, due to a bad control of the integrated resistances.

REFERENCES

1-H.Kroemer : Proc. IEEE. 70(1), 13 (1982)

2-P.Desrousseaux, C.Dubon-Chevallier, J.Dangla, D.Ankri : Eurocon (1986)

Self-Aligned Technology Using Refractory Ohmic Contacts for GaAs/GaAlAs Heterojunction Bipolar Transistors

K. Daoud-Ketata, C. Dubon-Chevallier, J. Etrillard, and J.F. Bresse

Centre National d'Etudes des Télécommunications, Laboratoire de Bagneux, 196 Av. Henri Ravera, F-92220 Bagneux, France

In this paper, we present a new self-aligned technology using refractory metals to contact the emitter layer, in order to afford the annealing process subsequent to the p type implantation, as the contact also acts as a mask for the p type implantation. GeMoW has been chosen for this purpose, GeMo to form the ohmic contact and W the implantation mask. The influence of the annealing conditions on the specific contact resistivity has been investigated. An As overpressure proved to be necessary for the formation of an ohmic contact.

I - INTRODUCTION

A self-aligned technology is necessary to process Heterojunction Bipolar Transistors (HBT's) in order to decrease the parasitic elements (base access resistance, collector capacitance). Recent works have been reported on self-aligned processes for HBT's /1,2/. In the first one, the same photoresist mask is used for the emitter layer etch and for the base contact deposition, thanks to the etch undercut. The technique used in the second one requires a dual level patterning which seems rather sophisticated. The new self-aligned technology presented here combines the use of refractory metals for the emitter ohmic contact and for the p type implantation mask.

II - A NEW SELF-ALIGNED TECHNOLOGY

This self-aligned technology is dedicated to the HBT's classical multilayers structure grown by MBE or MOCVD, which is convenient for the realization of small-scale digital integrated circuits /3/. The basic parameters are presented in table 1.

The emitter ohmic contact (GeMoW) is deposited first and delineated using reactive ion etching. The obtained profile after etching is a T shape structure where the undercut determines the distance between the emitter ohmic contact and the p type implantation. The p type implantation is then performed through the emitter contact which acts as a mask. The following annealing step permits both to form the ohmic contact and to activate the p type dopant. Then a light wet etch is achieved to remove the lateral diode between the n^+ cap layer and the p type implantation. A deeper wet etch is performed down to the collector layer. Then a high energy B^+, H^+ implantation takes place in order to obtain the electrical isolation between the transistors. AuGeNi delineated by lift-off and annealed up to 450°C is classically used for the collector ohmic contact, while AuMn/4/ is deposited to form the base ohmic contact. The final structure is presented in fig. 1. This processing technology is quite simple, does not require any critical registration and permits to

This work was partly supported by Esprit project 971.

Table 1 : epitaxial structure of HBT's

layer	thickness/µm/	type	doping/cm^{-3}/	Al composition
5	0.1	n^+	$4\,10^{18}$	0
4	0.35	N	$3\,10^{17}$	0.30
3	0.15	p^+	$5\,10^{18}$	0
2	0.4	n	$5\,10^{16}$	0
1	0.5	n^+	$2\,10^{18}$	0
substrate		SI		

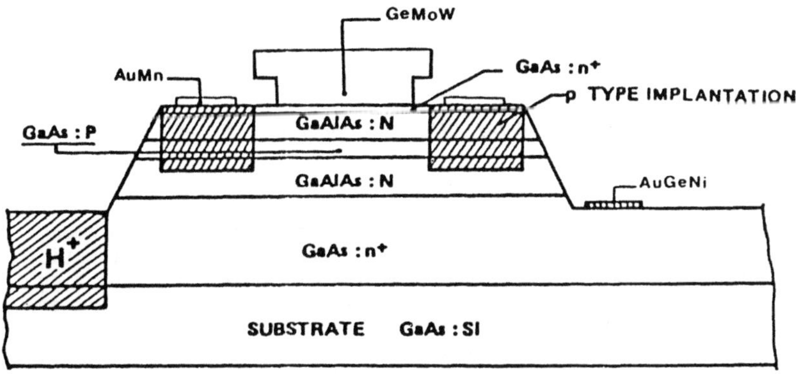

Fig. 1 : Schematic cross-section of a self-aligned HBT

decrease the base access resistance and the collector capacitance. However, different elemental steps have to be developed and made compatible to form a reliable process.

III - ELEMENTARY STEPS STUDY

III.1 - Refractory ohmic contact

A 150 Å Ge layer was deposited first using e-beam evaporation with a deposition rate of 10 Å/s, then a 150 Å Mo layer was RF sputtered with a deposition rate of 2 Å/s. Two different annealing techniques have been investigated to optimize the contact resistivity : rapid thermal annealing and classical annealing with a sealed ampoule under an As overpressure. It was found that the contact was ohmic only in the second case and the obtained contact resistivity was $3\ 10^{-6}\ \Omega cm^2$. Auger Electron Spectroscopy and Transmission Electron Microscopy were carried on, showing that the formation of the contact was due both to a n^+ layer created at the MoGe/GaAs interface by the Ge diffusion (fig. 2) and to the formation of a Mo_5Ge_3 stable phase. It was thought that in the first case (RTA), compensation mechanisms occur due to As loss.

III.2 - Reactive ion etching patterning

Mo, Ge and W can be etched selectively on GaAs using a SF_6 plasma discharge. The important point was to determine a convenient etching mask.

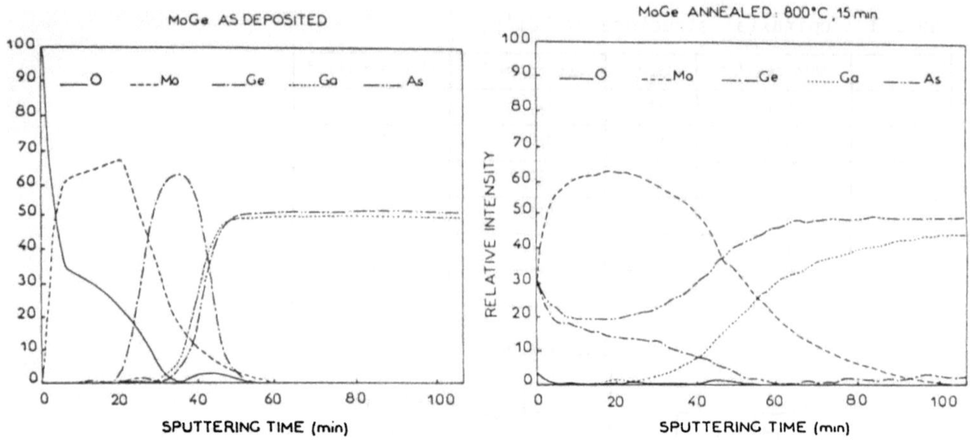

Fig. 2 : Auger Electron Spectroscopy profiles obtained a) before annealing, b) after annealing

Metallic films were found to be more suitable than photoresist, and among the metallic films Pt was found to be the most convenient. A systematic study of the etch rates and selectivities as a function of the operating conditions (pressure, RF power) has been carried on. It was found that an anisotropic profile (fig 3.a) was obtained when operating at low pressure (30 mT) and high power (50 W). In order to obtain a "T shape" (fig 3.b) two steps were necessary, the first one at low pressure (30 mT) and high power (50 W) and the second one at high pressure (60 mT) and low power (30 W).

III.3 - Implantation mask

The metal acting as an implantation mask depends on the emitter contact structure. With an anisotropic profile, W acts as the implantation mask, but with the "T shape", the patterning Pt acts as the implantation mask.

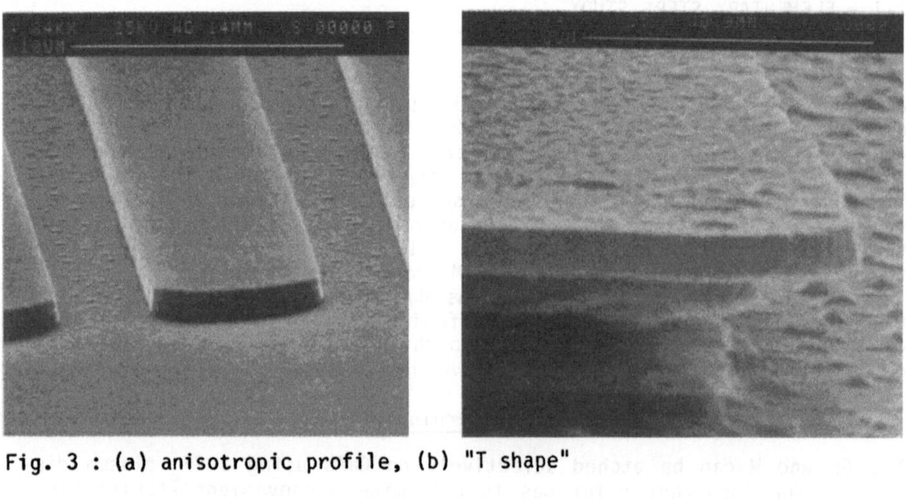

Fig. 3 : (a) anisotropic profile, (b) "T shape"

Both Pt and W have a high stopping power and the required thickness does not exceed half the penetration depth in GaAs.

III.4 - Transistor processing

A multilayer structure has been processed along the different steps of this new technology showing the feasibility of the process. It proved particularly the compatibility of the thermal treatments, as the classical annealing with a sealed ampoule under an As overpressure permits to obtain a convenient dopant profile.

IV - CONCLUSION

We have developed a new self-aligned technology which permits to define the base implant with the emitter contact metallization. A layer has been processed to prove the feasibility of this new technology.

REFERENCES

1 - R. Fisher, J. Klem, J.S. Gedymin, T. Henderson, W. Kopp, H. Morkoc : Proc IEDM (1985)

2 - M.F. Chang, P.M. Asbeck, D.L. Miller, K.C. Wang : Dev. Lett. EDL7(1) (1986)

3 - C. Dubon-Chevallier, P. Desrousseaux, A.M. Duchenois, C. Besombes, J. Dangla, C. Bacot, D. Ankri : High-Speed Electronics Conference (1986)

4 - C. Dubon-Chevallier, M. Gauneau, J.F. Bresse, A. Izrael, D. Ankri : J. Appl. Phys. 59(11) (1986)

Novel Cryoelectronic Device Concept Based on Magnetically Controlled Current Flow in Bulk Semiconductors

J. Mannhart, J. Parisi, and R.P. Huebener

Physikalisches Institut II, Universität Tübingen,
Auf der Morgenstelle 14, D-7400 Tübingen, Fed. Rep. of Germany

Magnetic control of the filamentary current flow during avalanche breakdown in bulk semiconductors at low temperatures offers an interesting new principle for the design of a novel device family. Power losses are minimized by using superconducting lines for the generation of the magnetic control fields and for the interconnections. Based on such a hybrid concept implementing p - germanium and niobium for the semiconducting and superconducting components, respectively, we have experimentally verified analog and digital cryoelectronic device approaches including magnetic field effect transistors, logic gates, and memory devices.

1. Introduction

Electrically driven into the regime of low-temperature avalanche breakdown, homogeneously doped semiconductors have been demonstrated to display spatially inhomogeneous current flow via high-conductivity filament channels generated by impurity impact ionization /1,2/. If an external magnetic field is applied perpendicular to the current direction, the current filaments are spatially shifted due to the Lorentz force. The corresponding shift of the current-voltage characteristic (magneto-resistance) leads to a "cooling effect" of the magnetic field on the hot charge carriers. Particularly in p-germanium at liquid-helium temperatures where the carrier mobility and, hence, the Lorentz force is high, a relatively large magneto-resistance is obtained with applied magnetic fields as small as only a few Gauss. The sensitive response to the magnetic control field suggests an interesting new principle for the design of a novel cryoelectronic device family.

Recently, we have experimentally demonstrated the concept of a new magnetic field effect transistor /1,3/. Our experiments were carried out at 4.2 K on homogeneously doped p-type Ge crystals with shallow ($\sim$ 10 meV) acceptor levels and properly arranged ohmic Al contacts. The magnetic control of the filamentary current flow was achieved by a small magnetic field (10 - 100 G), generated by superconducting Nb lines placed on the top surface of the Ge crystal. In this way a three-terminal device is obtained, based on a hybrid concept implementing semiconducting and superconducting components. Beyond analog operation as magnetic field effect transistor, this paper mainly focuses on the experimental demonstration of digital cryoelectronic device applications including logic operation and data storage. The digital operating modes are based on the magnetically induced switching behavior of current filaments between two stable states of different conductance (logic gates) or two stable channels of high conductance (memory device).

2. Transistor

Having typical dimensions of 0.1 x 5 x 6 mm^3, the sample configuration of the magnetic field effect transistor is sketched in Fig. 1, indicating the evapo-

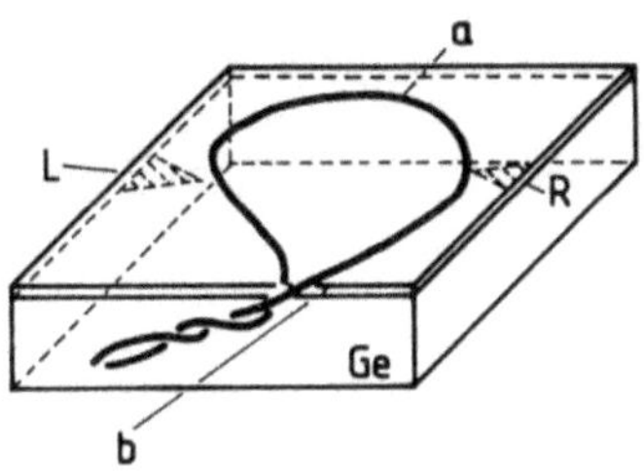

Fig. 1: Scheme of the magnetic field effect transistor

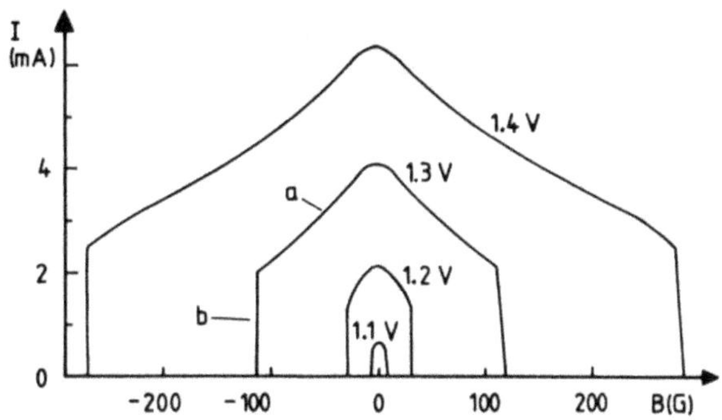

Fig. 2: Sample current vs applied magnetic field at different bias voltages

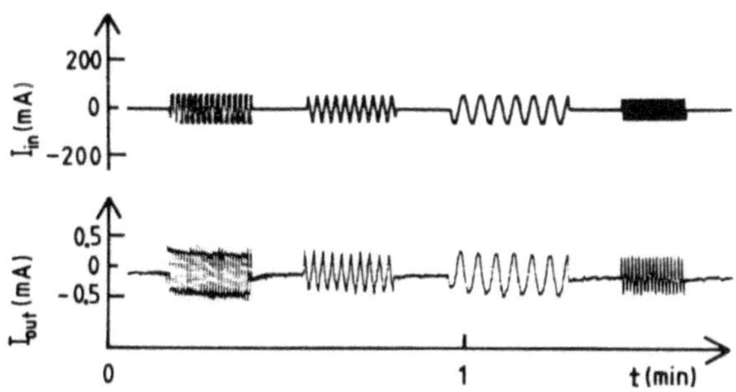

Fig. 3: Transistor current I_{out} controlled by the input current I_{in} at a slow time scale

rated ohmic Al contacts (L and R) and the superconducting Nb loop (a) for generating the magnetic control field. A thin capton foil (b) ensures electric insulation and magnetic coupling between the Ge crystal surface and the overlay Nb loop. Figure 2 clearly illustrates the magneto-resistive behavior of the above sample configuration. Further details on this behavior can be found elsewhere /1/.

Representing a magnetic analog to the electric field effect transistor, the corresponding operation principle of our transistor approach is demonstrated in Fig. 3 at a relatively slow time scale. The upper trace gives the temporal structure of the input current applied to the superconducting loop, while the lower trace shows the resulting output current of the voltage-biased semiconducting crystal. Due to the superconducting input circuit, the input impedance is nearly zero, and the operation is current controlled. We further note that the frequency range of the present transistor configuration extends up to about 1 MHz /1/.

3. Logic Gates

The operating regimes of the analog transistor and the digital logic and memory devices are indicated in Fig. 2 by a and b, respectively. So the realization of the digital operation modes discussed in the following can be based on one logic state associated with the sample current exceeding a certain value and the other state with the sample current switched off.

As one example, the sample configuration of a NOR-gate is sketched in Fig. 4. Here the input currents I_{in1} and I_{in2} of two electrically separated super-

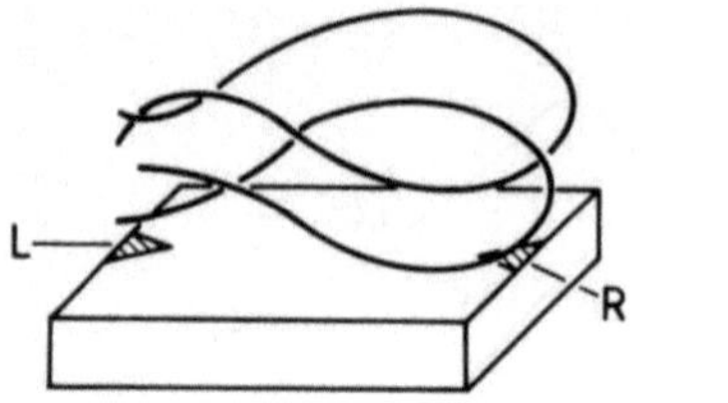

Fig. 4: Scheme of the NOR gate

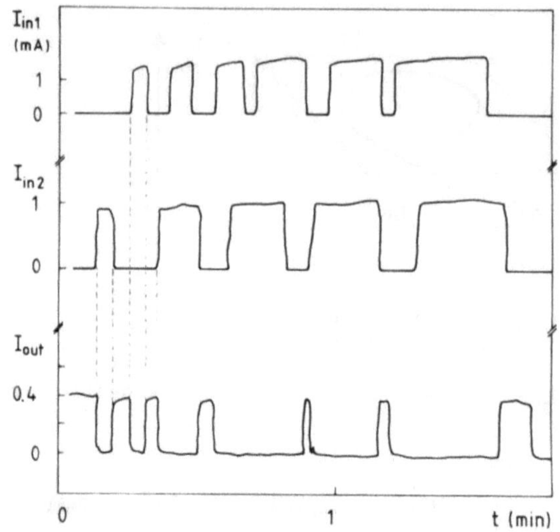

Fig. 5: NOR gate current I_{out} controlled by the two input currents I_{in1} and I_{in2} at a slow time scale

conducting loops control the electric current I_{out} flowing through the semiconductor between the ohmic contacts L and R. Figure 5 clearly demonstrates the basic operating principle that the magnetic control field generated by the input current of either loop is capable of switching the sample current I_{out} off. For plotting the measured temporal structures on an X-Y recorder, all traces were again taken at a relatively slow time scale.

4. Memory Device

In order to verify the digital operation mode based on magnetically induced switching behavior of current filaments between two stable channels of high conductance, the sample configuration of the memory device consists of three ohmic contacts (L,R, and R') evaporated onto the surface of the semiconductor crystal, as outlined in Fig. 6. Applying proper bias voltages to the three ohmic contacts (say, -2.4 V to R, -2.5 V to R', and 0 V to L), a stable current flow between the contacts L and R' (current $I_{LR'}$) is obtained, while the sample still remains insulating between the contacts L and R (current I_{LR}). By further applying a time-dependent transverse magnetic field B (generated by a superconducting loop) of the profile shown in the upper trace of Fig. 7, the sample current I_{LR} can be switched on and off in a well-defined way. It is evident that this magnetically controlled switching process occurs between the two stable channels of high conductance, established between the contacts L and R and the contacts L and R' (of course, the sample always remains insu-

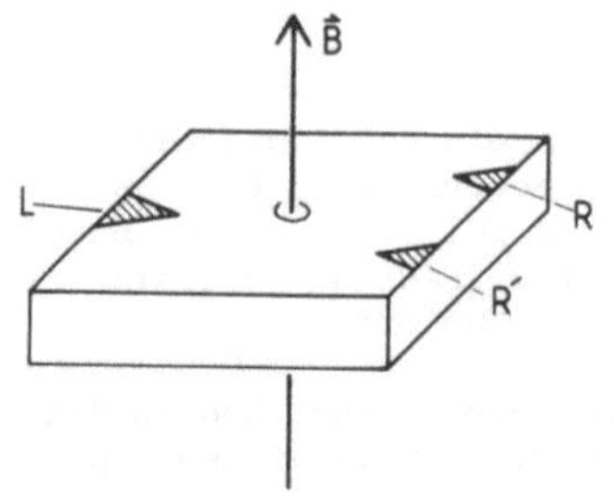

Fig. 6: Scheme of the memory device

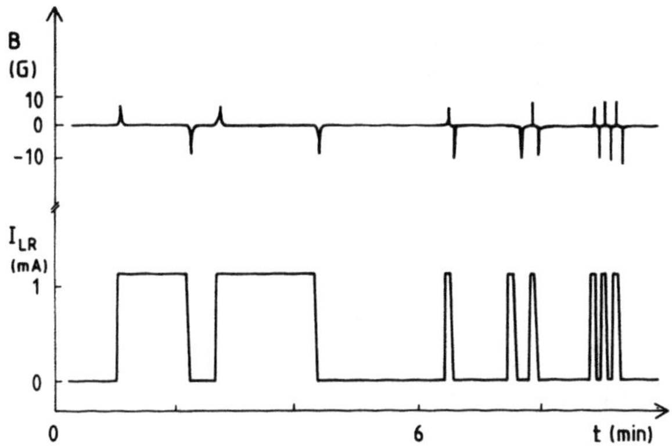

Fig. 7: Memory device current I_{LR} controlled by the transverse applied magnetic field B at a slow time scale

lating between the r.h.s. contacts R and R'). The stability of the current channel I_{LR} is shown to be preserved up to the onset of an additional magnetic field pulse having opposite polarity.

5. Discussion and Conclusions

Analog and digital cryoelectronic applications of a novel hybrid device family combining both semiconducting and superconducting elements have been demonstrated experimentally. Particularly attractive features of the new hybrid configuration are its simplicity, the utilization of well-established semiconductor technology, and its robust nature promising a high production yield. Using superconducting lines for generation of the magnetic control field and for all interconnections allows small dimensions and minimizes the need for multi-layer wiring in complex integrated systems. Further, deterioration caused by electromigration can be avoided, and signal propagation without dispersion and losses is achieved. Consequently, our device configuration represents an interesting concept for the development of ultrafast electronic circuits requiring high packing density and correspondingly low power dissipation.

Evaluating the most important performance characteristics /4/, the limitations due to the time-of-flight per length of the carriers and due to the power losses per distance between the ohmic contacts are estimated to about 50 ps/µm and 0.5 µW/µm, respectively. Assuming for the minimum distance between the ohmic contacts a still reasonable value of about 10 µm, operation up to frequencies in the GHz regime appears feasible. The magnetic field needed for magnetic control is found to be inversely proportional to the carrier mobility of the semiconductor. Therefore, highly pure semiconducting materials such as Ge and Si as well as operation at liquid-helium temperatures where the mobility becomes large appear favorable for the proposed device concept.

1. J. Mannhart, R.P. Huebener, J. Parisi, and J. Peinke: Solid State Commun. 58, 323 (1986)
2. K.M. Mayer, R. Gross, J. Parisi, J. Peinke, and R.P. Huebener: Phys. Rev. Lett. (to be published)
3. R.P. Huebener, J. Mannhart, and J. Parisi: German Patent Application P 3541290.9 from 22. Nov. 1985
4. J. Mannhart and R.P. Huebener: J. Appl. Phys. (to be published)

InGaAlAs/InGaAs/InGaAlAs NnpnN Double Heterojunction Bipolar Transistors: Experimental Characteristics and Monte-Carlo Interpretation

J.L. Pelouard[1], *P. Hesto*[1], *J.P. Praseuth*[2], *and L. Goldstein*[2]

[1]Laboratoire de Microstructures et de Microélectronique (CNRS)
[2]Laboratoire de Bagneux (CNET)
These laboratoires are within the Groupement Scientifique CNET/CNRS, 196, Av. H. Ravera, F-92220 Bagneux, France

Heterojunction Bipolar Transistors allow one to obtain an emitter efficiency very close to unity even when using a heavily doped base. Moreover the symmetric Double Heterojunction Bipolar Transistor (DHBT) is of interest for I^2L and ECL digital circuits. Nevertheless the coincidence between the metallurgical heterojunction and the electrical junction leads to large Generation-Recombination currents in the Space Charge Layers (SCL) and the spikes restrict both the electron injection from the emitter and the electron collection at the collector /1/. These effects have been avoided by introducing gradual heterojunctions or thin n type layers (spacers) at the emitter-base and collector-base junctions /2-4/. In this paper the influence of spacers on the transistor characteristics will be reported and the results will be interpreted using Monte-Carlo simulations.

Figure 1 shows the schematic cross-section of the transistors. Layers are grown by Molecular Beam Epitaxy. The low gap material (the base) was $In_{.53}Ga_{.47}As$ on account of its very promising properties for optoelectronic applications. The large gap semiconductor (the emitter and the collector) was $In_{.53}Ga_{.28}Al_{.19}As$. This quaternary compound is of a much better crystalline and electrical quality than $In_{.52}Al_{.48}As$ /5/. The MESA structures were processed with a classical technique similar to what is done with GaAlAs/GaAs devices /6/. Studies concerning transistors with a spacer at the base-collector junction (SHBT) and transistors with spacers both at the base-collector junction and at the emitter-base junction (DSHBT) have been carried out.

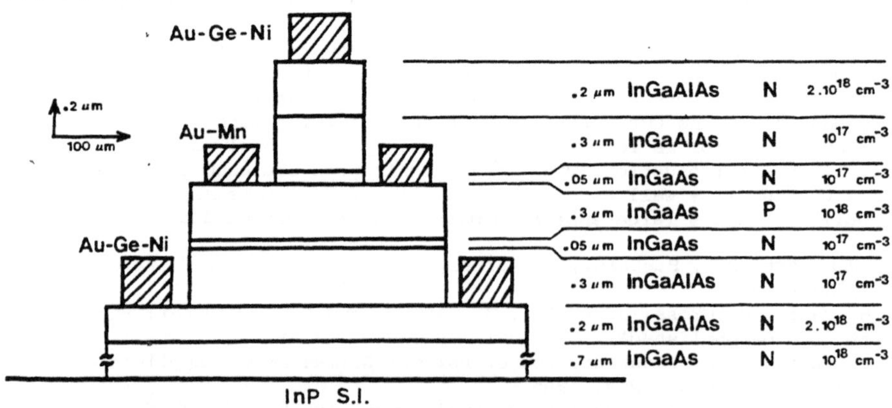

Fig.1. Schematic cross-section of the transistors

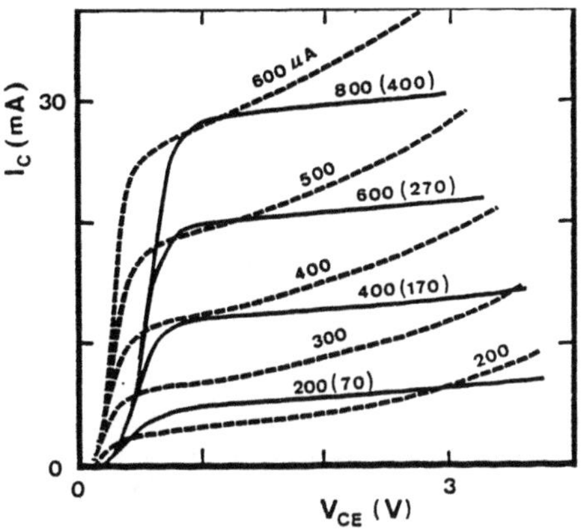

Fig.2. $I_C(V_{CE})$ characteristics
dashed line: DSTBH
solid line: STBH

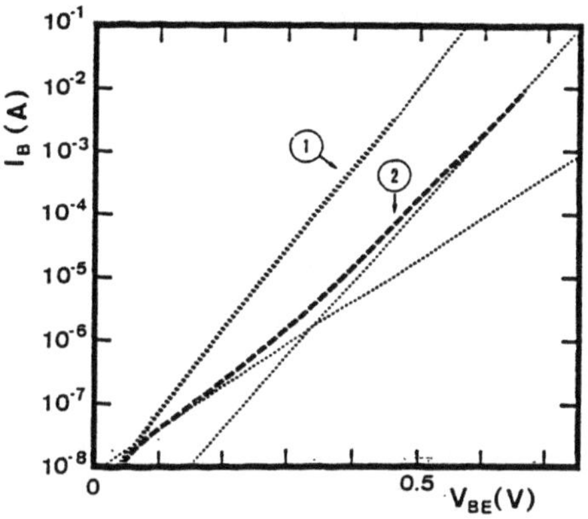

Fig.3. $I_E(V_{BE})$ characteristic
dotted line: with spacer
dashed line: no spacer

The $I_C(V_{CE})$ characteristics for the SHBT and the DSHBT are given in fig.2. The shapes of these characteristics are very different. For the SHBT, the collector current is virtually independent of V_{CE} while it drastically increases with V_{CE} for the DSHBT. The lack of collector current saturation in the SHBT can be explained by the Early effect : the Early voltage has been computed to be equal to 50 V, a realistic value for a 0.3 µm base width with a doping level equal to 10^{18} cm^{-3}. On the other hand the large increase of I_C with V_{CE} for the DSHBT cannot be explained by the Early effect nor by the multiplication factor at the collector junction. As a matter of fact, if the DSHBT collector junction exhibits a carrier multiplication, this must also occur in the SHBT because of their identical structure. An explanation of the collector current increase will be given later on, using Monte-Carlo simulations. We also considered the base current in the two structures. For a given collector current, the total base current is larger for the SHBT than for the DSHBT. But it is well known that two effects contribute to this current: the electron recombination in the base and the electron recombination in the emitter-base Space Charge Layer. In fact this SCL recombination current is very dependent upon the emitter-base junction configuration. Figure 3 shows the $I_E(V_{BE})$ characteristics for an emitter junction with a spacer (curve 1) and without spacer (curve 2). Curve 1 exhibits a linear variation over 5 decades with an ideality coefficient equal to 1.3 which is typical of a very low SCL recombination. On the other hand two linear variations can be noticed in the case of the junction without spacer (curve 2). The first one, at high voltages corresponds to an ideality factor equal to 1.45 and the second one, at lower voltages corresponds to an ideality factor equal to 2.5 which is characteristic of the existence of a SCL recombination. In order to compare the electron behavior of both transistors, we have reparametered the $I_C(V_{CE})$ characteristics of fig.2 by the base current due only to the recombination in the base (in brackets in fig.2). It can now be noticed that the base recombination current is lower for the SHBT. This must obviously be considered from a qualitative point of view. Finally as could be expected, the offset voltages are larger for the SHBT asymmetrical transistor (185 meV) than for the DSHBT symmetrical one (70 mV).

We will now interpret the DSHBT characteristics by studying the electron behaviour using Monte-Carlo simulations /7/. The base-emitter vol-

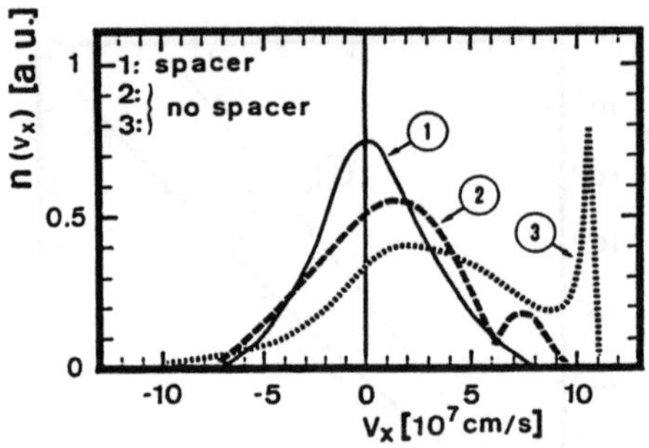

Fig.4. Velocity electron distribution in the base at 0.1µm from the emitter. 1: spacer, 2: no spacer, ΔE_c=0.13eV, 3: no spacer ΔE_c=0.3eV

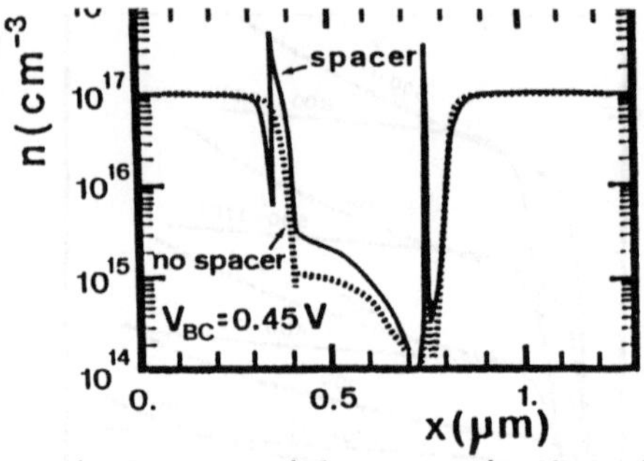

Fig.5. DTBH with spacer in the collector. Dotted line: no spacer in the emitter , solid line: spacer in the emitter

tages have been chosen so that the simulated emitter current is about the same for both transistors. Taking V_{BE}=0.633 V for the SHBT and V_{BE}=0.573 V for the DSHBT the emitter current is of the order of 1500 A/cm².

The distributions of the velocity along the device axis for the electrons are drawn in fig.4. These spectra are taken in the base at 0.1 µm from the emitter. Curve 1 corresponds to an emitter-base junction with spacer. The distribution is symmetrical and centered around a very low velocity. We can conclude an electron population in thermal equilibrium at the lattice temperature and then a diffusive current (the electric field in the base remains a constant). Curve 2 corresponds to an emitter-base junction without any spacer and a conduction band discontinuity equal to 130 meV. There are 2 peaks. The first one is nearly symmetrical and centered around a velocity equal to 1.5 10^6 cm/s. So it corresponds to relaxed velocity electrons having experienced impurity or alloy scatterings which randomize the electron velocity direction. The second peak is centered around a high velocity (7.5 10^7 cm/s) and corresponds to quasiballistic electrons having experienced only few interactions with polar optical phonons (during these interactions the electron velocity direction remains approximately constant). The quasiballistic electrons are 10% of the total electron population but carry 50% of the total current. For comparison curve 3 corresponds to the same structure but with a conduction band discontinuity equal to 300 meV. The quasiballistic phenomenon is about the same but the quasiballistic peak is better defined on account of the higher electron energy. For the structures without any spacer one can notice there are no electrons with a negative velocity high enough to retrodiffuse in the emitter. Thus the injection from the emitter corresponds to a thermionic injection /8/. Finally in the case of an injection from a junction without spacer the electron population is at a higher temperature than in the case of an injection from a junction with a spacer and a large part of the current is due to quasiballistic electrons.

Let us turn now to the simulated electron concentration along the devices. These concentrations are given in fig.5. One can notice the electron accumulation in the notches and the differences between the SHBT (dashed line) and the DSHBT (solid line). Three mains results can be pointed out. The density of the accumulated electrons near the collector notch, the gradient of the electron population in the base and the mean electron density in the base are much larger in the DSHBT than in the SHBT. These differences can be due to the higher electronic temperature for the electron population in the base of the SHBT than of the DSHBT. In particular the mean electron density is 2 times greater for the DSHBT (1.7×10^{15} cm^{-3} instead of 8×10^{15} cm^{-3}). If we assume the recombination current in the base to be proportional to the mean electron density the base current for the DSHBT must be 2 times greater at a given voltage. Experimental results, taking into account the correction due to the recombination current in the emitter-base SCL, give a factor of 1.5 (fig.2). We can then conclude that the higher electron temperature gives better collection properties to the collector in the SHBT which reduces the electron density and then the recombination current in the base. Moreover for the SHBT the mean electron density in the base is a constant with respect to V_{CB} while it decreases by 15% when V_{CB} increases from 0 to 0.5 V for the DSHBT (1.9×10^{15} cm^{-3} instead of 1.7×10^{15} cm^{-3}). This phenomenon can explain the increase of I_C with V_{CE} in the case of the DSHBT. If we assume once more a base current proportional to the mean electron density in the base, the variation of I_B with V_{CE} given by the Monte-Carlo simulations (15%) is comparable with the measured variations (7%).

Acknowledgments

We acknowledge G.TREMBLAY and A.ARQUEY for technical support and the members of CNET Microelectronics division for fruitful discussion on DHBT technology and characterisation. The authors are indebted to R.CASTAGNE from IEF for helpful discussions and criticisms. This work was supported by the DAII.

References

1. H.KROEMER: Proc.IEEE 17,13(1982)
2. R.J.MALIC,J.R.HAYES,F.CAPASSO,K.ALAVI,A.Y.CHO: IEEE Elect.Dev.Lett. EDL4,383(1983)
3. M.CHAUDHURY,K.TABATABAI-ALAVI,C.G.FONSTAD: IEEE Elect.Dev.Lett. EDL5,251(1984)
4. L.M.SU,N.G.GROTE,R.KAUMANNS,H.SCHROETER: Appl.Phys.Lett. 47,28(1985)
5. L.GOLDSTEIN,J.P.PRASEUTH,J.L.PELOUARD,P.HESTO: 4th International Conference on Molecular Beam Epitaxy,YORK(1986)
6. C.DUBON,A.M.DUCHENOIS,J.F.BRESSE,D.ANKRI: Elect.Lett. 21,614(1985)
7. These algorithms were developed in the Institut d'Electronique Fondamentale. See P.HESTO,J.F.PONE,M.MOUIS,J.L.PELOUARD, R.CASTAGNE: Nasecode IV DUBLIN,315(1985)
8. J.L.PELOUARD,P.HESTO,R.CASTAGNE: 10th IEEE/Cornell Conference,52(1985)

Lattice-Strained Double Heterojunction InGaAs/GaAs Bipolar Transistors

L.P. Ramberg[1*], *P.M. Enquist*[1], *Y.K. Chen*[1], *F.E. Najjar*[1], *L.F. Eastman*[1], *E.A. Fitzgerald*[2], *and K.L. Kavanagh*[2]

[1]School of Electrical Engineering, Cornell University, Phillips Hall, Ithaca, NY 14853, USA

[2] Department of Materials Science and Engineering, Cornell University, Bard Hall, Ithaca, NY 14853, USA

Abstract: A study of a series of NpN GaAs/InGaAs/GaAs lattice-strained heterojunction bipolar transistors, in which the indium mole fraction has been varied, shows that it is possible to obtain good dc current gain and a high cut-off frequency in this materials system despite relatively small band offsets and the presence of interface lattice irregularities.

Introduction: While a perfect lattice-match between different constituents was long considered essential for heterostructure device applications, improved performance has recently been shown by devices that contain a certain amount of mismatch introduced by the presence of indium. Following the work of ROSENBERG et al.[1], HENDERSON et al.[2] have explored the properties of the InGaAs/AlGaAs modulation doped field effect transistors (MODFETs), while ENQUIST et al.[3] have researched the use of the same materials system in the heterojunction bipolar transistor (HBT) context.

For a given layer thickness, only a finite amount of mismatch can be accommodated by coherent strain. If more indium is added, or a thicker layer is grown, a relaxation will eventually take place, after which some of the mismatch will be accommodated by misfit dislocations and the remainder by strain. The onset of this mechanism is reasonably well defined; non-relaxed material that is essentially free from misfit dislocations will be termed pseudomorphic.

In the case of HBTs, an indium-rich alloy is generally preferred. At least up to a point, device performance will respond advantageously to an indium content increment owing to improvements in the injection efficiency. Simultaneously, the base transport factor will benefit from a rise in the saturated velocity of the minority carriers. Taken to some extreme compositionwise, the InGaAs/AlGaAs materials system could even provide the band gaps and offsets, necessary for Auger transistor operation, as discussed by KROEMER [4]. However, the bane of many extrapolated expectations can be the misfit dislocations. In the case of InGaAs/AlGaAs MODFETs, these have been observed to have a highly detrimental effect upon the two-dimensional electron gas. For HBTs, recombination at the mismatched emitter-base interface may increase considerably as dislocations start to form, as elaborated by KRESSEL [5]. On the other hand, YAMAKOSHI et al.[6] did not find

* Permanent address: Department of Electron Physics I, Chalmers University, S-412 96 Gothenburg, Sweden

any close correspondence between misfit dislocations and recombination in InGaAsP/InP diodes.

Insofar as the dislocations in the InGaAs/AlGaAs system remain relatively harmless, it would permit the design of HBTs containing little or no aluminium while still retaining sufficient leeway for bandgap engineering. Abandoning Al altogether would cater for a significant reduction of the emitter trap density, a lower attainable emitter resistance, and less opportunities for reactivity-related problems during epitaxial growth.

Experiments: In an effort to illuminate the device physics discussed above, we have fabricated and evaluated a series of large area as well as microwave HBTs in the InGaAs/GaAs materials system. The structure of the four layers is displayed in Table I. The growth method was molecular beam epitaxy, the growth rate was 1.0 µm/h, and the substrate temperatures were 550°C and 600°C for InGaAs and GaAs, respectively. The indium mole fractions aimed for were 0%, 2%, 5%, and 8%. No attempt was made to compositionally grade the emitter-base junction.

Table I. Layer structures

material	purpose	thickness	doping
GaAs	cap	2000Å	n^+ 2×10^{18} cm^{-3}
GaAs	emitter	2000Å	n 2×10^{17} cm^{-3}
InGaAs	Be setback	200Å	π -
InGaAs	base	800Å	p^+ 1×10^{18} cm^{-3}
InGaAs	grading of In	500Å	n^- 2×10^{16} cm^{-3}
GaAs	collector	4500Å	n^- 2×10^{16} cm^{-3}
GaAs	coll. contact	5000Å	n^+ 2×10^{18} cm^{-3}

The unprocessed wafers were subject to studies by Nomarski interference microscopy, Rutherford backscattering spectroscopy (RBS), ion channeling, transmission electron microscopy (TEM), and cathodoluminescence (CL) as will be discussed later. To minimize the obscuring effects of surface recombination, a large junction area was chosen for devices that would be dc characterized only. Specifically, the emitter-base junction area is 1.2×10^{-4} cm^2 and the base-collector area is 3.7×10^{-4} cm^2. Alloyed Ni/AuGe/Ag/Au formed contacts to emitter and collector, while Ti/Pt/Au contacted the base. The etchant used throughout the processing was H_3PO_4:H_2O_2:H_2O (3:1:50). High-frequency characterization was carried out using an HP8510 system on devices with dual 8 µm x 30 µm emitter stripes, a 2 µm emitter mesa to base contact separation, and a 1.7×10^{-5} cm^2 base-collector junction area.

Results and discussion: Nomarski microscopy showed a cross-hatched pattern on the surface of the wafer with a 8% In base layer; remaining wafers were featureless. After that the RBS had confirmed the nominal mole fractions, TEM carried out on the cross-hatched layer revealed the presence of 40,000 dislocations per cm. With no cross-hatches observed on the 0%, 2%, and 5% layers, and with reference to the work by FRITZ et al.[7], we expect these la-

Table II. Common emitter current gains

Indium mole fraction	0%	2%	5%	8%
Current gain	3.5	8	21	340

yers to be virtually free from misfit dislocations, i.e. pseudomorphic. Ion channeling studies did not indicate any significant de-channeling in either layer.

The common emitter current gains of the large area transistors are shown in Table II.

The initial increase in the gain with increased indium composition is exponential. With a ratio of five between base and emitter doping levels, the injection efficiency will, given a low indium content, be quite poor and thus it will govern the gain. More precisely, the logarithm of the gain will depend linearly upon the the difference between the heights of the energy barriers that holes and electrons face at the emitter-base heterojunction. Consequently, this barrier, which according to simple theory should be equal to the valence band offset KROEMER [4], can be deduced from the slope of the initial part of the gain vs. composition curve. In our case it amounts to 9.7 meV/%In.

While the gain increases strictly exponentially up to 5% In, the outstanding gain of the $In_{0.08}Ga_{0.92}As$ base device represents a superexponential growth. Furthermore, in contrast to the lower composition devices, the highest composition device has a gain that is dependent on the collector current I_c. Thus at I_c=95 mA, the gain reaches β=650.

Another fundamental difference between the pseudomorphic devices and the one that is partially relaxed is manifested by the gain vs. temperature curves. If the minority carriers diffuse across the base, the gain should at first increase as the temperature is lowered from RT due to an increased minority carrier lifetime. However, at some point the minority carrier liftime will no longer increase. As the value of the diffusion constant evolves inversely with temperature, the gain will then start to fall as described by ITO et al.[8]. This sort of behaviour is exhibited by the $In_{0.05}Ga_{0.95}As$ base device. The gain of the $In_{0.08}Ga_{0.92}As$ base device on the other hand increases monotonically upon cooling, reaching β=3000 at 50K. Such behaviour is compatible with the presence of an electric or pseudo-electric field in the base, as well as with base transport mechanisms that rely on high-energy injection. In conclusion, the non-pseudomorphic device displays a superior dc current gain and a different base transport mechanism as compared to its pseudomorphic counterparts, albeit the exact nature of the transport mechanism has yet to be established.

Much as a high dc gain is desirable, the rf properties will still provide the ultimate test of device feasibility. Mismatched material may e.g. contain potentially ruinous traps not apparent from dc measurements. This being the case, we have investigated the properties of a series of transistors from the four layers mentioned above. Table III presents some relevant device data.

The cut-off frequency f_T improves monotonically with increased indium content, rapidly at first as the injection efficiency improves, but gradually slower as composition-independent para-

Table III. Microwave device data

Indium mole fraction	0%	2%	5%	8%
dc current gain	6	8	25	300
Cut-off frequency f_T (GHz)	2.2	4.2	6.6	7.0

sitical effects start to dominate. Since f_T increases smoothly with increased indium content, and reaches f_T=7GHz (quite acceptable considering the size of the present device), we conclude that there are no major deleterious effects on the speed of the devices caused by the large mismatch and the concomitant lattice irregularities. This is further corroborated by the values of the components in a small signal model that we fitted to the S-parameters of the $In_{0.08}Ga_{0.92}As$ base device. They reveal that, as should be the case, the emitter-base junction charging time provides the speed limit.

The same kind of transistors were also tested in switching mode. Five-stage ring oscillators fabricated out of the different layers showed, as expected, monotonically decreasing switching times with increased indium composition.

To complement the partly surprising device results, we decided to take a closer look at the recombination properties of the various layers. The spectrally integrated CL technique chosen for the purpose revealed that elongated centres of elevated recombination, appearing as dark line defects (DLDs), had started to form in the pseudomorphic $In_{0.02}Ga_{0.98}As$ base device. Progressively denser networks of DLDs are found in the layers with 5% and 8% indium; the 0% In layer appeared featureless. Interestingly, the DLDs not only form before the misfit dislocations do, but the two are also quite differently spaced when co-existing. In the 8% In case, typical DLD separations and misfit dislocation separations are 5μm and 0.25μm, respectively. We conclude that neither DLDs, nor misfit dislocations are necessarily very influentcial on device performance, and, contrary to e.g. GOURLEY et al.[9], that the two are not strongly correlated.

<u>References</u>:

1. J.J. Rosenberg, M. Benlamri, P.D. Kirchner, J.M. Woodall, and G.D. Pettit: IEEE Electr.Dev.Lett. vol EDL-**6**, no 10, p.491 (Oct. 1985)
2. T. Henderson, J. Klem, C.K. Peng, J.S. Gedymin, W. Kopp, and H. Morkoç: Appl.Phys.Lett.**48** (16), p.1080 (21 April 1986)
3. P.M. Enquist, L.P. Ramberg, F.E. Najjar, and L.F. Eastman: to appear in Appl.Phys.Lett.
4. H. Kroemer: Proc.IEEE, <u>70</u>(1), p.13 (1982)
5. H. Kressel: J.Electr.Mat. vol **4**, no 5, p.1081 (1975)
6. S. Yamakoshi, M. Abe, O. Wada, S. Komiya, and T. Sakurai: IEEE J.Quant.Electr. vol QE-**17**, no 2, p.167 (Feb. 1981)
7. I.J. Fritz, S.T. Picraux, C.R. Dawson, T.J. Drummond, W.D. Laid ig, and N.G. Anderson, Appl.Phys.Lett.**46**(10), p.967 (1985)
8. H. Ito, T. Ishibashi, and T. Sugeta: Japn.Jour.Appl.Phys. **24**(4) p.L241 (April 1985)
9. P.L. Gourley, R.M. Biefeld, and L.R. Dawson: Appl.Phys.Lett. **47**(5), p.482 (1 Sept. 1985)

Cryogenic GaAs Integrated Circuits Using a Lightly Doped GaAs FET Structure

J. Kamdem[1], *M. Henry*[1], *J.P. Klein*[2], *F. Alexandre*[3], *and M. Gloanec*[4]

[1]C.N.E.T. Lannion, BP 40, F-22301 Lannion Cedex, France
[2]C.N.E.T. Grenoble, F-38031 Grenoble, France
[3]C.N.E.T. Bagneux, F-92220 Bagneux, France
[4]Thomson-CSF Orsay, F-91405 Orsay, France

INTRODUCTION

Several papers in recent years about GaAs MESFETs with low doping levels have shown that lowering the operating temperature of this device (to around 100K) [1,2], can improve the carrier transport properties inside the active layer. One observes a rise of electron mobility and saturation velocity which leads to an increase of transconductance and saturation drain-source current. Cooling GaAs integrated circuits provides a means of raising the charge/discharge current inside the parasitic capacitances; thus it may be possible to combine either the high switching speed of conventional MESFETs with a low power dissipation or a higher logic gate speed capability with no increase of consumption. This work presents the performance of cooled high speed logic gates, based on a special MESFET structure, designed for low temperature operation.

DEVICE STRUCTURE AND PROCESS

The low temperature GaAs FET structure we adopted (Fig.1) is made up of a double N^+/N layer, grown by MBE and separated from the semi-insulating substrate by a buffer layer 2 µm thick. The N^+ region with a doping level around 2×10^{17} cm^{-3} was used to lower the ohmic contact resistance. Device isolation was performed by firstly a mesa recess and secondly a B^+ implantation, while a gate recess up to 1500 Å was employed to attain the desired value of pinchoff voltage.

LOGIC CELL DESIGN

Prior to the technological process, logic cell behavior was simulated at 77K, in the time domain, using a non-linear GaAs MESFET model which had previously been validated at 300K [3,4]. Algebraic equations for non-linear elements in the model (Fig. 2) were kept according to the assumption that no ballistic carrier transport phenomenon would occur at 77K in our structures for L = 1 µm gate length.

At this temperature, the mean free distance $L_m = (3kTm^*/q^2)^{1/2}$ between carrier collisions in GaAs with a 10^{16} cm^{-3} doping level, is equal to 0.135 µm for a low-field mobility μ_0=15000 cm²/V.s; so we have $L_m < L$.

Two logic gate configurations, one with and the other without a diode in the voltage shifter branch, have been designed and simulated in the time domain analysis.

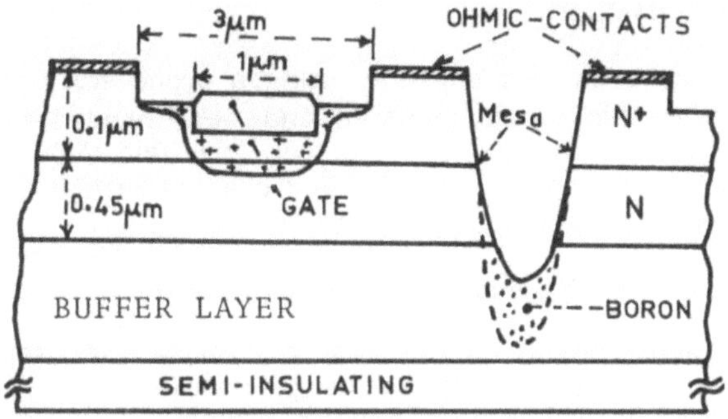

Fig.1.Low-temperature GaAs FET structure

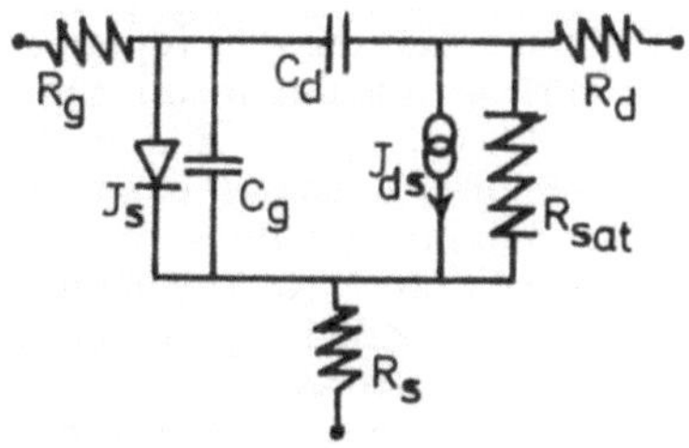

Fig.2. Non-linear GaAs FET model

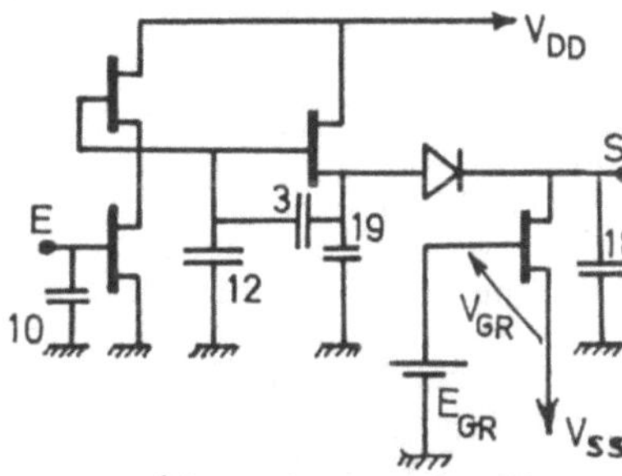

(Capacitances in fF)

Fig.3.Ga As MESFET logic gate

We have introduced a voltage source E_{gr} in the driver shifter branch in order to control the high and low logic levels simultaneously (Fig. 3). Five different types of 13-stage ring oscillator (RO) circuits with width of switch FET equal to 10 μm, 30 μm and 50 μm have been designed.

A 3.8 mm x 3.8 mm master field made of RO circuits and test devices (FETs, isolated logic gates), in 500 μm x 500 μm chip size, was formed and fabricated in a 6 x 4 array arrangement. Figure 4 shows a microphotograph of a RO chip.

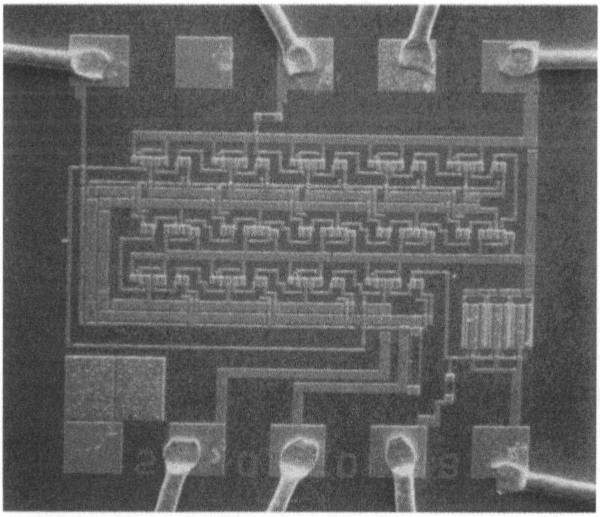

Fig.4. Microphotograph of a 13-stage RO chip.

EXPERIMENTAL RESULTS AND COMMENTS

After fabrication, the circuits were evaluated on slice, using a probe station; the wafer was then diced and the chips were mounted and were bonded to alumina chip carriers. Carrier concentration n(x) and electron low-field mobility $\mu_0(x)$ profiles were determined from capacitance C_{gs} and channel resistance R_c measurements versus gate-source voltage V_{gs}. The effects of Debye length $L_d(x) = [kT\varepsilon/q^2 n(x)]^{1/2}$ on R_c and C_{gs} were taken into account during the automatic measurement process by comparing the step increase Δx of the measured active layer thickness $x(V_{gs})$ to the width X_t of the transition region between the depletion layer and the channel

$$X_t = L_d(x)\ln\left[\ln\left(\frac{1}{(1-\lambda)}\right)^2\right] \quad \text{with} \quad 0.05 < \lambda < 0.1$$

n(x) and $\mu_0(x)$ were then computed from the classical flat doping profile relations. While the device was cooled to 77K, the n(x) curves remained close, with an average doping level equal to 10^{16} cm^{-3} in 0.4 µm mean depth (Fig.5). We noted a shift by 200 mV of the threshold voltage of the FETs, while the built-in potential increased by 250 mV. Figure 6 shows examples of low field mobility profiles at 300K and 77K. The maximum values reached are 4660 $cm^2/V.s$ at room temperature and 11400 $cm^2/V.s$ at 77K. The measured parasitic drain and source resistances were then divided by a factor of around 2 while the ohmic contact resistance remained unaltered.

Figure 7 presents an example of RO output waveforms, and Table 1 displays part of the data summary for RO measurements. The maximum frequency observed is 1013 MHz at 77K, which corresponds to a propagation delay time T_{pd} = 38 ps per gate and a power delay product F = 1pJ; at 300K the same circuit oscillates at 500 MHz for T_{pd} = 78 ps, and F = 1.1 pJ. From 123 K to 300 K we observed a quasi constancy of F, and a linear dependency of the power consumption per logic gate (Fig.8); so the relation $F = P.\ T_{pd}$, leads to a linear variation of $1/T_{pd}$ with temperature.

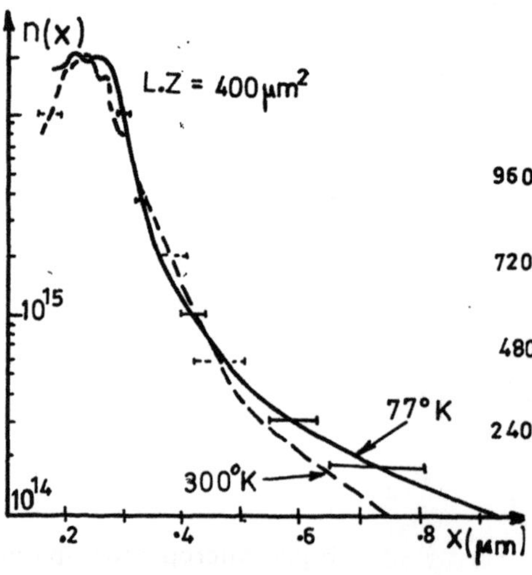

Fig 5 Carrier concentration profiles

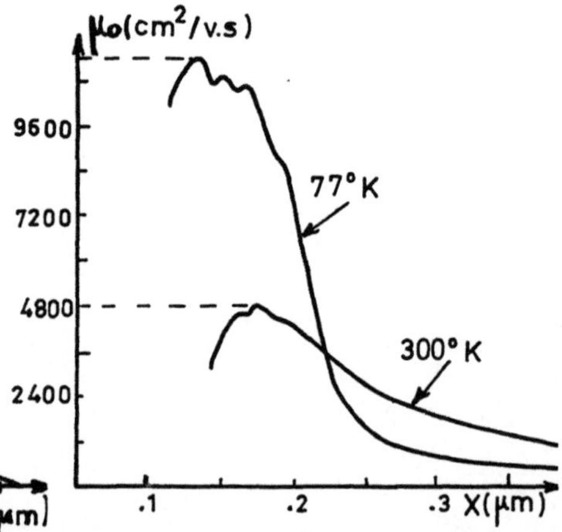

Fig 6 Low-field mobility profiles

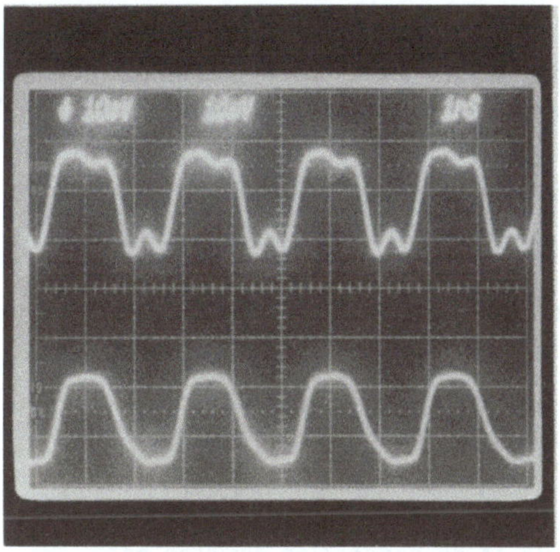

Fig 7 RO waveforms - 1nS/div.

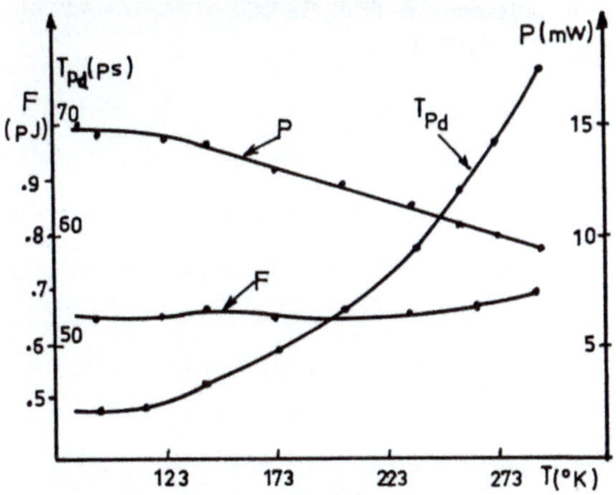

Fig 8 Temperature dependency of Tpd, F, P.

Table 1. Summary of RO performances.

No. of CIRCUIT	SP1007/1613		OPT150/1913		OPT150/1105	
Switch FET gate width(μm)	10	10	50	50	50	50
Temp.(K)	300	77	300	77	300	77
Freq.(MHz)	443	702	520	945	500	1013
T_{pd}(ps)	87	58	74	40.7	78.2	38
P(mW)	5.1	6.7	12,8	20,7	14,7	27
F(pj)	0.45	0.37	0.95	0.8	1.1	1

CONCLUSION

High-speed logic gates optimised for low-temperature operation have been characterised. Large average mobility levels have been obtained for 1-2 10^{16} cm^{-3} doped active layer, cooled to 77 K. A minimum propagation delay time of 38 ps/gate was measured with a power delay product of 1 pJ.

[1] Ress H, Low-temperature FET for low-power high-speed logic, Electron. Lett. 13, 156 (1977)

[2] Pone J.F, Castagne R, Kamdem J, Klein J.P, Henry M, Design of 3-5 devices for logic applications at 77K, E.S.S.D.E.R.C Conf. 1984, Lille (Fce)

[3] Rossel P, Azizi C, Graffeuil J, Tranduc H, Sautereau J.F, Modèles du transistor à effet de champ a l'Arseniure de Gallium, Revue polytechnique 1372, 335 (1978)

[4] Kamdem J, Maimuni R, Azizi C, Graffeuil J, Rossel P, Méthodologies d'étude experimentale et de conception assistée par ordinateur des amplificateurs de puissance à TEC GaAs, Revue de Physique appliquée 18, 213 (1983)

GaAs-on-Insulator Structure Prepared by Heteroepitaxy of Fluorides and GaAs

T. Asano, K. Tsutsui, H. Ishiwara, and S. Furukawa

Graduate School of Science and Engineering, Tokyo Institute of Technology, 4259 Nagatsuda, Midoriku, Yokohama 227, Japan

GaAs-on-insulator(GaAs-SOI) structure has been prepared on Si or GaAs substrates by the molecular beam epitaxy of $(Ca,Sr)F_2$ and GaAs. Microstructure of the GaAs films has been investigated by cross-sectional TEM. It has been found that microtwins and dislocations exist in the films on (100) oriented substrates, while only dislocations exist on (111) substrates. The origin of these defects is discussed and related to electrical properties of the GaAs films. MESFETs fabricated in the GaAs-SOI film have shown a g_m value of about quarter of the value obtained from homoepitaxial GaAs.

1. INTRODUCTION

GaAs electronic devices are usually fabricated on semi-insulating GaAs substrates to make electrical isolation between devices. However, particularly for fabrication of future very high speed ICs having very small geometry devices, it will be required to realize the electrical isolation by dielectric materials with low dielectric constants in order to reduce parasitic capacitances and leakage currents. From this point of view, the GaAs-SOI(semiconductor-on-insulator) structure is one of ideal structures for GaAs ICs. Moreover, if the GaAs-SOI structure is formed on Si substrates, it will bring further advantages for IC fabrication because Si provides mechanically strong and high thermal conductive substrates and well-controlled processing techniques.

Recently we have demonstrated that a GaAs-SOI heteroepitaxial structure can be formed on GaAs or Si substrates by use of $(Ca,Sr)F_2$ fluoride insulators.[1-3] The use of the fluorides is very attractive because their widely variable lattice constants(0.546-0.580nm)[4,5] and low dielectric constants(6.5-6.8).

In this paper, we report details of the crystalline quality of the GaAs-SOI observed by cross-sectional transmission electron microscopy(TEM) and electrical properties obtained from MESFETs.

2. EXPERIMENT

The GaAs-SOI structures were prepared by using two separate growth systems. One is a high vacuum system for the growth of fluorides and the other is an MBE system for the growth of GaAs. CaF_2 was grown on (100) and (111)Si substrates at 600°C and 700°C respectively. $(Ca,Sr)F_2$ mixed fluorides were grown on both (100) and (111)GaAs at 450°C. The growth of GaAs layers on top of the fluoride layers was performed at substrate temperature from 350°C to 650°C. Details of the growth process have been described in Refs. 1 and 2. For microstructural characterization, cross-sectional TEM samples were prepared by dicing, polishing and ion-milling with 5 keV Ar. TEM observations were carried out at an acceleration voltage of 200 kV. For electrical characterization, Hall effect measurements were carried out by using the Van der Pauw method. MESFETs having a Au gate electrode were also fabricated on the GaAs/(Ca,Sr)F_2/GaAs structure.

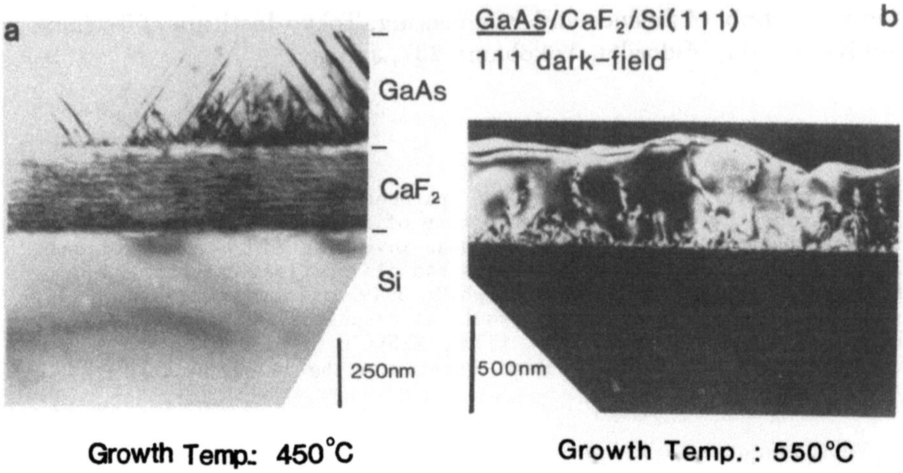

Fig. 1 Cross-sectional TEM micrographs of GaAs/CaF_2/Si(100)(a) and (111)(b) structures. The GaAs films were grown at 450°C and 550°C respectively.

3. MICROSTRUCTURE OF GaAs-SOI

Figure 1 shows cross-sectional TEM micrographs for GaAs films grown on (100) and (111) oriented CaF_2/Si structures. In the GaAs film on the (100) substrate, both microtwins along {111} planes and dislocations exist. On the contrary, only dislocations are observable in the GaAs film on the (111) oriented substrate. These results clearly demonstrate that GaAs films of better quality can be grown on (111) substrates than on (100) substrates. However, the doping of impurities is difficult for (111)GaAs owing to the compensation effect[6]. Therefore we mostly investigated the formation of GaAs films on (100) oriented substrates.

In Fig. 2, depth distributions of the density of microtwins and dislocations in the GaAs film on the (100) substrate are shown. Both defects are relatively high in density near the GaAs/CaF_2 interface region. The density of microtwins rapidly reduces toward the surface. However the dislocation density is almost constant throughout the film except in the interface region.

These two types of defects have been observed in GaAs films grown at different growth temperatures. But the dependence on the growth temperature of the defect density was different between the two as shown in Fig. 3. Although the dislocation density reduces with elevating growth temperature, the microtwin density increases. Similar characteristics have been obtained for GaAs films grown on (Ca,Sr)F_2/GaAs structures. From these results we presume origins for the two types of defects as follows. The microtwins are produced by a slip deformation of the GaAs crystal induced by a compressive stress due to the difference in the thermal expansion coefficient between GaAs(6×10^{-6}/deg) and fluoride(19×10^{-6}/deg). The dislocations extending to the surface are produced at low-angle grain

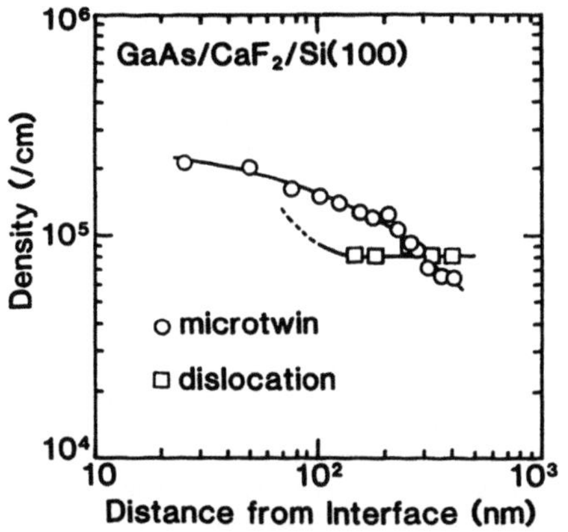

Fig. 2 Depth distributions of the density of microtwins and dislocations in the GaAs film grown on the CaF_2/Si(100) at 450°C.

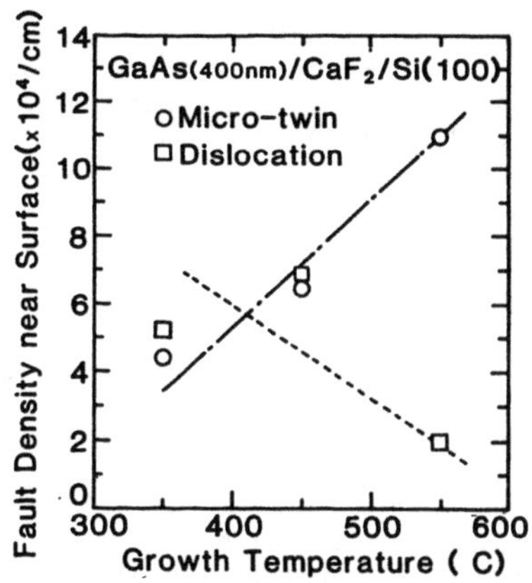

Fig. 3 Growth temperature dependence of the microtwin and dislocation densities at the surface of 400nm thick GaAs films on CaF_2/Si(100).

boundaries which originate from the three-dimensional nucleation of GaAs on fluoride. In fact, ion channeling measurements have shown that the spread of the crystallite orientation of GaAs films is 0.4-0.7° wider than that of a bulk GaAs crystal.

4. ELECTRICAL PROPERTIES OF GaAs-SOI

Figure 4 shows Hall electron mobilities in GaAs films grown on $(Ca,Sr)F_2$/GaAs(100) structures as functions of the thickness of the undoped buffer layer and the growth temperature. The growth at higher temperatures, for example 550°C, resulted in high resistivity. The GaAs

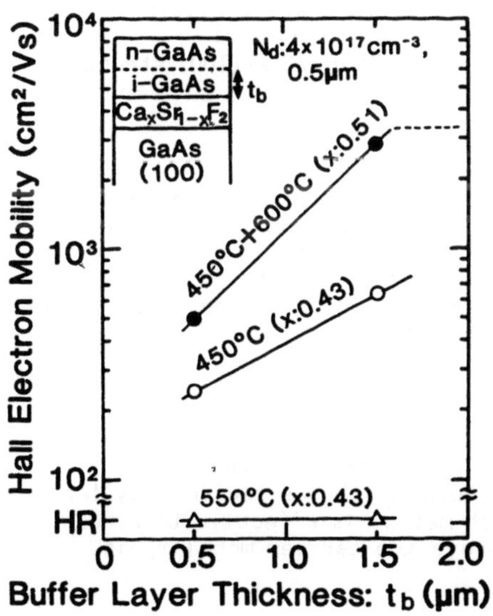

Fig. 4 Hall electron mobility in GaAs on $(Ca,Sr)F_2$/GaAs(100) as functions of growth temperature and the buffer layer thickness.

film grown at 450°C showed 250-630 cm^2/Vs depending on the thickness of the buffer layer. The increase of the buffer layer thickness resulted in higher mobility. These results and the above microstructural investigation suggest that the mobility is mainly limited by the microtwin defect.

It is well known that substrate temperatures higher than 600°C are required to grow high quality GaAs films by MBE. In order to improve the quality of GaAs-SOI films by growing at higher temperatures without increasing the microtwin density, we have developed a two-step growth method.[1] That is, an about 100nm thick GaAs layer is first grown on the fluoride layer at 450°C and it is followed by the growth of GaAs at 600°C. In this method, the GaAs layer grown at the lower temperature is expected to reduce the thermal stress applying to the GaAs layer grown at the higher temperature, since the relative atomic spacing at the GaAs/fluoride interface is fixed at the state of the lower temperature. Actually the density of microtwins of a 400nm thick GaAs film grown on a $(Ca,Sr)F_2/GaAs(100)$ by the two-step method was about 8×10^4/cm at the surface, which was even lower than that of a GaAs film grown at 450°C(1.3×10^5/cm).

The mobility is improved by using the two-step growth method as shown in Fig. 4. A mobility value of 2700 cm^2/Vs was obtained on a 1.5µm thick buffer layer, which corresponded to about 80% of the mobility of a homoepitaxial GaAs.

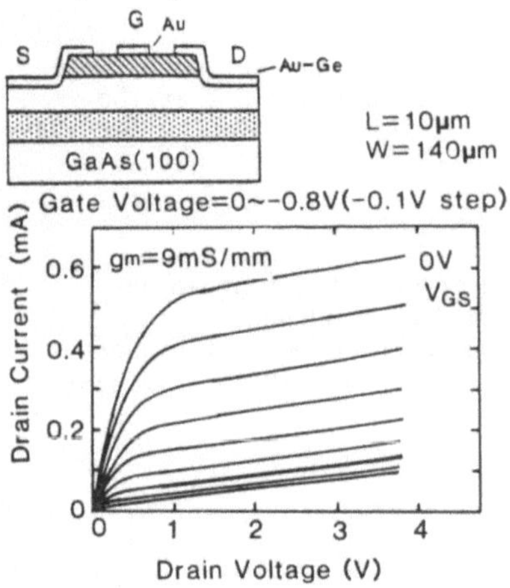

Fig. 5 I_d-V_d characteristic of a MESFET fabricated on a $GaAs/(Ca,Sr)F_2/GaAs(100)$ structure.

Figure 5 shows a characteristic of a MESFET having a gate length of 10µm fabricated on a $GaAs/(Ca,Sr)F_2/GaAs(100)$ structure. A g_m value of 9 mS/mm is obtained at the zero gate voltage. Although this g_m value is lower and the leakage current is higher than that of a MESFET fabricated on homoepitaxial layer(g_m=42mS/mm), the result indicates the potential of the GaAs-SOI structure for device applications.

5. CONCLUSION

GaAs-SOI can be formed on GaAs or Si substrates by the heteroepitaxial growth of fluoride and GaAs, which has feasibility for device applications. The two-step growth method is effective to reduce the microtwin densities, and thus to improve the carrier mobility in GaAs-SOI films.

ACKNOWLEDGMENT

The authors are thankful to Prof. A. Satoh for the use of the TEM. This work was partially supported by the 1986 Grant-in-Aid for the Special Distinguished Research (No. 59060002) from the Ministry of Education, Science and Culture of Japan.

REFERENCES

1. K. Tsutsui, H. C. Lee, H. Ishiwara, T. Asano, and S. Furukawa: Inst. Phys. Conf. Ser. No. 79, Chapter 2 (Adam Hilger Ltd, 1986) p. 109, Paper presented at Int. Symp. GaAs and Related Compounds, Karuizawa, Japan, 1985
2. T. Asano, H. Ishiwara, H. C. Lee, K. Tsutsui, and S. Furukawa: Jpn. J. Appl. Phys. 25(1986)L139
3. K. Tsutsui, H. Ishiwara, and S. Furukawa: Appl. Phys. Lett. 48(1986)587
4. E. G. Chernevskaya and G. V. Ananeva: Soviet Phys. Solid State 8(1966)169
5. H. Ishiwara and T. Asano: Proc. 14th Conf. Solid State Devices, Tokyo, 1982, Jpn. J. Appl. Phys. 22(1983) Suppl. 22-1, p. 201
6. W. I. Wang: Collected Papers of the 2nd Int. Conf. Modulated Semiconductor Structures, Kyoto, 1985, p.257

Part IV

High-Speed Opto-Electronics

Optoelectronic Generation of Very High Speed Electromagnetic Transients

D.H. Auston, M.C. Nuss, and P.R. Smith

AT & T Bell Laboratories, Murray Hill, NJ 07974, USA

Optoelectronic techniques for generating extremely short electromagnetic transients are reviewed with emphasis on the fundamental properties of photoconducting and electro-optic materials. Circuit considerations will also be discussed, including the use of radiative methods such as the electro-optic Cherenkov effect and photoconducting Hertzian dipoles.

1. INTRODUCTION

As the development of new high-speed electronic devices continues at a rapid pace, it is essential that experimental techniques be available which are capable of determining the speed of response of these new structures and the materials from which they are made. Optoelectronics has proven to be an effective approach which satisfies this need/1/. It derives a speed advantage from the use of short pulse dye and solid state lasers which can have durations as short as 30 fs today/2/. In addition, optoelectronics has a capability for relatively sensitive measurements which are virtually jitter-free and in some cases can be made non-invasively without perturbing the structure or operation of the circuit or device under test/3/.

An essential aspect of this approach is the use of high-speed transducers to couple optical and electronic signals. Two specific material systems have been used for this purpose. These are photoconductors and electro-optic materials. In this paper, we address some of the key problems related to the use of these materials in the limit of very high speeds.

2. OPTICAL RECTIFICATION

Optical rectification is the property of electro-optic crystals which permits one to produce an electrical waveform which follows the envelope of an optical pulse. The effect was first observed by Ward and co-workers/4/ using millisecond pulses from a solid-state laser. The effect was extremely weak and required very energetic optical pulses to produce an observable electrical pulse. With the picosecond and femtosecond lasers available today, however, it is possible to generate relatively large amplitude electrical signals with extremely fast durations. For this reason, it is an attractive method of producing electrical transients for the measurement of high-speed devices and materials.

The basic physical mechanism of optical rectification comes from a second order nonlinear polarization which can be written as:

$$P_i(t) = \chi_{ijk} E_j(t) E_k(t) \; . \tag{1}$$

The nonlinear susceptibility, χ_{ijk}, is related to the electro-optic tensor, r_{kij}, by the expression:

$$\chi_{kij} = -\frac{\epsilon_0}{4} n_i^2 n_j^2 r_{ijk} \; . \tag{2}$$

If the optical pulse has a rapidly varying envelope, as in the case of picosecond and femtosecond pulses, the nonlinear polarization, P(t), will also be rapidly varying and will have a very broad frequency spectrum which can extend from dc up through microwaves, millimeter waves and even into the far-infrared. This poses a major difficulty for extracting this signal and applying it to a specific problem. The actual electrical signal that is produced depends strongly on its environment, i.e. the circuit embodiment. We will discuss three specific embodiments which have been used for coupling the nonlinear polarization to an external circuit. These are (1) a voltage source, (2) a current source , and (3) a radiation source.

2.1 Electro-optic Voltage Source

When an electro-optic crystal is used as a voltage source, it is treated as a lumped capacitor which drives a large resistive load. This is the situation initially explored by Ward and his colleagues. If we simplify the geometry to a cube of dimension a, and place electrodes of the polar faces (3 direction) of the crystal, the voltage developed across these electrodes is given by the expression:

$$V(t) = \frac{1}{\epsilon_{dc}} 30\pi n^3 r_{33} I(t) a \; , \tag{3}$$

where I(t) is the intensity envelope of the optical pulse. The remarkable feature about this result is that in spite of the fact that the signal is extracted in a capacitor, the duration of the voltage waveform can be extremely short. A typical value for the magnitude of the peak voltage given by expression (3) is 10^{-7} V/W/cm. This assumes an electro-optic coefficient of 3×10^{-11} and a dc dielectric constant of 42 (e.g. lithium tantalate). Although this sensitivity figure seems small, with ultrashort optical pulses it is possible to have relatively large peak intensities for moderate pulse energies and average powers. For example, a $1\mu J$ optical pulse having a duration of 1ps in a lithium tantalate crystal of dimensions $10\mu m$ will produce a peak voltage of approximately 1 V.

It is important to note that the peak voltage is proportional to the ratio of the electro-optic coefficient to the dc dielectric constant. Since these usually scale together, it is not necessarily advantageous to use materials with large r coefficients, since they also have very large ϵ_{dc}. Expression (3) is valid only when the size of the crystal is very small compared to the spatial extent of the optical pulse(e.g. less than 100 μm for a 1 ps pulse). Also, the voltage source configuration is not useful for driving low impedance loads at high speed.

2.2 Electro-optic Current Source

A more useful configuration is to use the electro-optic crystal as a current source. In this case, the crystal is mounted on a low impedance electrical transmission line, or may even form a part of the line. This provides a tight coupling between the crystal and the line and current flows readily down the line to produce the signal. Assuming a crystal of roughly cubic dimensions, a^3, the current waveform is given by the expression:

$$i(t) = \frac{1}{4c} n^3 r_{33} \frac{\partial W}{\partial t}, \tag{4}$$

where W(t) is the instantaneous optical power in the electro-optic crystal. An important distinction between the current source and voltage source is that the current waveform is proportional to the first derivative of the envelope of the optical pulse, whereas in the voltage source it is directly proportional to the envelope. This means that the current waveform will be a bipolar signal, having both positive and negative lobes, whereas the voltage is a unipolar signal. Another important distinction is that the peak current is inversely proportional to the square of the optical pulse duration for a given pulse energy, whereas the peak voltage is inversely proportional to the first power of the pulse duration. In both cases the efficiency of optical rectification improves greatly with the use of very short pulses. For example, a $1\mu J$

pulse of 1 ps duration will produce a peak current of approximately 0.3 A in lithium tantalate.

2.3 Electro-optic Cherenkov Radiation

When the pulse duration is less than one picosecond, it becomes more difficult to design circuits which can handle the extremely large bandwidths associated with these pulses. A solution to this problem which we have recently demonstrated/5/ is to let the polarization current radiate directly into a uniform dielectric medium, rather than attempt to couple it into a transmission line. In this case, the size of the crystal is not critical and it can have dimensions that are substantially larger to provide an extended interaction. The polarization source is not stationary, however, but moves with the group velocity of the optical pulse. If the optical pulse is tightly focussed, its spatial extent can be only a few microns,so that it has the appearance of a moving dipole moment. Since it has rapidly varying temporal components, it will radiate in a broad range of frequencies from radio to microwave, millimeter waves, and into the far-infrared. Due to the additional contribution to the low-frequency dielectric response from the infrared lattice vibrations, the velocity of the source exceeds the radiation velocity. This produces a characteristic cone of radiation in the form of a shock wave, analogous to the classical Cherenkov radiation from a charged particle moving at a velocity that exceeds the radiation velocity. The temporal profile of the radiation pulse is approximately a single cycle having a period comparable to the envelope duration of the optical pulse. This technique produces electrical pulses that are shorter than any other method. Effective durations as short as 300 fs have been measured.

Recently, it has been possible to substantially improve the amplitude of the electro-optic Cherenkov pulses by using optical pulses which have been amplified to an energy of approximately $0.1 \mu J$. This has resulted in electric field strengths as high as 7 kV/cm, which corresponds to a peak power of approximately $1 MW/cm^2$.

An interesting aspect of the electro-optic Cherenkov radiation technique is that the peak electric field depends on the inverse 5/2 power of the optical pulse duration(the peak power depends on the inverse 5th power of τ_p). Consequently the efficiency is such that femtosecond optical pulses are essential.

The spectral content of these pulses extends as high as 5THz, making them valuable sources of coherent radiation in a region of the electromagnetic spectrum where tunable coherent sources are not available. When combined with phase-sensitive coherent detection, a complete far-infrared spectroscopy system results which has been used to measure the momentum relaxation times of free carriers in semiconductors/6/ and the dynamics of lattice vibrations /7/.

3 ULTRAFAST PHOTOCONDUCTORS

Photoconductors are also extremely effective sources of very short electrical pulses/8/. When illuminated by ultrafast optical pulses they have been used to generate electrical transients as fast as 0.5 ps/9/. Usually, ultrafast photoconductors are mounted in transmission line structures which are used to propagate the signals to a test system or measurement device. At moderate illumination levels they are current sources and have excellent sensitivity, requiring only nanojoules to produce signal levels of a few hundred milliamps. At higher excitation levels, their resistance can be reduced to values which are only a few ohms, much less than the characteristic impedance of the transmission lines they are driving. In this case they function as voltage sources and produce a signal which is determined only by the bias conditions and show a saturated behavior with respect to light intensity.

3.1 Photoconducting Hertzian Dipoles

An alternative arrangement which we have demonstrated/10/ is to design photoconductors which can radiate directly into free space as a dipole antenna. Their radiation properties can be described by a classical Hertzian dipole. As a first approximation, the radiation field can be determined from the Hertz vector potential of a time-varying dipole moment p(t), of infinitesimal spatial extent. The electric field at a distance r, and angle θ relative to the dipole axis, is then equal to

$$E(r,\theta,t) = \frac{1}{4\pi\epsilon}\left\{\frac{1}{r^3}\ p(t) + \frac{n}{cr^2}\ \dot{p}(t) + \frac{n^2}{c^2 r}\ddot{p}(t)\right\}\sin\theta\ , \qquad (5)$$

where $n = \sqrt{\epsilon/\epsilon_0}$ is the index of refraction of the medium between the two dipoles We immediately recognize in expression (1) three terms: the static field, the near-field, and the far field, each having different spatial and temporal variations. Since the far field (last) term varies as r^{-1}, it will dominate as we go farther from the source. The distance where this occurs depends on the pulse duration, τ_p, and can be estimated by the relation $r > \tau_p c/n$. If τ_p is extremely short, this distance is also short (e.g. 0.1 mm for 1 ps).

The magnitude of the radiated electric field can be estimated from the fact that the first derivative of the dipole moment is equal to the product of the photocurrent and the length of the illuminated region of the photoconductor. The radiated electric (far) field is proportional to the first derivative of the current pulse. If the photocurrent has a rapid rise time due to excitation of free carriers by a subpicosecond optical pulse, and also decays rapidly due to their capture at traps, then the radiated electrical field will have a bipolar shape with a positive pulse having an amplitude proportional to the photocurrent rise time followed by a negative pulse proportional to the decay time. A novel feature of these photoconducting Hertzian dipoles is their reciprocal property which enables them to be used equally effectively as receiving antennas as well as transmitting antennas. This permits a complete measurement system to be made which can be used to determine their speed of response by electrical autocorrelation techniques/10/. Autocorrelation widths as fast as 0.9 ps have been measured/11/.

3.2 Large Aperture Photoconducting Antennas

We have recently extended this approach to large aperture photoconducting antennas. These are photoconductors whose radiating apertures are greater than the pulse length of the electromagnetic signal they generate. They can be described as an array of Hertzian dipoles. In principle these large aperture photoconducting antennas can have appreciable radiation efficiencies. This requires sufficient optical excitation to produce a photoconductance having a sheet resistance that is comparable to or lower than the free-space characteristic impedance. The electric field of a large aperture photoconducting antenna in the region near the radiating surface (near field) can be related to the surface photoconductivity, σ_s, by the expression:

$$E = \left[\frac{\sigma_s Z_o}{2 + \sigma_s Z_o} \right] E_{dc} \,, \qquad (6)$$

where Z_o is the characteristic impedance of the surrounding medium, and E_{dc} is the applied bias electric field. From equation (6) it can be seen that when the sheet resistance due to the photoconductivity (i.e. σ_s^{-1}) becomes comparable to the free-space characteristic impedance, Z_o, the radiated electric field can be comparable to the bias field which can be very large. The large-aperture feature of these antennas also gives them a directional property. As with the individual Hertzian dipole antennas, they can also be used as receiving antennas. Recent experiments have confirmed the novel features of these photoconducting antennas.

REFERENCES

[1] For a review, the reader is referred to: *Picosecond Opto-electronic Devices,* ed. C. H. Lee, Academic Press (1984).

[2] J. A. Valdmanis, R. L. Fork, and J. P. Gordon, Opt. Lett., **10**, 11 (1985).

[3] B. H. Kolner, K. J. Weingarten, M. J. W. Rodwell, and D. M. Bloom, *Picosecond Electronics and Optoelectronics,* Ed. G. A. Mourou, D. M. Bloom, and C. H. Lee, pp50-53, Springer-Verlag Series in Electrophysics, No.21(1985).

[4] M. Bass, P. A. Franken, J. F. Ward, and G. Weinreich, Phys. Rev. Lett. **9**, 446 (1962).

[5] D. H. Auston, K. P. Cheung, J. A. Valdmanis, and D. A. Kleinman, Phys. Rev. Lett., **53**, 1555 (1984).

[6] D. H. Auston and K. P. Cheung, J. Opt. Soc. Am. B, **2**, pp. 606-612 (1985).

[7] K. P. Cheung and D. H. Auston, Phys. Rev. Lett., **55**, 2152 (1985).

[8] D. H. Auston, "Picosecond Photoconductors: Physical Properties and Applications," in *Picosecond Optoelectronic Devices,* ed. C. H. Lee, Academic Press, New York (1984).

[9] M. B. Ketchen, D. Grischkowsky, T. C. Chen, C-C. Chi, I. N. Duling, N. J. Halas, J-M. Halbout, J. A. Kash, and G. P. Li, Appl. Phys. Lett., **48**, 751 (1986).

[10] D. H. Auston and K. P. Cheung, "Picosecond Photoconducting Hertzian Dipoles," Appl. Phys. Lett., **45**, pp. 284-286 (1984).

[11] P. M. Downey and J. R. Karin, *Picosecond Electronics and Optoelectronics,* Ed. G. A. Mourou, D. M. Bloom, and C. H. Lee, pp201-204, Springer-Verlag Series in Electrophysics, No.21(1985).

Picosecond Electro-Optic Sampling

G.A. Mourou

Laboratory for Laser Energetics, University of Rochester,
250 East River Road, Rochester, NY 14623, USA

The electro-optic sampling technique is a new characterization technique based on short optical pulses (picosecond or femtosecond) and the electro-optic effect. With this technique, electrical waveforms have been characterized with temporal resolution of less than 300 fs, corresponding to a 1 THz bandwidth. Since its inception in 1982, this method has been adapted to a variety of measuring situations, leading to a large number of embodiments. The electro-optic sampling was first used to study the dispersion of picosecond electrical pulses on transmission lines. Using a superconducting transmission line, it has been recently shown that conduction losses can be defeated and single picosecond pulses propagated over a few millimeters. Additionally, Ultrafast transistors have been tested with single picosecond resolution that is, the transistor response was time-resolved as the gate was pulsed using a step function with a picosecond rise time. MESFET, TEGFET, and permeable base transistor (PBT) responses have been studied, and switching times as low as 5 ps have been directly measured for the PBT. Because GaAs exhibits the Pockels effect, the GaAs components and circuits can be directly probed in a noninvasive way. We will review the principle of electro-optic sampling in its various embodiments and applications to device and circuit characterization, as well as the direct investigation of velocity overshoot and ballistic and resonant tunneling transport.

As switching speeds of microelectronic circuits increase, new problems arise in characterizing the real-time responses of these circuits. It is now typical for a new device to be faster than conventional electronic techniques can measure.

Figure 1(a) illustrates the evolution of the sampling system for characterization of fast transients over the years. The first technique is based on a tunnel diode and was introduced by Tektronix in the late 1960s. By using Josephson junctions and cryoelectronics, FARIS [1], WOLF [2], signals could be sampled with 2-ps gating time. With the advances in ultrashort laser pulses, new techniques were developed using laser-induced photoconductivity, AUSTON et al. [3-5], and the electro-optic effect. The technique of electro-optic sampling, with its few terahertz potential bandwidth, will be the subject of this paper.

This electro-optic sampling technique is based on short optical pulses and the electro-optic effect, i.e. induced birefringence in a noncentro-symmetric crystal by the signal under analysis. Figure 1(b) exhibits the general embodiment of the technique. Excellent temporal resolution has been demonstrated down to several hundred femtoseconds. Because extremely short pulses, down to 30-fs, have been demonstrated, the speed limitation of this technique is mainly due to the lattice absorption appearing in the terahertz range, according to the material under study. The sensitivity is on the order of 50 $\mu V/(Hz)^{1/2}$. A typical waveform produced by a photoconductive switch triggered by 100 fs pulses is shown in Fig. 2.

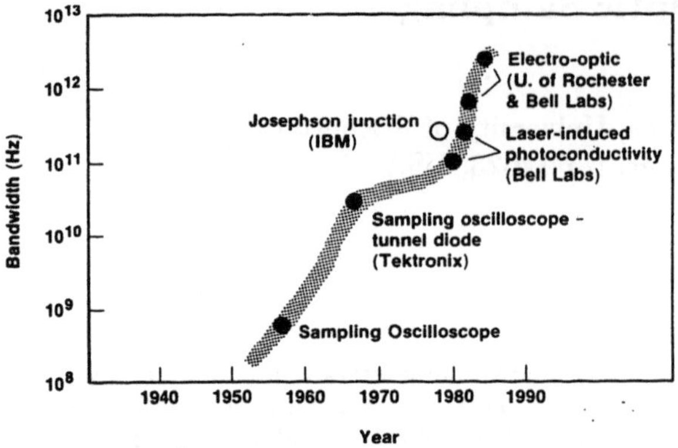

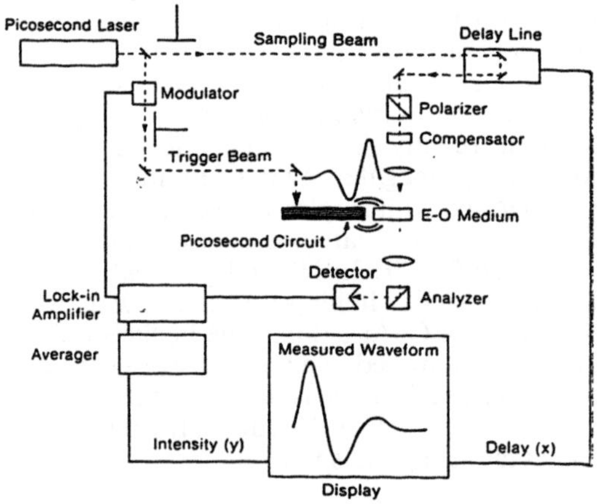

Figure 1 (a) The evolution of sampling systems for the characterization of electrical transients, shown here against a scale based on temporal resolution, has resulted in an electro-optic system having a temporal resolution of nearly 200 fs. (b) General configuration of the electro-optic sampling system. A high repetition rate (~ 100 MHz) train of picosecond pulses is split into a trigger beam and a sampling beam. The trigger beam initiates the generation of the electrical signal to be characterized which propagates through an electo-optic medium. The sampling beam travels through a scanning delay line and then through the Pockels cell modulator arrangement. The intensity of the sampling beam is measured by a lock-in amplifier and then plotted as a function of delay, resulting in an equivalent time representation of the picosecond electrical signal

Since the early work performed on balanced strip lines, the electro-optic technique has been adapted to a wide variety of measurement situations. In doing so, many different embodiments for electro-optic modulation have evolved. Figure 3 depicts simple cross-sectional views of six different modulator configurations that each utilize some form of traveling wave electrode geometry either on the electro-optic crystal itself or in the circuit being measured, KOLNER et al. [6].

Fig. 2

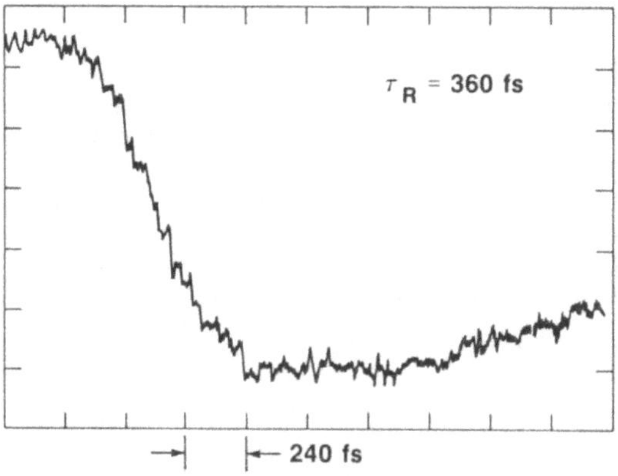

Fig. 3

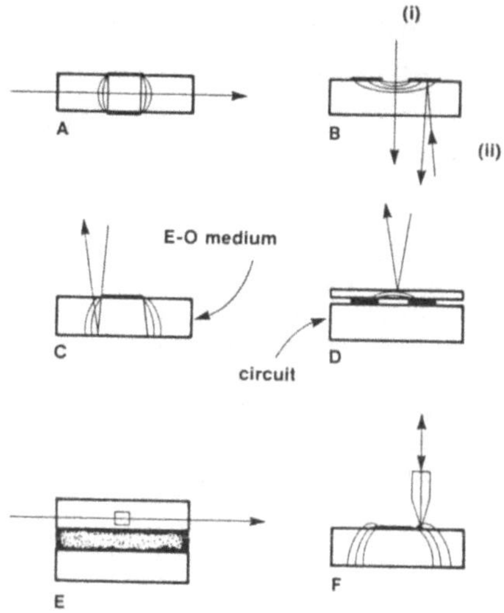

Figure 2 A typical waveform produced by a photoconductive switch is captured using electro-optic sampling. The temporal resolution is less than 300 fs

Figure 3 The various embodiments of electro-optic modulators (depicted with electric field lines) used in electro-optic sampling include: (a) transverse balanced line modulator; (b,I) coplanar line modulator, transverse line geometry; (b,II) coplanar line modulator, longitudinal back-side sampling geometry for use in noninvasive integrated circuit sampling where the substrate is also electro-optic; (c) longitudinal microstrip modulator which has also been implemented for noninvasive sampling of integrated circuits; (d) transverse superstrate modulator; (e) horizontal electrodeless probe modulator; and (f) vertical electrodeless probe modulator. The shaded areas represent electronic device substrates, typically a semiconductor, while the clear areas represent dielectric, electo-optic media

I. Electrical Dispersion in Strip Line Electrodes

(a) Transmission Line at Room Temperature, VALDMANIS et al. [7]

Dispersive effects in the open transmission line electrode geometry can be considerable for high-frequency signals. Dispersion is due to a frequency-dependent, relative effective dielectric constant, ε_{eff}, and the specific electrode geometry. In typical microstrip electrode geometries there exists a critical frequency, f_c, at which the onset of significant dispersive effects appears, due to the excitation of spurious modes in the form of surface wave TM and TE modes. The critical frequency at which coupling occurs to the fundamental quasi-TEM mode is given by

$$f_c = v_o / 4h\sqrt{\varepsilon_{rel} - 1} \quad , \tag{1}$$

where v_o is the velocity of light in vacuum, and h is the electrode separation. The critical frequency occurs in the regime where the electrical wavelength in the dielectric is of physical dimensions comparable to the transmission line cross-sectional dimensions. Below the critical frequency, ε_{eff} assumes a constant nondispersive lower value, ε_{tem}, and above the critical frequency ε_{eff} is equal to a then higher relative dielectric constant, ε_{rel}, and is also nondispersive. Hence, the dispersion curve is basically two-valued, with a smooth transition regime in between.

The electro-optic sampling system provides a unique method, VALDMANIS et al. [7], of measuring these dispersive effects in an electrical signal as it propagates along a transmission line. By translating the sampling beam by a known amount with respect to the stationary modulator it is possible to interrogate the modulator aperture at ever increasing distances from the electrical signal source. This configuration is depicted in Fig. 4. Sampling at each successive point provides a progression of pulse shapes that displays the accumulating effects of the dispersive strip line on the electrical pulse. Figure 5 shows a photoconductive detector as it propagates along a 250-μm thick modulator.

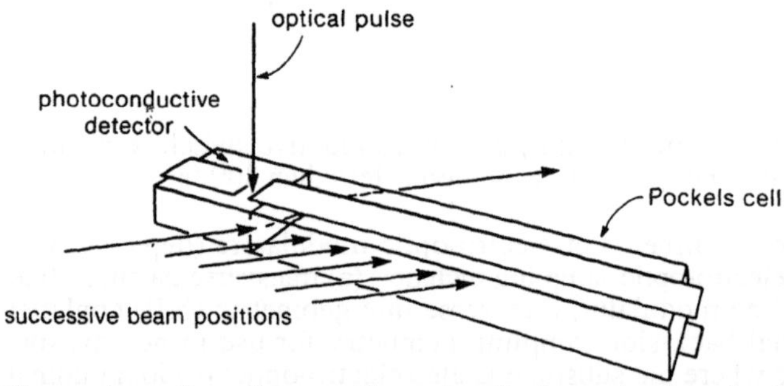

Figure 4 Sampling configuration for investigating the effects of dispersion. The photoconductive detector generates a large bandwidth electrical pulse which is then sampled progressively as it propagates along a balanced line modulator.

When a signal contains frequencies in the regime of the critical frequency for a particular geometry, as in this case, the dispersive characteristics of the strip line conceptually divide the signal bandwidth into two "packets," one containing the lower frequencies and the other the higher frequencies. The lower frequencies will experience a lower effective dielectric constant, ε_{eff}, and hence travel at higher velocity along the strip line. Higher frequencies, in turn, travel more slowly owing to their increased ε_{eff}. Thus as the pulse propagates, its higher frequency components appear later in the waveform. Figure 5 illustrates the effect of dispersion produced on a picosecond electrical pulse at different points along a 250-μm thick balanced microstrip transmission line.

From Eq. (1) it can be seen that the dispersive characteristics of a microstrip transmission line are sensitive to the product of the electrode separation and the square root of the substrate dielectric constant. Figure 6(a) compares the magnitude of the dispersion for several different strip line electrode geometries by plotting the rise time of a test pulse as a function of a propagation distance along the strip line. Curve A plots the rise times of a pulse traveling along a 500-μm thick balanced transmission line on $LiTaO_3$ with $\varepsilon_{rel} = 43$. Significant dispersion is observed

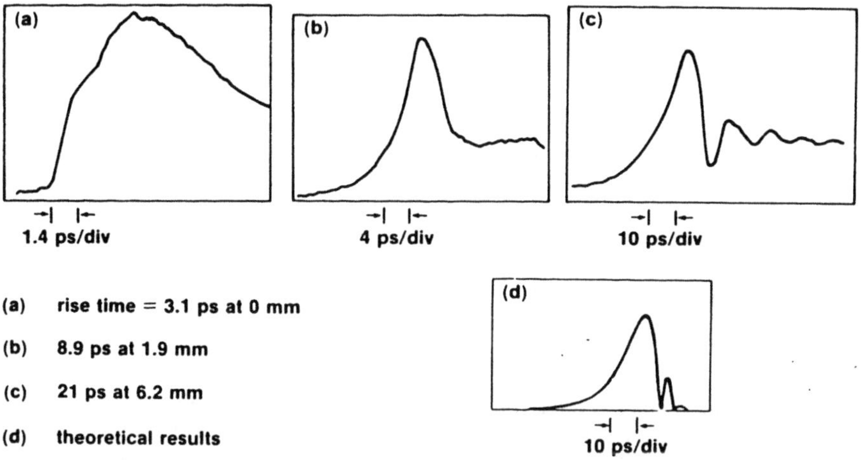

Figure 5 Dispersion of a picosecond electrical signal traveling along a balanced strip line modulator. (a) shows the signal with a rise time of 3.1 ps at the beginning of the modulator. As the pulse propagates, dispersion delays high frequencies relative to lower frequencies and extends the rise time. After 6.2 mm (c) the initial portion of the signal is dominated by low frequencies and has a rise time of 21 ps. Delayed high frequencies now sharpen the trailing side of the pulse and also give rise to a theoretically predicted oscillation.

because of the relatively large electrode separation and high dielectric constant. To decrease the electrode separation, Mourou et al, [8], used a coplanar line on $LiTaO_3$. Curve B plots the dispersion of such a line with 50 μm electrode separation. As predicted, three is a noticeable improvement over the balanced line geometry.

Recently, Kryzak et al, [9], have implemented a similar embodiment of this electro-optic sampling technique to study and successfully develop new transmission lines that can propagate picosecond electrical signals with less dispersion than has been previously achieved. Their design is an air-spaced microstrip geometry where the upper electrode is supported by a 125-μm thick glass superstrate 500 μm above the ground plane. The electrode width is 750 μm. Measurements were done by sampling a small piece of $LiTaO_3$ inserted into the electrode air gap. Due to the elimination of a high ε dielectric between the electrodes, the electric field exists predominantly in the uniform air gap,thus significantly reducing dispersion. Curve 6(c) shows the improvement of nearly two orders of magnitude.

(b) Superconducting Transmission Lines, DYKAAR et al. [10]

We have shown that the electro-optic sampling could be used at less than 2 K. As shown in Fig. 6(b), a $GaAs/LiTaO_3$ hybrid structure was fabricated. In operation, the entire unit was placed in a superfluid helium ($T < 2$ kelvins) environment.

The use of superfluid helium allowed the laser pulses to propagate into and out of the cryogenic environment, undistorted by fluid bubbling. An early version of the

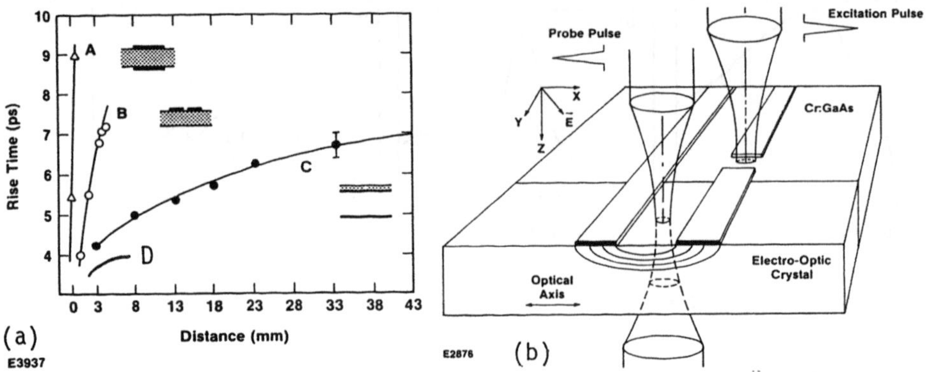

Figure 6(a) The dispersive characteristics of four transmission lines are compared. Cross-sectional views of the electrode structures are also included: (a) balanced strip line on $LiTaO_3$ with 500 μm electrode separation; (b) coplanar strip line on $LiTaO_3$ with 50 μm electrode separation; (c) inverted, air-spaced strip line -- the top electrode is supported by a thin glass superstrate 500 μm above the ground plane (note the vastly improved dispersive character of this geometry); (d) superconducting coplanar transmission line. The line dimensions are 50 μm. Figure 6(b) Superconducting sampling configuration. Electrodes were evaporated directly across a GaAs/$LiTaO_3$ interface

sampler, using a standard Pb-alloy technology for superconducting electronics showed a 5-fold increase in sampler performance to 1.0 ps, when cooled from room temperature. A rise time of 360 fs was obtained using indium electrodes. As shown in Fig. 6, the rise time dispersed to only 2 ps after 3 mm of propagation. These results were modeled using a technique previously reported, HASNAIN et al [11], with the only change in the model being to set the absorption of the transmission lines to zero, suggesting that only modal dispersion is present.

II. High-Speed Transistor Characterization

(a) MODFET Characterization

Three-terminal electrical devices, such as ultrafast transistors, can also be characterized using the electro-optic system. Instead of direct optical excitation, however, an intermediate photodetector is employed to generate a jitter-free electrical trigger signal. Meyer et al, [12], have used a hybrid transverse coplanar modulator to measure the response of a modulation-doped field effect transistor, otherwise known as a MODFET or TEGFET.

The experimental configuration is similar to that employed by Smith et al, [13], for photoconductive sampling measurements of GaAs FETs. Figure 7(a) depicts the experimental arrangement. A GaAs photoconductive switch illuminated by a short (100-fs) optical pulse provides a fast input signal to the gate superimposed on the dc gate bias. The source is grounded, and the drain is connected to the coplanar electro-optic sampler, which also provides the drain bias. The response of the MODFET is a current pulse out of the drain that propagates onto the modulator.

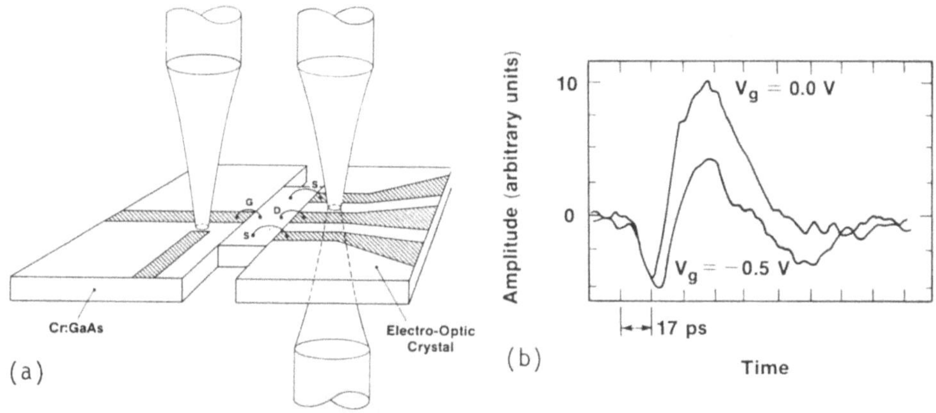

Figure 7 (a) The coplanar sampling geometry used for characterizing a MODFET. The transistor is sandwiched between a GaAs photoconductive detector used to trigger the gate and a coplanar lithium tantalate modulator for sampling the current pulse output from the drain. (b) MODFET response to a short electrical pulse produced by a GaAs photoconductive detector. The traces show the response with the gate pinched off (V_g = -0.5 V) and with the gate on (V_g = 0.0 V). The rise time with the gate on is 16 ps

Some experimental results are shown in Fig. 7(b). With a drain bias of +1.0 V and a gate bias of -0.5 V, the gate is pinched off and some of the input pulse is capacitively coupled through to the output. When the gate bias is increased to 0.0 V the device turns on and gain is apparent. The measured 10% - 90% rise time of the TEGFET response is 16 ps which agrees well with the theoretical rise time based on the calculated cut-off frequency of 23 GHz.

Measurements of this type are traditionally performed with several devices cascaded together to form a ring oscillator. By calculating the delay time for the entire chain, the average time for a single element could be deduced. This electro-optic technique provides the time response of a single element directly.

(b) Permeable Base Transistor (PBT)

Conceived at Lincoln Laboratory in 1979, BOZLER et al. [14], the PBT presents unique problems in its manufacture. As shown in Fig. 8, a PBT is arranged similarly to a vacuum tube triode, except that here the "grid" has submicrometer dimensions. By interrupting the GaAs epitaxy process to fabricate the grating, contaminants are introduced, keeping yields low. As of this writing, only single devices have been fabricated, precluding the construction of ring oscillators. In addition, analog measurements have proven difficult because of the very high gain of the PBT. However, extrapolated frequency-gain measurements show a maximum frequency of oscillation in excess of 200 GHz, HOLLIS et al. [15]. The speed of this device results from advances in fabrication techniques that allow for extremely small dimensions in both the horizontal and vertical directions. These advances have led to a device that, until now, remained completely uncharacterized in the time domain.

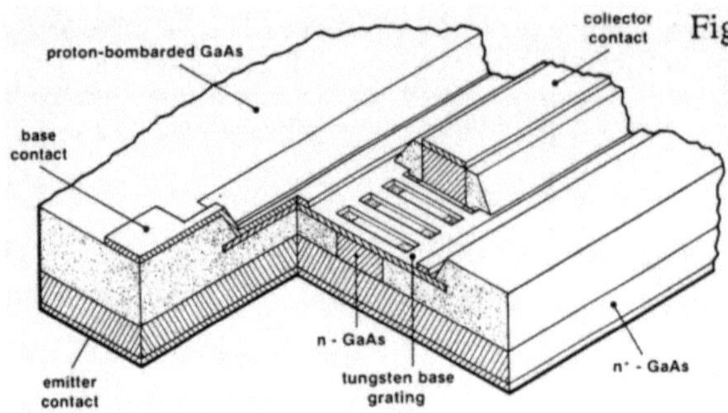

Figure 8 Geometry of PBT

For electro-optic characterization, the PBT was built into the sampling geometry shown in Fig. 9(a). This geometry permits dc biasing as well as high-speed operation. The CPM laser-based sampling system measured the rise time, as shown in Fig. 9(b). Electromagnetic radiation from the input connection was thought to cause the negative prepulse. This feature did not scale in amplitude with PBT gain and has also been present in earlier measurements on TEGFET and MESFET devices. This measurement identifies, to the best of our knowledge, the currently fastest three-terminal room-temperature device.

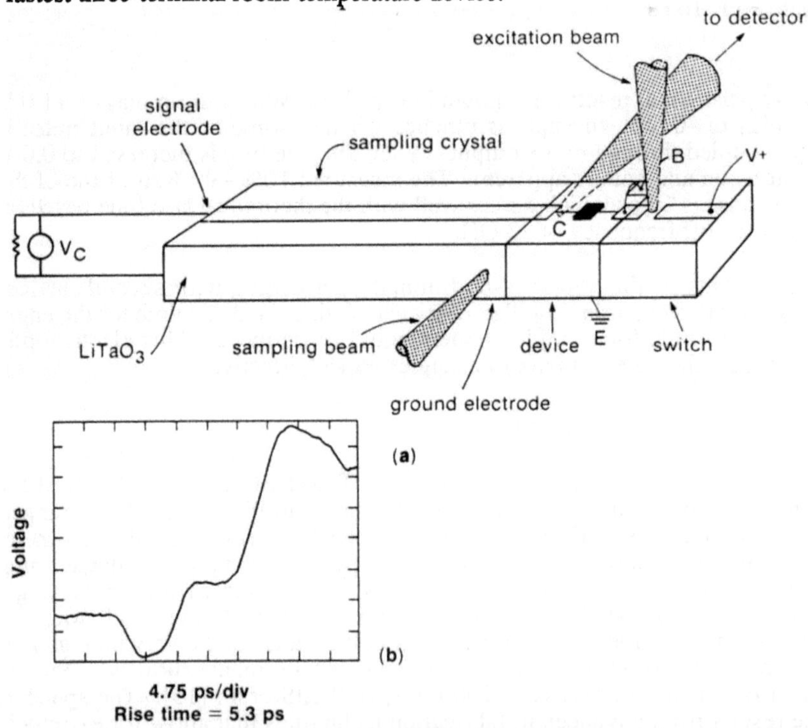

Figure 9(a) Sampling geometry for PBT characterization. Looping wire bonds were used for bias leads to prevent loading of high-speed signal paths. (b) Step response of PBT as measured by the electro-optic sampling system

In conclusion, the electro-optic sampling is presently the best technique to look at waveforms in a nonevasive way with few hundred femtosecond resolution. This technique has been used to study the response of circuits and devices with unprecedented resolution. It should play a key role in helping the electronics to move from the gigahertz to the subterahertz regime.

Acknowledgment

This work was supported by the Laser Fusion Feasibility Project at the Laboratory for Laser Energetics which has the following sponsors: Empire State Electric Energy Research Corporation, General Electric Company, New York State Energy Research and Development Authority, Ontario Hydro, Southern California Edison Company, and the University of Rochester. Such support does not imply endorsement of the content by any of the above parties.

References

1. S.M. Farris: In Appl. Phys. Lett 36 (12) pp. 1005-1007 (1980)
2. P. Wolf: In Proc. Conf. on Picosecond Electronics and Optoelectronics, March 1985, ed. by G. Mourou et al., pp. 238 (1985)
3. D.H. Auston et al: In Appl. Phys. Lett. 37 (4) pp. 371-373 (1980)
4. D.H. Auston and P.R. Smith: In Laser Focus 18 (4) pp. 89-93 (1982)
5. D.H. Auston, K.P. Cheung, and P.R. Smith: In Appl. Phys. Lett. 45 (3) pp. 284-286 (1984)
6. B.H. Kolner and D.M. Bloom: In Elect. Lett 20 , pp. 818-819 (1984)
7. J.A. Valdmanis and G.A. Mourou: In Picosecond Optoelectronic Devices, ed. by C.H. Lee (Academic Press, Orlando 1984), pp. 249-270
8. G.A. Mourou and K.E. Meyer: In Appl. Phys. Lett. 45 (5) pp. 492-494 (1984)
9. C.J. Kryzak et al.: In "Transmission Line Designs With a Measured Step Response of 3 ps per Centimeter," Proc. Conf. on Picosecond Electronics and Optoelectronics, ed. by G. Mourou et al (Springer-Verlag, Berlin, Heidelberg 1985), pp. 244-248
10. D.R. Dykaar, R. Sobolewski, J.F. Whitaker, T.H. Hsiang, G.A. Mourou, M.A. Hollis, B.J. Clifton, K.B. Nichols, C.O. Bolzer, and G.A. Murphy: In Proc. Conf. on Ultrafast Phenomena, Snowmass CO, June 1986 (to be published in Springer-Verlag)
11. G. Hasnain, A. Dienes, and J.R. Whinnery: In IEEE Trans. on Microwave Theory and Techniques MTT 34 , pp. 738-741 (1986)
12. K.E. Meyer, D.R. Dykaar, and G.A. Mourou: In Picosecond Electronics and Optoelectronics, ed. by G.A. Mourou, D.M. Bloom, and C.H. Lee (Springer-Verlag, Berlin, Heidelberg, New York, and Tokyo 1985), pp. 54-57
13. P.R. Smith, D.H. Auston, and W.M. Augustyniak: In Appl. Phys. Lett 39 (9) pp. 739-741 (1981)
14. C.O. Bozler, G.A. Alley, R.A. Murphy, D.C. Flanders, and W.T. Lindley: In Proc. 7th Bien. Cornell Conf. on Active Microwave Devices, pp. 33 (1979)
15. M.A. Hollis, K.B. Nichols, R.A. Murphy, R.P. Gale, S. Rabe, W.J. Piacentini, C.O. Bozler, and P.M. Smith: In International Electron Devices Meeting Technical Digest , pp. 102-105 (1985)

High-Speed Integrated Circuit Testing by Time-Resolved Photoemission

R. Clauberg, H.K. Seitz, A. Blacha, J.A. Kash, and H. Beha
IBM Zurich Research Laboratory, CH-8803 Rüschlikon, Switzerland

The growing complexity of integrated circuits (IC) and the rapid development in VHSICs require contactless testing methods for logic state analysis and waveform measurements on internal nodes of VLSI chips. Considering the fast evolution of VLSI/VHSIC technologies, a contactless testing method has simultaneously to provide challenging features like high-voltage resolution (mV), high spatial resolution (sub-micrometer), and high time resolution (ps) in a short testing time. In addition to these requirements, it is desired that the method should be applicable to all technologies, e.g., Si as well as GaAs-based ones.

In this paper we present a novel method for contactless internal node testing of VLSI chips and VHSICs. The new method is based on some IBM patent applications [1] where it was named "Integrated Circuit Laser Testing" (ICLT). The fundamentals of the method and its attractive features are explained. First experimental results are shown and discussed.

1. BASIC PRINCIPLE

The ICLT approach exploits time-resolved photoemission for voltage measurements. In principle, there are two ways to induce photoemission. The first uses a single-photon process where the photon energy exceeds the work function of the illuminated material. The second uses a multi-photon process, where the photon energy is below the work function, and the sum of two or three photons is required to excite photoemission. Here, the single-photon process is described. The basic principle is illustrated in Fig. 1. A pulsed laser beam is focused on a metal line with voltage V_s. The induced photoelectrons have a distribution of kinetic energies between 0 eV and $\Delta E = h\nu - \Phi$, the energy by which the photon energy $h\nu$ exceeds the work function Φ. Between the emission point and the detector, a retarding grid is installed. If the retarding voltage is higher than V_s all electrons will be accelerated towards the grid, and are detected. On the other hand, if the potential energy $e(V_s - V_{ret})$ at the emission point exceeds the maximum kinetic energy ΔE of the photoelectrons, no electrons can reach the retarding grid, and a zero current is measured. Hence, the detected signal as a function of the retarding voltage is an S-like curve which shifts linearly with V_s [Fig. 1(b)]. The voltage measurement is performed by a feed-back loop system which keeps the detected current at a constant value I_0.

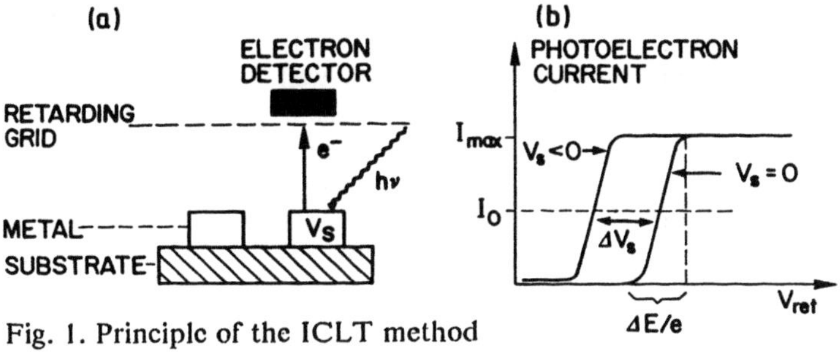

Fig. 1. Principle of the ICLT method

2. VOLTAGE, SPATIAL, AND TIME RESOLUTION

The voltage resolution is limited by shot-noise. Therefore, the minimum detectable voltage is

$$\Delta V_s = \frac{\Delta E}{e \times C} \times \sqrt{\frac{2e \times \alpha}{(I_0/\alpha) \times \tau_{int}}} \times (S/N),$$

where τ_{int} is the integration time of the measurement, $C \approx 0.9$ is a constant defining the quality of the retarding field analyzer, and $I_0 = \alpha \times I_{ph} \times e \times Y$ $(0 < \alpha < 1)$ is the detected current at the operating point of the S-curve. The important parameters are the photon current I_{ph}, the photoyield Y, and the width ΔE of the energy distribution. For clean Al, it is $Y = 10^{-4}$ electrons/photon at 5 eV photon energy. A laser power of $I_{ph} = 5 \times 10^{14}$ photons/s leads to $\Delta V_s = 150\ \mu V$ for $\tau_{int} = 10$ ms, and $S/N = 1$ or $\Delta V_s = 10$ mV for $\tau_{int} = 2 \mu s$.

Spatial resolution is determined by the wavelength λ of the light and the numerical aperture of the focusing optics. For $\lambda < 250$ nm ($h\nu > 5$ eV), a resolution better than 0.5 μm is achievable with state-of-the-art optics. Since ICLT is not limited to a minimum wavelength, in principle it is not limited in spatial resolution.

The time resolution is determined by the width of the laser pulses and the transit time t_{Tr} of the electrons; t_{Tr} is the time needed by the emitted electrons to reach a well-defined potential which screens them against the potential changes at the emission point. For macroscopic structures, this distance is given by that of the first grid from the surface of the sample. Ideally, this is an accelerating grid which increases the velocity of the electrons. Electrons with 0 eV starting energy and an accelerating voltage V_{acc} at a distance d, have a transit-time $t_{Tr} = d \times (2m/eV_{acc})^{1/2}$. For $V_{acc} = 1000$ V and $d = 100\ \mu$m, one finds $t_{Tr} \approx 10$ ps . For microstructures, the first-defined potential may already occur a few micrometers above the emission point leading to a transit time of only a few ps [2,3].

In conclusion, the ICLT method allows simultaneously a high-voltage resolution, a sub-μm spatial resolution and a pico-second time resolution in a short testing time. In addition, it is non-invasive for laser powers which still allow a short testing-time, and it is applicable to Si as well as GaAs-based technologies.

Comparing with the other known methods for contactless signal measurement, none of them offers these attractive features simultaneously. For e-beam, it is well-known that spatial, time, and voltage resolution as well as testing time are strongly interrelated. Testing times with pico-second e-beam pulses exceed those with ICLT by some orders of magnitude when the same voltage resolution is required [4]. Electro-Optical Sampling is restricted to electro-optically active materials like GaAs. It is also limited to photon energies smaller than the band-gap in the substrate material ($h\nu < 1.4$ eV, i.e., $\lambda > 0.88$ μm for GaAs) [5] and therefore also in spatial resolution.

3. EXPERIMENTAL RESULTS

The experiments were performed with a special laser system and second harmonic generation producing 1-2 ps laser pulses with 5 eV photon energy at a repetition rate of 4 MHz.

Two S-curves measured from air-exposed Al on a Si wafer are shown in Fig. 2(a). The UV power was $I_{ph} = 10^{13}$ photons/s, the yield $Y = 10^{-6}$ electrons/photon, $\Delta E \approx 1$ eV. The experiments showed that the minimum detectable voltage change under these conditions was $\Delta V_s \approx 10$ mV measured within only 12.5 ms. In comparison, the shot-noise equation gives $\Delta V_s = 4$mV for S/N = 1 and the values stated.

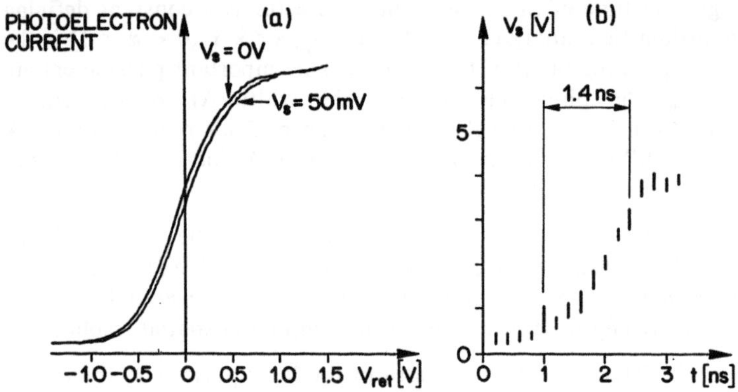

Fig. 2. Experimental results: (a) S-curves measured for $V_s = 0$ V and $V_s = 50$ mV. (b) Rise time measured by the sampling technique with ICLT.

Figure 2(b) shows a rise time measured with the stroboscopic sampling technique. The accelerating grid was installed about 100 μm above the sample at $V_{acc} = 1000$ V, and the input signal rose from 0 V to 5 V in 2 ns. The agreement demonstrates the validity of our theoretical estimate, which gives a time resolution of about 10 pico-seconds.

4. SUMMARY

The fundamentals and key features of the ICLT method with time-resolved photoemission have been presented and the experimental verification of these features has been demonstrated. The method has the potential to evolve into a major technique in advanced contactless testing of VLSI chips and VHSICs.

Acknowledgement

We should like to thank G.W. Rubloff for technical support and fruitful discussions. We are also grateful to B. Reihl and H. Salemink for supporting our work.

1. H. Beha, R.W. Dreyfus, G.W. Rubloff: US patent (pending)
2. H. Fujioka, K. Nakamae, K. Ura: J. Phys. D 18, 1019 (1985)
3. R. Clauberg: to be published
4. E. Menzel, E. Kubalek: Scanning 5, 103 (1983)
5. G.A. Mourou, K.E. Meyer: Appl. Phys. Lett. 45, 492 (1984); B.H. Kolner, D.M. Bloom: Electronics Lett. 20, 818 (1984)

Overview of Optical Switching and Bistability

N. Peyghambarian[1], *H.M. Gibbs*[1], *D. Hulin*[2], *A. Antonetti*[2], *A. Migus*[2], *and A. Mysyrowicz*[2]

[1]Optical Sciences Center, University of Arizona,
Tucson, AZ 85721, USA
[2]Laboratoire d'Optique Appliquée, Ecole Polytechnique – ENSTA,
F-91120 Palaiseau, France

Abstract

In this paper, recent advances in optical bistability and gating of semiconductors will be reviewed. Nonlinear mechanisms responsible for switching will be described. The advantages of nonlinear etalons for massive parallelism and current problems in performing such operations will be mentioned.

Optical bistable devices and logic gates are of current interest for their potential applications in future optical computers and signal processing systems.[1,2] These devices take advantage of the nonlinearity of optical materials and combine it with a feedback mechanism for operation. The most common device consists of a nonlinear medium inside two partially reflecting mirrors, forming a nonlinear etalon. Depending on whether the feedback is provided electronically or optically, the bistability is hybrid[3] or all-optical,[4] respectively. In usual dispersive or absorptive optical bistability, the feedback is external (e.g. external Fabry-Perot mirrors). Induced absorption or increasing absorption bistability occurs when the feedback is intrinsic (or internal).[5]

Optical gating has also attracted much attention. Optical bistable devices may perform gating operations under different operating conditions. More than one input beam is employed for optical gating. For example, to demonstrate a NOR gate, the transmission of a probe pulse is monitored in the presence or absence of two input pulses. The probe transmission, which is the gate output, is adjusted (in the absence of any inputs) to the "high" state by matching its peak wavelength to one of the Fabry-Perot peaks. The input pulse changes the index of refraction of the material, shifts the Fabry-Perot peak away from the probe wavelength and thereby forces the gate output to the "low" state. The second input pulse shifts the Fabry-Perot peak even further and the probe transmission stays in the "low" state. Various gating operations such as AND, OR, NOR, etc. have been demonstrated in bulk GaAs and GaAs-AlGaAs MQWs.[6,7]

Semiconductors are attractive materials for this application due to their large nonlinearity, rapid response time, room-temperature operation and small size. Various nonlinear mechanisms in semiconductors have been employed. Free-exciton saturation in GaAs-AlGaAs multiple-quantum-wells,[8] screening of the continuum states and bandfilling in bulk GaAs,[9] biexciton generation in CuCl,[10,11] bound-exciton saturation in CdS,[12] and bandfilling in InSb,[13] InAs[14] and CdHgTe[15] and thermal nonlinearity in ZnS and ZnSe interference filters[16,17,18] have been successfully used to demonstrate bistability. The operating parameters, and the origins of the nonlinearities in various semiconductors are summarized in Table 1. Such tables should be used with care because the reported parameters are usually not the optimized values and sometimes the switching speed may not correspond to the switching power indicated. Tables like this should be only used as a guide for selection of

Table 1

MATERIAL	TEMP.(K)	λ(μm)	SWITCHING INTENSITY	PULSE ENERGY (GATING)	SPEED	NONLIN.	REFERENCES
GaAs bulk	300	0.88	5 mW 5 kW/cm^2	3 pJ	30 ps	Bandfilling and Screening	9
GaAs MQW	300	0.82–0.88	5 mW 5 kW/cm^2	3 pJ	--	Free Exciton Saturation	8
InSb	77	5	8 mW 40 W/cm^2	--	500 ns	Bandfilling	13
InAs	77	3	3 mW 75 W/cm^2	--	200 ns	Bandfilling	14
CdHgTe	300	10.6	500 $\frac{kW}{cm^2}$	--	400 ns	Bandfilling	15
CdS	2	0.48	0.75 mW 330 W/cm^2	--	2 ns	Bound Exciton	12
CuCl	77	0.4	10 $\frac{MW}{cm^2}$	--	100 ps	Biexciton	10,11
InGaAsP/ InP	300	1.3	--	20 pJ(elect.) 1 fJ(opt.)	1 ns	Hybrid	26
SEED	300	0.84–0.86	0.67 μW to 3.7 mW	1 nJ 18 $fJ/\mu m^2$	1.5 ms–400 ns	Hybrid	27
ZnS,ZnSe	300	0.5–0.7	5 mW	--	1 ms–10μs	Thermal	16,17,18

the proper material for the desired wavelength and temperature and gives an overall range of switching powers and speeds.

One of the recent advances in this field has been the understanding of the origin and measurement of the magnitude of optical nonlinearity of bulk GaAs and multiple quantum wells of GaAs-AlGaAs at various wavelengths in the vicinity of the bandedge.[8,9] It is now well documented that in bulk GaAs, the contribution to the nonlinear index, Δn from the free-exciton saturation is quite small, and the main index change at room temperature comes from bandfilling and plasma screening of Coulomb enhancement of continuum states. In contrast, for intensities required for optical bistability, the nonlinearity is mainly excitonic for GaAs-AlGaAs MQWs with well sizes, W, of 50 Å, 100 Å and 152 Å which are either smaller or slightly larger than the exciton radius a. For 299-Å MQW, ($W \approx 2a$), where MQW behaves more like bulk, but still has distinct excitonic features, the nonlinearity is due to the combination of exciton saturation, bandfilling and screening of continuum states. In addition, the measured nonlinear indices are rather similar for all these samples. The 50-Å MQW has a slightly smaller value of nonlinear index because the nonlinearity which in this sample is due to saturation of the exciton resonance occurs at higher excitation intensities due to larger exciton binding energy.

Optical gating in GaAs has also progressed considerably. One picosecond switching-on time of a GaAs MQW gate is demonstrated.[19] This time indicates how fast the etalon peak shifts in response to an input pulse with an above bandgap frequency. Recovery of the gate requires removal of the carriers produced by the input pulse. This recovery takes more than 10 ns in the usual GaAs and MQW etalons due to the long lifetime of carriers. The recovery time has been shortened to 200 ps[20] (detector limited) using surface recombination in an etalon with no AlGaAs outside layers (the AlGaAs windows are normally used to stop etching of the GaAs substrate). A 30-ps recovery and 70-ps cycling time has also been demonstrated in similar etalons.[21]

If the input pulse frequency is tuned below the exciton, in the transparency region of the material, then the electric field of the laser may be used to switch the etalon. The advantage of this optical-Stark-effect switching is that the carrier lifetime is no longer a limitation as carriers are not excited in this scheme. We have demonstrated a 1-ps recovery using this effect as shown in Fig. 1.[22] It shows NOR-gate switch-on and off times of 1 ps.

For optical signal processing and computing applications, many switches or gates should operate simultaneously. GaAs NOR-gate arrays are the most promising for this parallel processing implementation. Assume 10^4 pixels on 1 cm^2 requiring 10 pJ per bit operation and operating once every nanosecond. This requires 100 W of laser power and results in 100 W/cm^2 heat load. It yields 10^{13} bit operations per second, more than that performed by a CRAY. Much work is required to convert the array's capability for many operations in parallel into a programmed or programmable system able to make useful computations, to recognize patterns, or to learn. However, the first few steps have been already taken as displayed in Figure 2 which shows an array that has exhibited NOR-gate and bistable operations.[23] This was a 100 x 100 array of pixels, each 9 μm x 9 μm in size, formed by reactive-ion etching.

For implementation of nonlinear gates and bistable switches in commercial systems, several problems have to be solved: uniformity of gate characteristics anywhere on the array, cascadability, fan-out capability, and addressability to independent pixels on an array, to mention just a few.

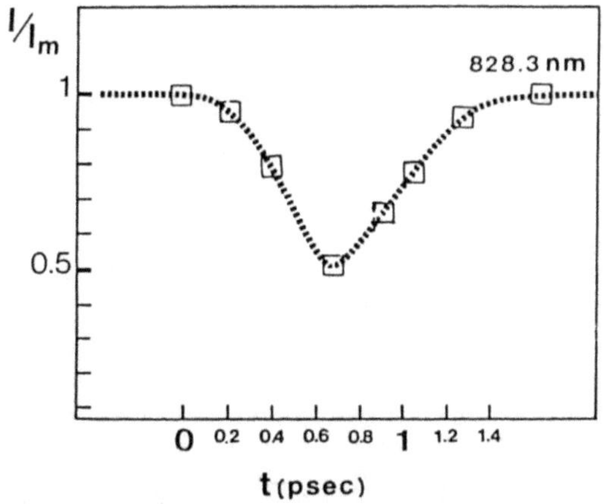

Fig. 1 Picosecond switching of a GaAs optical logic gate. Optical Stark effect has been employed to obtain these switching speeds.

Fig. 2 A two-dimensional array of optical logic gates, etched in GaAs.

However, as our knowledge about the physics of the nonlinearity and operation of these devices grows, we should be able to model accurately the operation of a single device, thereby optimizing the operating parameters. This should lead to realization of low-power operation, cascading, and gain. Materials with large nonlinearities should be sought and implemented to reduce the power requirement. Fly's eye lens arrays, holograms, and spatial light modulators should allow beam divisions and addressing. Already, a 2 x 3 version of Fig. 2 was defined by a fly's-eye lens array and a simple pattern was recognized.[24] New algorithms and architectures that take advantage of nonlinear etalons should be designed.[25]

We gratefully acknowledge support by the Air Force Office of Scientific Research, the Army Research Office, the National Science Foundation, and the Optical Circuitry Cooperative.

References

1. H. M. Gibbs: Optical Bistability: Controlling Light with Light, Academic Press, 1985.
2. H. M. Gibbs, P. Mandel, N. Peyghambarian, and S. D. Smith, eds.: Optical Bistability III, Springer-Verlag, Berlin, 1986.
3. P. W. Smith: Opt. Eng. 19, 456 (1980); P. W. Smith and W. J. Tomlinson: IEEE Trans. Spectrosc. 18, 16 (1981).
4. N. Peyghambarian and H. M. Gibbs: Physics Today 39, No. 1, S-57 (1986); N. Peyghambarian and H. M. Gibbs: Optics News 11, No. 12, 7 (1985).
5. Y. Toyozawa: Solid State Commun. 28, 533 (1978); E. M. Epshtein: Sov. Phys. Tech. Phys. 23, 983 (1978); K. Bohnert, H. Kalt, and C. Klingshirn: Appl. Phys. Lett. 43, 1088 (1983); H. Rossmann, F. Henneberger, and J. Voigt: Phys. Stat. Sol. b115, K63 (1983); H. X. Nguyen and R. Zimmermann: Phys. Stat. Sol. b124; 191 (1984); D. A. B. Miller: J. Opt. Soc. Am. B1, 857 (1984); H. E. Schmidt, H. Haug, and S. W. Koch: Appl. Phys. Lett. 44, 787 (1984); J. A. Goldstone and E. Garmire: Phys. Rev. Lett. 53, (1984); J. W. Haus, C. C. Sung, C. M. Bowden, and J. M. Cook: Optical Materials Conference, Gaithersberg, Maryl, May 7-9, 1985, NBS Publication no. 697, ed. A. Feldman; H. M. Gibbs, G. R. Olbright, N. Peyghambarian, H. E. Schmidt, S. W. Koch, and H. Haug: Phys. Rev. A 32, 692 (1985).
6. N. Peyghambarian and H. M. Gibbs: Opt. Eng. 24, 68 (1985).
7. J. L. Jewell, Y. H. Lee, M. Warren, H. M. Gibbs, N. Peyghambarian, A. C. Gossard, and W. Wiegmann: Appl. Phys. Lett. 46, 918 (1985).
8. J. F. Morhange, S. H. Park, N. Peyghambarian, A. Jeffrey, H. M. Gibbs, Y. H. Lee, A. Chavez-Pirson, S. W. Koch, A. C. Gossard, J. H. English, M. Maselink, and H. Morkoc: The Annual Meeting of the Optical Society of America, Seattle, Washington, Oct. 19-24, 1986.
9. S. W. Koch, Y. H. Lee, A. Chavez-Pirson, H. M. Gibbs, J. Morhange, S. H.Park, T. Carty, A. Jeffrey, N. Peyghambarian, B. Batdrof, L. Banyai, A. C. Gossard, and W. Wiegmann: postdeadline paper PD14, IQEC, 86, San Francisco, CA, June 9-13, 1986.
10. N. Peyghambarian, H. M. Gibbs, M. C. Rushford, and D. Weinberger: Phys. Rev. Lett. **51**, 1692 (1983).
11. R. Levy, J. Y. Bigot, B. Honerlage, F. Tomasini, and J. B. Grun: Solid State Commun. **48**, 705 (1983).
12. M. Dagenais and H. G. Winful: Appl. Phys. Lett. **44**, 574 (1984).
13. D. A. B. Miller, S. D. Smith, and A. Johnston: Appl Phys. Lett. **35**, 658 (1979); A. K. Kar, J. G. H. Mathew, S. D. Smith, B. Davis, and W. Prettl: Appl Phys. Lett. **42**, 334 (1983); C. T. Seaton, S. D. Smith, F. A. P. Tooley, M. E. Prise, and M. R. Taghizadeh: Appl. Phys. Lett. **42**, 131 (1983); D. Sarid, R. S. Jameson, and R. K. Hichernell: Opt. Lett. **9**, 159 (1984).
14. C. D. Poole and E. Garmire: Appl. Phys. Lett. **44**, 363 (1984).
15. A. Miller, G. Parry, and R. Daley: IEEE J. Quantum Electron. **QE-20**, 710 (1984).
16. F. V. Karpushko and G. V. Sinitsyn: J. Appl. Spectrosc. (USSR) **29**, 1323 (1978); Appl. Phys. B **28**, 137 (1982).
17. G. Olbright, N. Peyghambarian, H. M. Gibbs, A. Macleod, and F. Van Milligen: Appl. Phys. Lett. **45**, 1031 (1984); S. D. Smith, J. G. H. Mathew, M. R. Taghizadeh, A. C. Walker, B. S. Wherrett, and A. Hendry: Opt. Commun. **51**, 357 (1984); R. Jin, L. Wang, R. W. Sprague, H. M. Gibbs, G. Gigioli, H. Kulcke, H. A. Macleod, N. Peyghambarian, G. R. Olbright, and M. Warren: in Optical Bistability III, ed. H. M. Gibbs, P. Mandel, N. Peyghambarian, and S. D. Smith (Springer-Verlag, Berlin, 1986).
18. S. D. Smith, J. G. H. Mathew, M. R. Taghizadeh, A. C. Walker, B. S. Wherrett, and A. Hendry: Opt. Commun. **51**, 357 (1984); M. R. Taghizadeh, F. A. P. Tooley, and J. G. H. Mathew: in Optical Bistability III, ed. by H. M. Gibbs, P.
19. A. Migus, A. Antonetti, D. Hulin, A. Mysyrowicz, H. M. Gibbs, N. Peyghambarian, and J. L. Jewell: Appl. Phys. Lett. **46**, 70 (1985).

20. Y. H. Lee, M. Warren, G. Olbright, H. M. Gibbs, N. Peyghambarian, T. Venkatesan, J. S. Smith, and A. Yariv: Appl. Phys. Lett. **48**, 754 (1986).
21. J. L. Jewell, Y. H. Lee, J. F. Duffy, A. C. Gossard, W. Wiegmann, and J. H. English: in Optical Bistability III, eds. H. M. Gibbs, P. Mandel, N. Pey-
22. A. Migus, D. Hulin, A. Antonetti, A. Mysyrowicz, H.M. Gibbs, N. Peyghambarian, M. Masselink, H. Morkoç: CLEO 1986 postdeadline paper ThU9, San Francisco, CA, June 9-13.
23. T. Venkatesan, B. Wilkins, Y.H. Lee, M. Warren, G. Olbright, H. M. Gibbs, N. Peyghambarian, J. S. Smith, and A. Yariv: Appl. Phys. Lett. 48, 145 (1986) ; Y. H. Lee, M. Warren, G. R. Olbright, H. M. Gibbs, N. Peyghambarian, T. Venkatesan, J. S. Smith, and A. Yariv: Appl. Phys. Lett. 48, 754 (1986).
24. H. M. Gibbs and N. Peyghambarian : Proceedings of The International Optical Computing Conference, Jerusalem, Israel, July 6-11, 1986.
25. A. Huang: IEEE, 1983, 10th International Optical Computing Conference, pp. 13-17.
26. W. F. Sharfin and M. Dagenais: Appl. Phys. Lett. 46, 819 (1985).
27. D. A. B. Miller, D. S. Chemla, T. C. Damen, A. C. Gossard, W. Wiegmann, T. H. Wood, C. A. Burrus: Appl. Phys. Lett. 44, 16 (1984).

Monte Carlo Investigation of High-Speed GaAs Schottky Barrier Photodiode*

M.A. Osman, U. Ravaioli, and D.K. Ferry

Center for Solid State Electronics Research, Arizona State University, Tempe, AZ 85287, USA

The progress in sub-picosecond laser pulse techniques has opened a new horizon in high-speed optical communication and logic circuits. Consequently the development of high-speed photodetectors such as Schottky barrier photodiodes capable of providing the link from these ultrashort optical signals to electrical signals has intensified in recent years. The speed of response of these devices depends on how fast the photoexcited carriers are swept outside the high field region and the diffusion of the minority carriers generated outside the depletion region. The transit time is usually estimated by representing the transport of the electrons and holes by two sheets of charges drifting at the saturated velocity in a constant electric field. Although this approach provides some qualitative understanding of the shape of the impulse response, considerable discrepancy exists between the calculated transit time and the experimentally measured values. Besides, in picosecond applications the electric field distribution inside the device is determined by both the built-in field and the self-consistent corrections due to the distribution of the photoexcited carriers, whose density can exceed the background carrier density. Because the highly excited electron-hole plasmas are generated on very short time scales the field is not modified instantaneously due to the fact that the space charges of electrons and holes are initially neutralized by their equal densities. The space charge develops as the electrons and holes drift apart in the field resulting in regions of unequal density. During this period, the carrier-carrier interaction might influence the response of the device.

We have investigated the ultrafast dynamics of the photoexcited carriers generated by sub-picosecond laser pulses inside a GaAs n+/n/metal submicron Schottky diode, using an Ensemble Monte Carlo approach. In this model the interaction between electrons and holes is included in addition to carrier-phonon interactions and ionized impurity scattering, which were the only scattering mechanisms considered in our previous models [1,2,3]. The recombination process which occurs at much slower rates, is neglected in this model. The excess energy of the excited electrons and holes was calculated using a non-parabolic conduction band and two parabolic valence bands [1]. The generated carriers were distributed nonuniformly within the device according to the following relationship:

$$x_r = -\{\ln[1-r(1-e^{-\alpha W})]\}/\alpha , \qquad (1)$$

where α is the absorption coefficient, r is a random number uniformly distributed between 0 and 1, and W is the length of the device. The time evolution of the electric field is determined by solving Poisson's equation using an accurate orthogonal collocation method [2]. Thus, this

*Supported by the U.S. Office of Naval Research.

model takes into account the self-consistent corrections to the electric fields due to the nonuniformly generated electron-hole pairs.

The influence of the electron-hole interaction on the separation of the electrons and holes generated in the depletion region was investigated for an excitation level of 5×10^{16} cm^{-3} in a background doping of 5×10^{16} cm^{-3}. The spatial distribution of the electrons generated by a 1.55 eV laser pulse within the high-field region is shown in Fig. 1. The curve denoted by 1 in the figure is the distribution of the electrons immediately after the excitation, while curves 2 and 3 are the distributions at 50 and 500 fs after the excitation. Identical curves were obtained when electron-hole interaction was included, reflecting the fact that the electron-hole interaction has not significantly affected these distributions. This is because the density of the photoexcited carriers is low, so that the electron-hole scattering rates are not high enough to influence the energies of the electrons and holes. Besides, the rate at which electrons gain energy from the fields in the depletion region exceeds the rate of energy loss through electron-hole interaction.

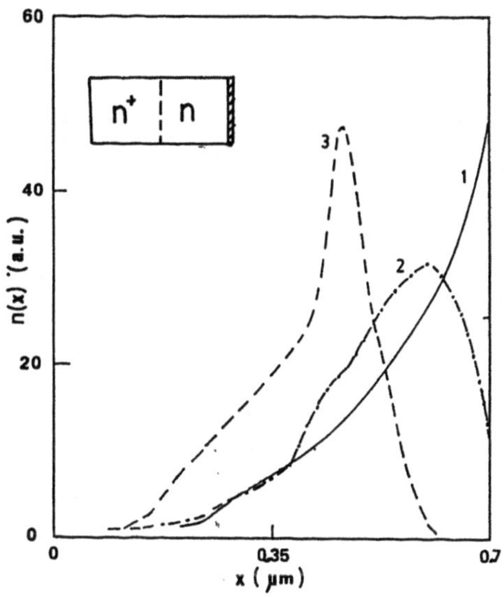

Fig. 1 - Spatial distribution of photoexcited electrons inside a Schottky photodiode at 1) t=0, 2) t=50 fs, 3) t=500 fs after excitation through Schottky contact.

Since almost all the carriers within the high-field region are photoexcited carriers, we have investigated a model where all the carriers are generated by a laser pulse in the presence of a constant electric field. The response of the photogenerated carriers to high applied fields of the same order of magnitudes as that present in the depletion region, was investigated with and without including electron-hole interaction effects. The model was then investigated at fields ranging between 1 kV/cm and 5 kv/cm for excitation levels of 10^{17} and 10^{18} cm^{-3}. Figure 2 shows that the electron-hole interaction has negligible effect on the instantaneous velocity of the electrons at low carrier concentrations, which correspond to the densities of photoexcited carriers in photodetectors. This explains why no significant difference was obtained in the evolution spatial distribution of the carriers in Fig. 1. At an

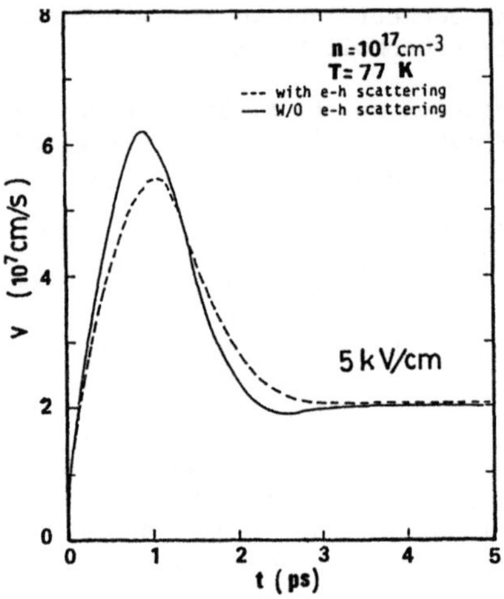

Fig. 2 - Plot of velocity vs time for electrons photoexcited by a 1.55 eV Laser pulse in GaAs.

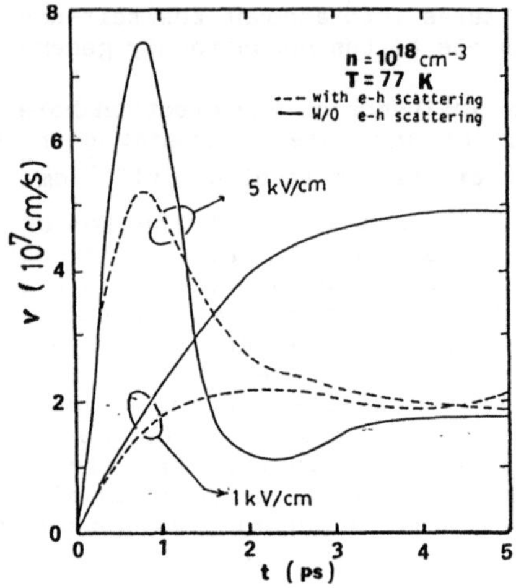

Fig. 3 - Plot of velocity vs time for electrons photoexcited by a 1.55 eV Laser pulse in GaAs.

excitation level of 10^{18} cm^{-3} the velocity is reduced more significantly at low applied fields than at higher fields as shown in Fig. 3 for 1 kV/cm and 5 kV/cm fields. The fraction of the electrons in the upper valleys is plotted in Fig. 4 for applied fields of of 1 kV/cm and 2 kV/cm in the presence and absence of electron-hole interaction. From this figure it is obvious that the presence of electron-hole interaction reduces the fraction of electrons that transfer to the upper valleys more significantly at lower electric fields. This is because the strong electron-hole interaction transfers a fraction of the energy gained by the electrons from the electric field to the holes as can be seen in Fig. 5.

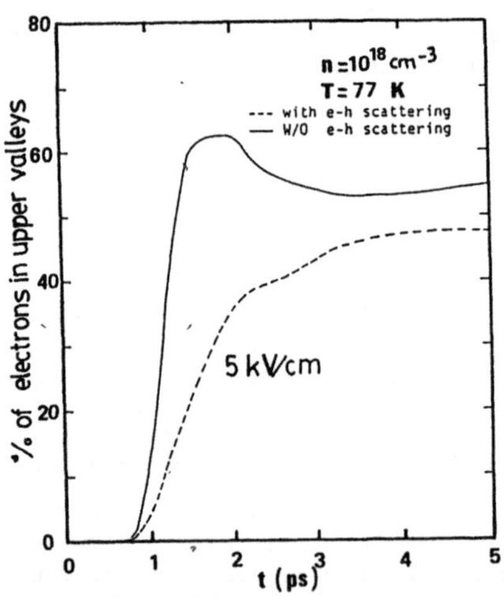

Fig. 4 - Plot of fraction of electrons in the upper valleys vs time for the situation plotted in Fig. 3.

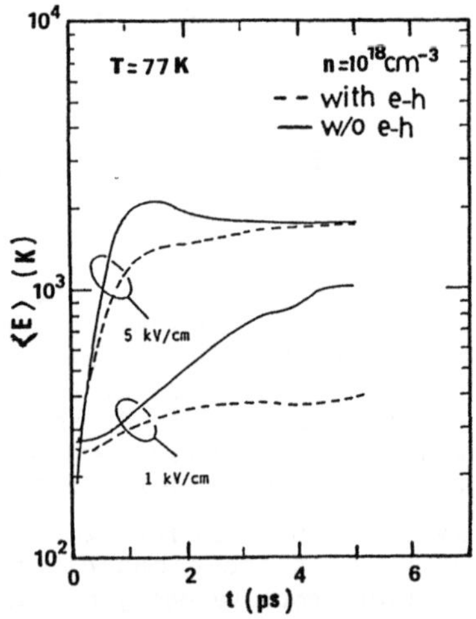

Fig. 5 - Plot of average energy of electrons in the central valleys vs time for the situation plotted in Fig. 3.

This effect is more significant at low fields where the rate at which electrons gain energy from the field is small, so that the energy transferred to the holes through electron-hole interaction is a significant fraction of the electron's energy. On the other hand, at high fields the rate at which electrons gain energy from the field is much larger than the rate at which the energy is transferred to the holes.

References

1. M.A. Osman, U. Ravaioli, and D.K. Ferry, in "Picosecond Electronics and Optoelectronics", Ed. by G.A. Mourou, D.M. Bloom, and C.-H. Lee (Springer, Berlin, 1985), p. 97.
2. U. Ravaioli, P. Lugli, M.A. Osman and D.K. Ferry, IEEE Trans. Electron Devices, ED-32, 2097 (1985).
3. P. Lugli, U. Ravaioli ,and D.K. Ferry: "Monte Carlo simulation of contacts in submicron devices", AIP Conference Proceedings no. 122, The Physics of VLSI, p. 162, Palto Alto (1984).

A High-Speed Au/$In_{0.53}Ga_{0.47}As$/InP Schottky Barrier Photodiode for 1.3–1.65 μm Photodetection

J.H. Kim and S.S. Li

Department of Electrical Engineering, University of Florida, Gainesville, FL 32611, USA

1. Introduction

With the advent of femtosecond laser pulses and light modulation in the tens of GHz frequency range, the development of high-speed photodetectors is necessary for microwave optical fiber communications at 1.30 - 1.65 [µm]. High-speed photodetectors have been fabricated mainly on GaAs using Schottky barrier structure for 0.80 - 0.90 [µm] and on $In_{0.53}Ga_{0.47}As$ for 1.30-1.65 [µm] using p-i-n structure. In this paper, we propose a novel high-speed long wavelength Au/p^+-n-$In_{0.53}Ga_{0.47}As$/n^+-InP Schottky barrier structure to enhance the effective barrier height. The proposed Au/p^+-n-$In_{0.53}Ga_{0.47}As$/n^+-InP Schottky barrier photodiode structure requires the use of a very thin surface layer of p^+-$In_{0.53}Ga_{0.47}As$ on the n-$In_{0.53}Ga_{0.47}As$ epitaxial layer grown on n^+-InP substrate in order to enhance the effective barrier height. The significance of this new structure lies on its ability to increase the effective barrier height , and hence to overcome the problem of large dark current encountered in such a detector [1-2]. This approach for barrier height enhancement has been demonstrated previously in GaAs [3] and InGaAs Schottky diodes [4], and is used in our Au/$In_{0.53}Ga_{0.47}As$/InP system for photodetector applications.

2. Schottky Barrier Contacts on $In_{0.53}Ga_{0.47}As$

Schottky barrier contacts on n-$In_{0.53}Ga_{0.47}As$ usually yield very low barrier height (Φ_{Bn} = 0.2-0.3 eV), which makes Schottky contacts too leaky to be useful for photodetector applications. Therefore, the effective barrier height needs be enhanced in order to reduce the dark current in the metal/ n-$In_{0.53}Ga_{0.47}As$ Schottky barrier diode. The barrier height enhancement can be achieved by growing a very thin p^+-$In_{0.53}Ga_{0.47}As$ epilayer on top of the n-$In_{0.53}Ga_{0.47}As$ epilayer with a selected dopant density and layer thickness. The thickness and dopant density of the p^+-$In_{0.53}Ga_{0.47}As$ surface layer, which must be completely depleted in thermal equilibrium, can be related to the effective barrier height, Φ'_{Bn} by:

$$\Phi'_{Bn} = (q/2\epsilon_o\epsilon_r N_A)(N_A W_p - N_D W_n)^2 . \quad (1)$$

The enhanced barrier potential will reach a maximum value at $x = x_m$ inside the p^+-$In_{0.53}Ga_{0.47}As$ surface layer provided that $N_A W_p >> N_D W_n$. Thus, the effective barrier height Φ'_{Bn} obtained at $x = x_m$ is given by:

$$\Phi'_{Bn} = \Phi_{Bn} + E_m x_m - qN_A x_m^2/2\epsilon_o\epsilon_r \quad (2)$$

For $N_A W_p >> N_D W_n$, Eq.(2) may be simplified to

$$\Phi'_{Bn} = \Phi_{Bn} + qN_A W_p^2/2\epsilon_o\epsilon_r . \quad (3)$$

It can be shown that Eq.(3) holds only for $\Phi_{Bn} >> V_D N_D/N_A$. Note that N_A and N_D denote the dopant density of the p^+- and n-$In_{0.53}Ga_{0.47}As$ layer, respectively. W_p is the thickness of the p^+-$In_{0.53}Ga_{0.47}As$ layer, and V_D is the built-in potential of the p^+-n junction. The thickness and dopant density of the $In_{0.53}Ga_{0.47}As$ layers are determined by the condition that $\Phi_{Bn} >> V_D N_D/N_A$. The barrier height for such a photodetector can be tailored to its optimum value via properly selected thickness and dopant density of the surface layer. Therefore, the effective barrier height equal to the bandgap energy of $In_{0.53}Ga_{0.47}As$ can in principle be achieved by the proposed structure.

3. Experimental Results and Discussion

High-speed Au/p^+-n-$In_{0.53}Ga_{0.47}As$/n^+-InP and Au/p-$In_{0.53}Ga_{0.47}As$/p^+-InP mesa Schottky barrier photodetectors have been fabricated using a photolithographic lift-off process. The structures of two different types of Au/$In_{0.53}Ga_{0.47}As$/InP Schottky barrier photodiodes are shown in Fig.1. The p^+-n-$In_{0.53}Ga_{0.47}As$ epitaxial layer was grown on n^+-InP substrate by MBE technique. The thickness of the p^+-$In_{0.53}Ga_{0.47}As$ epilayer is chosen to be 0.05 [μm] with a dopant density of 3.5×10^{17} cm^{-3}, and the thickness of an n-$In_{0.53}Ga_{0.47}As$ epilayer is 2 [μm] with a dopant density of 3.0×10^{15} [cm^{-3}] for the proposed structure. A 100 [Å] gold film was deposited on the p-$In_{0.53}Ga_{0.47}As$ to obtain a semi-transparent Schottky contact and Cr/Au (60/1,000 Å) was deposited for the bonding pad. For ohmic contact to n^+-InP, Au-Ge (88-12 %) alloy (1,500 Å) was deposited and alloyed at 400 [°C] for 30 [sec] in H_2-N_2 (5-95 %) forming gas ambient. For ohmic contact to p^+-InP, Mn/Au (100/900 Å) was deposited and alloyed at 460 [°C] for 30 [sec] in forming gas ambient. For mesa etch, $H_2SO_4:H_2O_2:H_2 = 3:1:1$ was used for the $In_{0.53}Ga_{0.47}As$ epilayer with an etch rate of 0.1 [μm/sec] at T = 300 [K]. To facilitate device characterization, the 50 [ohm] microstrip transmission line was fabricated on the Cr-Au metallized alumina (Al_2O_3) substrate [5].

The current-voltage (I - V) characteristics of the Au/p^+-n-$In_{0.53}Ga_{0.47}As$/n^+-InP Schottky diode grown by MBE technique shows a large reverse leakage current. However, the Au/p-$In_{0.53}Ga_{0.47}As$/p^+-InP Schottky diode shows a much smaller reverse leakage current, as is shown in Fig.2. The junction capacitance was

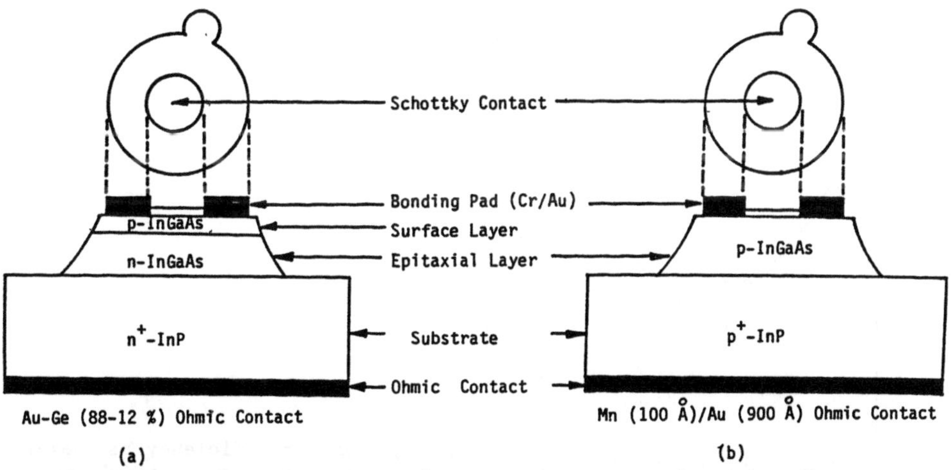

Fig.1 Structures of (a) Au/p^+-n-$In_{0.53}Ga_{0.47}As$/n^+-InP Schottky photodiode and (b) Au/p-$In_{0.53}Ga_{0.47}As$/p^+-InP Schottky photodiode.

found equal to 0.15 [pF] at V_R = 5 [V] for the Au/p^+-n-$In_{0.53}Ga_{0.47}As$/n^+-InP Schottky diode and 0.1 [pF] for the Au/p-$In_{0.53}Ga_{0.47}As$/p^+-InP Schottky barrier diode (with an area of 3.0×10^{-6} cm^2). From the a.c. admittance measurements, the series resistance was found to be 11.8 [ohm] for p^+-n-$In_{0.53}Ga_{0.47}As$ and 5.3 [ohm] for p-$In_{0.53}Ga_{0.47}As$ Schottky barrier photodiode. Fig.3 shows the quantum efficiency vs. wavelength for the Au/p-InGaAs/p^+-InP Schottky photodiode. Without antireflection coating, the maximum quantum efficiency was 68.5 [%] at 1.1 [μm] and 40.5 [%] between 1.3 to 1.65 [μm] for this detector. The impulse response of a Au/p-$In_{0.53}Ga_{0.47}As$/$p^\pm$-InP Schottky barrier photodiode to a laser pulse of 95 [ps] FWHM at 820 [nm] and a 10 [KHz] repetition rate measured by using a sampling scope (Tektronix 7S-12) with S-4 sampling head of a risetime of 25 [ps] (55 ps FWHM). The detector has a risetime of 110 [ps] and 490 [ps] FWHM. The measured RC time constant is around 0.56 [ps]. Therefore, further improvement in the packaging of the detector is needed in order to obtain a true response speed. Table.1 lists the design parameters for the Au/$p^\pm$-n-InGaAs/$n^\pm$-InP and Au/p-InGaAs/$p^\pm$-InP Schottky barrier photodiodes, and table. 2 summarizes the results of these photodiodes.

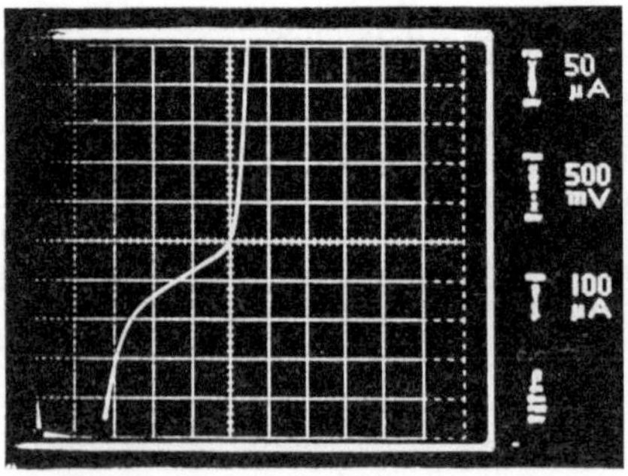

(a) Au/p^+-n-InGaAs/n^+-InP

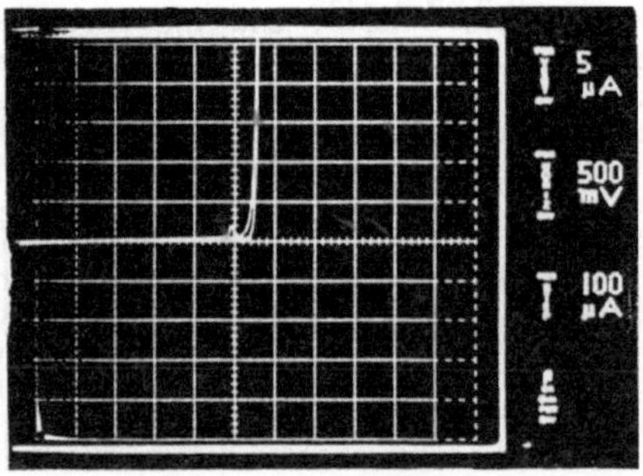

(b) Au/p-InGaAs/p^+-InP

Fig.2 The current-voltage characteristics for (a) Au/p^+-n-$In_{0.53}Ga_{0.47}As$/n^+-In and (b) Au/p-$In_{0.53}Ga_{0.47}As$/p^+-InP Schottky barrier photodetector.

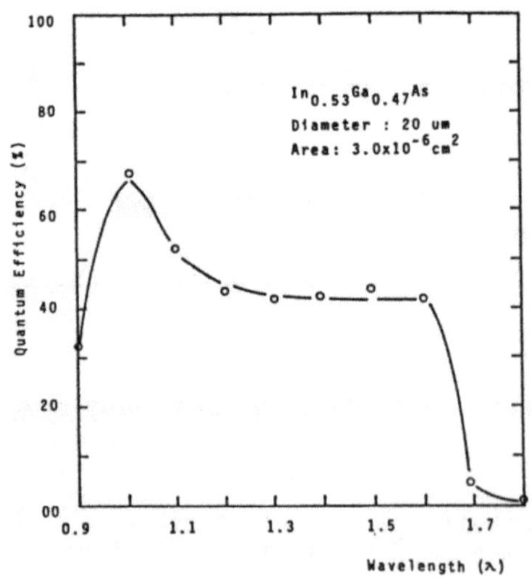

Fig.3 Quantum efficiency vs. wavelengt for the Au/p-$In_{0.53}Ga_{0.47}As$/p^+-InP Schottky barrier photodiode.

Table 1. Design Parameters for High-Speed Au/$In_{0.53}Ga_{0.47}As$/InP Schottky Barrier Photodetectors

Device Structure	p^+-n-InGaAs/n^+-InP	p-InGaAs/p^+-InP
Schottky Contact	Au (100 Å)	Au (100 Å)
Bonding Pad	Cr/Au (60/1,000 Å)	Cr/Au (60/1,000 Å)
p^+-InGaAs Surface Layer	0.05 - 0.15 µm 5.5×10^{16} - 3.5×10^{17} cm^{-3}	
Epitaxial Layer	n-$In_{0.53}Ga_{0.47}As$ 1.5 - 2.0 µm 3.0×10^{15}	p-$In_{0.53}Ga_{0.47}As$ 1.0 - 1.5 µm 1.0 - 5.0×10^{15}
Growth Technique	MBE	LPE
Substrate	Sn doped n^+-InP 3.0×10^{18} cm^{-3}	Zn doped p^+-InP 2.0×10^{18} cm^{-3}
Ohmic Contact	Au-Ge (88-12 %), 1,500 Å	Mn/Au (100/900 Å)
Alloy	400 °C for 30 sec	460 °C for 30 sec

Table 2. Test Results for High-Speed Au/$In_{0.53}Ga_{0.47}As$/InP Schottky Barrier Photodetectors

Device Structure	p^+-n-InGaAs/n^+-InP	p-InGaAs/p^+-InP
Diameter (µm)	20 - 50	20 - 50
Photodiode Area (cm^2)	0.3-2.0×10^{-5}	0.3-2.0×10^{-5}
Epilayer Doping (cm^{-3}) Thickness (µm)	3.0×10^{15} 1.5 - 2.0 µm	1.0 - 5.0×10^{15} 1.0 - 1.5 µm
Dark Current Density (A/cm^2) at V_R = 5 (V)		1.45×10^{-2}
Capacitance (pF) V_R = 5 (V)	<0.2	0.1
Series Resistance (ohm)	11.8	5.3
Quantum Efficiency (%) at 1.3 µm		40.8
Responsivity (A/W) at 1.3 µm		0.43
Bandwidth (GHz)	12.9	28.8
RC Time Constant (ps)	1.77	0.56
Risetime (ps)		100

4. Conclusions

High-speed Au/p^+-n-$In_{0.53}Ga_{0.47}As/n^+$-InP and Au/p-$In_{0.53}Ga_{0.47}As/p^+$-InP Schottky barrier photodetectors have been fabricated and characterized in this study. The results show that the intrinsic response speed for our photodetectors with an active area (3×10^{-6} cm^2) of 20 [µm] in diameter was estimated to be 13 [GHz] for Au/p^+-n-$In_{0.53}Ga_{0.47}As/n^+$-InP and up to 29 [GHz] for the Au/p-$In_{0.53}Ga_{0.47}As$ photodetector based on the RC time calculated from the measured values of the total capacitance (0.1-0.2 pF) and the series resistance (5.3-11.8 ohm). The true response speed can be further improved by improving our packging technique.

Acknowledgement

The authors wish to acknowledge Mr. R. Wagner of Los Alamos National Laboratory for the response speed measurements, Prof. P. Bhattacharya of the University of Michigan for the MBE growth of p±n-InGaAs epilayer, Prof. L. Figueroa of the University of Florida for providing a p-InGaAs epilayer grown by Dr. E. Rezek of TRW/EORC, and Drs. M. Timmons and P.K. Chiang of Research Triangle Institute for their assistance in device fabrication. This work was supported by U.S. Air Force Rome Air Development Center subcontracted through University of South Florida.

References

1. J. Selders, N. Emeis, and H. Beneking: IEEE Elec. Dev., 32 (3), 605 (1985).
2. N.Emeis, H.Schumacher, and H.Beneking: Electron. Lett., 21, 180 (1985).
3. S. S. Li: Solid State Electronics, 21, 435 (1978).
4. C. Y. Chen, A. Y. Cho et al.: Appl. Phys. Letts., 40 (5), 401 (1982).
5. J. H.Kim, S. S. Li, and J. J. Pan: SPIE Proc., 716, (1986).

High Sensitivity Picosecond Optical Pulse Detection by Semiconductor Laser Amplifiers Via Cross-Correlation

K. Ketterer, E.H. Böttcher, and D. Bimberg

Institut für Festkörperphysik I, TU-Berlin, Hardenbergstr. 36, D-1000 Berlin 12

A novel method to measure the temporal behaviour of ultrashort light pulses is presented. A semiconductor laser amplifier driven by short injection current pulses presents the light sensitive device in a cross-correlation experiment. The method is based on the picosecond temporal response of the gain of the semiconductor laser amplifier driven by short injection current pulses. Computer simulations show that the time resolution is better than 10 ps under appropriate injection conditions. This prediction is confirmed by experiments where a GaAlAs-twin-channel-substrate- mesa-guide laser amplifier is used to study ps-pulses of a current modulated multiple quantum well laser. Our results prove that the cross- correlation method is at least one order of magnitude more sensitive than the conventional method of detection by the combination of a fast photodiode and a sampling scope.

1. INTRODUCTION

The availability of simple and economic ps-light sources /1/ like directly modulated or Q-switched semiconductor lasers has recently rapidly increased and has created a demand for equally simple and sensitive methods to detect these pulses, unlike e.g. synchroscan streak cameras which are highly complex and expensive. A novel, accurate and very simple method to characterize the temporal behaviour of such ultrashort light pulses is presented here. This method utilizes the dynamics of the optical gain of a current-modulated laser diode /1/ which acts as the light-sensitive device in a cross-correlation experiment. The time resolution of the detection system is found to be better than 10 ps. The resolution is defined by the short time interval in which the laser diode operates as an efficient laser amplifier for the incident light. Light pulses of a pulse energy of 0.5 fJ can still be detected.

2. EXPERIMENTS

A sketch of the experimental setup is shown in Fig.1. The light source under test (ld1) is a GaAs/GaAlAs MQW laser driven by an avalanche generator. The light emitted from the test laser is focused on the active area of a second laser diode (ld2, in the present case a TCSM laser /3/), which can be considered as a sampling gate. This gate is opened by a short injection current pulse from a second avalanche generator (ag2). Both avalanche generators are triggered by a common

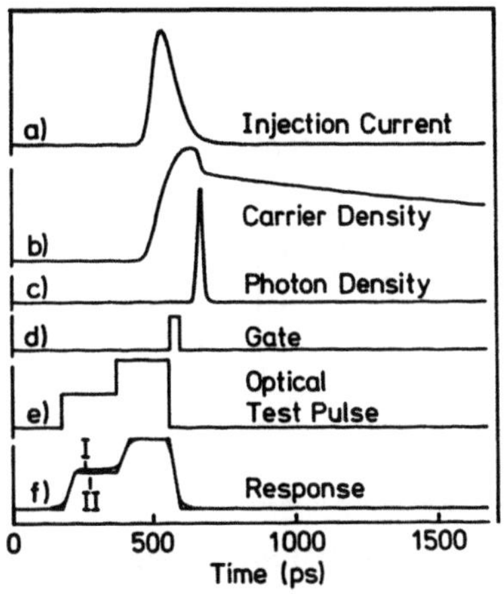

Fig.1. Schematic of the experimental setup. ld1 is the test laser diode and ld2 the light sensitive laser amplifier

pulse generator running at a repetition rate of 100 KHz. The gate ld2 reconstructs the temporal behaviour of the emission of ld1 by shifting the delay time between the injection current pulse of ld1 and ld2 using a variable electronic delay unit driven by a sawtooth voltage. A slow, integrating Si-photodiode monitors the light emitted from the rear mirror of the sampling laser. The photocurrent is recorded by a picoamperemeter and displayed versus the delay time on a x-y real time oscilloscope. The z-signal of the oscilloscope blanks out the return movement of the sawtooth voltage.

3. COMPUTER SIMULATIONS

In order to analyze the temporal response of the detection system theoretically we have numerically solved the coupled single mode rate equations for the charge carrier density and the photon density /1,4/ including gain saturation /5/. The simulations predict that the detection system represents an optical sampling system if the following excitation conditions are fulfilled:

1. The incident power of the laser under test is at least three orders of magnitude smaller than the light power of the first relaxation oscillation emitted by the sampling laser. This condition ensures that the temporal emission characteristics of the sampling laser like the turn-on delay or the emitted pulse shape are only little affected by the light pulses to be measured. This condition presents no restriction of the method because the power of the light source under test can be easily attenuated by means of neutral density filters.
2. The sampling laser is electrically pumped in such a way that only the first relaxation is emitted .

Under these circumstances the emission of ld2 is particularly sensitive to an external optical pumping in a narrow time interval which forms the sampling gate. This time interval is approximately determined by the dynamics of the effective signal gain G_{eff} of the laser amplifier/6,7/. We have calculated G_{eff} using the injected carrier density (Fig. 2 b) obtained by the solution of the rate equations. The dependence of the FWHM of G_{eff} -the gate- on the normalized current pulse

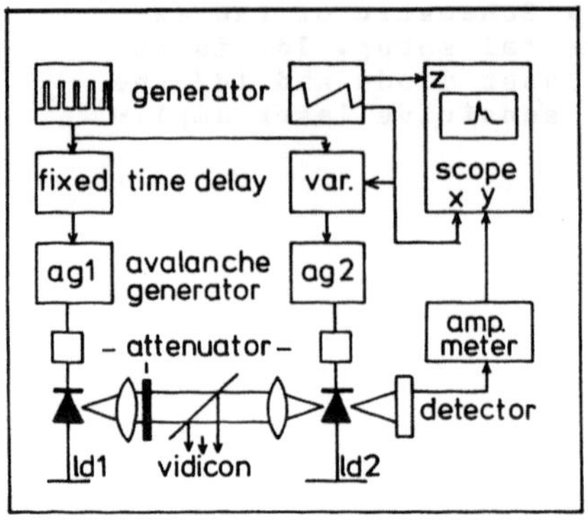

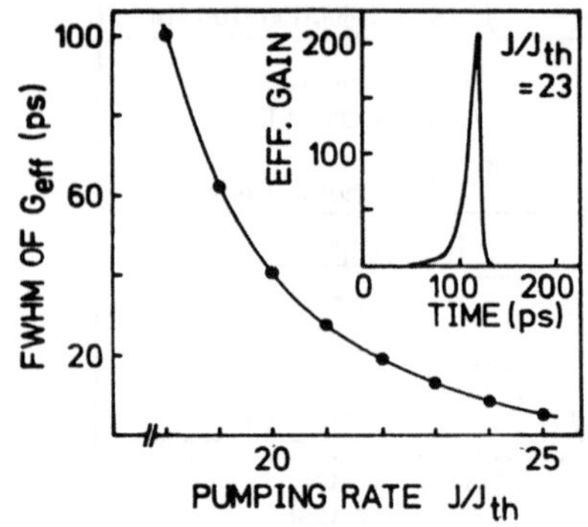

Fig.2. Calculated response (f) to a test pulse (e). Curve I: solution of the rate equations Curve II: convolution of the test pulse with the gate (d)

Fig.3. FWHM of the effective signal gain G_{eff} as a function of the pumping rate J/J_{th} The time evolution of G_{eff} at J/J_{th}=23 is shown in the inset

amplitude J/J_{th} is shown in Fig.3. In the insert of Fig.3 the time evolution of G_{eff} is shown for J/J_{th} =23. The calculated response of the optical sampling system to an optical test pulse (Fig.2e) is shown as curve I in Fig.2f. This particular shape of an optical test pulse has been chosen because it allows straightforward evaluation of the temporal response and linearity of the detection system. Each point of the 512 points of this curve represents the time integral of the photon density calculated by solving the rate equations. The curve II in Fig. 2f is the result of the convolution of the optical test pulse (Fig. 2e) with the gate pulse shown in Fig. 2d. The close correspondence of both curves justifies the treatment of ld2 as a sampling gate. In practice there has to be made a trade-off between sensitivity and time resolution. If the sampling laser is running at or just above the threshold the sampling gate is wide (as shown in Fig.3), the emitted light power is small and therefore the fractional increase of the correlated light signal is large. If ld1 is driven far above threshold the sampling gate becomes small but the light power emitted by the sampling laser increases and the correlated signal is superimposed on the strong light of ld2.

4. RESULTS

A typical experimental response obtained with the setup of Fig. 1 is shown in Fig. 4 and compared with a S1-synchroscan streak camera record. A nearly identical FWHM of 54 ps is obtained with both methods. The time resolution of the correlation method is therefore comparable to that of the synchroscan streak camera which is better than 10 ps. Curve I of Fig.5 shows the response of the detection system to MQW laser pulses driven at a higher pumping rate. For comparison, the

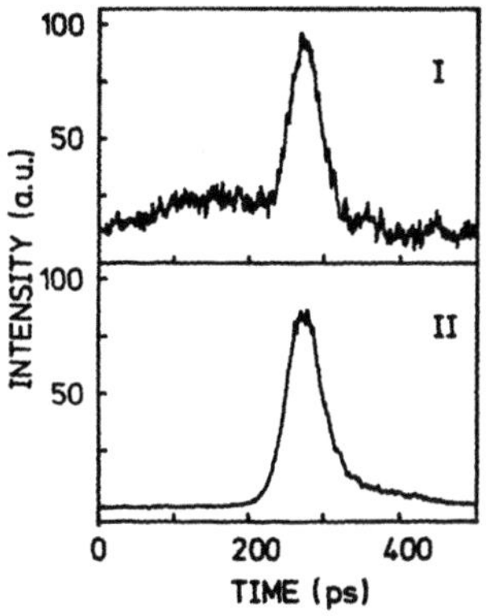

Fig.4. Correlation signal (curve I) measured with the setup of Fig.1 and synchroscan streak camera record (curve II) of the test laser pulse

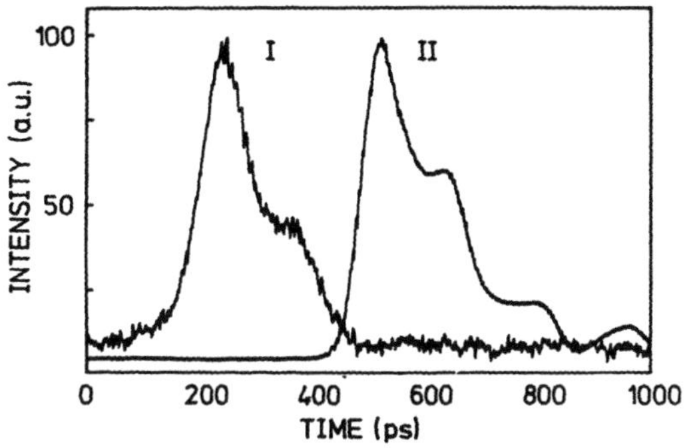

Fig.5. Correlation signal (curve I) and fast photodetector and sampling scope response (curve II) of the test laser emitting two relaxation oscillations

response of a fast GaAs-Schottkydiode and a fast sampling scope to these pulses is also shown in Fig.5 (curve II). The signals correspond closely. The first and second peak in the curves of Fig. 5 correspond to the first and second relaxation oscillation of the optical test pulse, respectively. If the photodetector was used, however, the test signal amplitude had to be increased by at least one order of magnitude above the signal level impinging on the laser amplifier.

5. CONCLUSION

In conclusion, it is demonstrated that the time-dependent gain of current modulated semiconductor lasers can be used as an optical gate for the investigations of picosecond optical pulses. A theoretical analysis of the transient gain was performed by solving numerically the coupled rate equations. Under appropriate electrical pumping conditions a time resolution of better than 10 ps is found. Experimental results confirm the theoretically predicted time resolution, and indicate further that this measurement technique is at least one order of magnitude more sensitive than the conventional technique based on fast photodetectors and sampling scopes.

6. ACKNOWLEDGEMENT

We are indepted to G.Trott and G.Weimann for the gift of the TCSM- and MQW- lasers, respectivly. We appreciate funding of part of this work by the Deutsche Forschungsgemeinschaft .

1. D.Bimberg, K.Ketterer, E.H.Böttcher, E.Schöll: Int. J. Electronics 60, 23 (1986)
2. P.-T.Ho:In Picosecond Optoelectronic Devices ed. by C.H.Lee, (Academic Press, New York 1984) pp.11

3. D.E.Ackley, G.Hom:Appl.Phys.Lett.42, 653 (1983)
4. M.Ikeda: IEEE J.Quant.Electron.QE-19, 157 (1983)
5. D.J.Channin: J.Appl.Phys.50, 3858 (1979)
6. T.Mukai, Y.Yamamoto, T.Kimura: In Semiconductors and Semimetals, ed. by W.T.Tsang, Vol.22 Part E (Academic Press, Orlando 1985) pp. 265
7. W.Lenth: Opt.Lett. 9, 396 (1984)

GaAs Photoconductors to Characterize Picosecond Response in GaAs Integrated Devices and Circuits

R.B. Hammond, N.G. Paulter, and A.J. Gibbs

Los Alamos National Laboratory, Los Alamos, NM 87545, USA

1. Introduction

Optoelectronic techniques that use femtosecond lasers and promise the precise measurement of transient response in high-speed electronic devices and circuits have been under study for several years. We have investigated on-wafer electrical-impulse generation and sampling using femtosecond-laser-excited GaAs photoconductors. This approach is applicable to any transmission line structure, it is directly integrable, noninvasive, jitter-free, and it is applicable to both microwave and digital circuits.

We have developed the capability to: a) generate electrical impulses in on-wafer GaAs transmission lines with 3-dB bandwidths in excess of 60 GHz, and b) sample the electrical impulse response of an on-wafer test structure without jitter and with a 3-dB measurement bandwidth of 103 GHz at a noise level of -67 dBm. We have studied dispersion in GaAs on-wafer coplanar-waveguide transmission lines.

2. Sample Preparation and Experiment

We prepared on-wafer photoconductor pulse generators and sampling gates in coplanar waveguide transmission lines on 625-μm-thick Cr-doped semi-insulating GaAs wafers. Center conductor widths were 67 μm and the spacing to the ground planes was 33 μm. This gave a characteristic impedance of approximately 50 ohms and a cut-off frequency of approximately 100 GHz. The photoconductor pulse generators consisted of 25 μm gaps in the center conductor. Photoconductor sampling gates were placed at 33, 400, and 1200 μm distances from the pulsers. The test structures were formed with evaporated AuGe alloy, patterned with lift-off photolithography, and annealed at 425°C for 2 minutes.

In order to produce photoconductors on GaAs with picosecond response we used ion-beam irradiation to introduce point defects into the semi-insulating substrate. This structural damage produces deep-level trapping sites that reduce carrier-relaxation time in the crystals. The samples were damaged with a dose of 5×10^{14} cm^{-2}, 6 MeV He ions. This dose produces a uniform point-defect density of 2×10^{19} cm^{-3} in the top 3 microns of the GaAs substrate. Increasing the ion-beam-damage dose above this level caused the material to become conductive.

Optoelectronic correlation measurements were performed on these test structures using a colliding-pulse modelocked dye laser producing 0.2 ps optical pulses at a wavelength of 620 nm, and a repetition rate of 125 MHz.

3. Experimental Results and Analysis

Figure 1 shows a measured optoelectronic correlation from an on-wafer GaAs photoconductor pulse-generator/sampler pair. A fit to the measured correlation using the photoconductor model of D. H. Auston [1] allows us

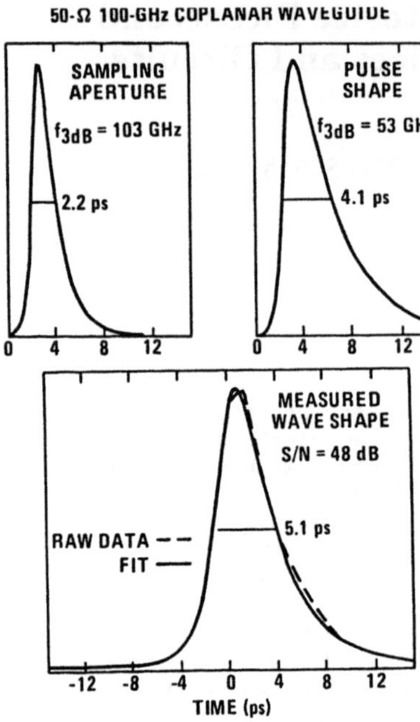

Fig. 1.
Measured optoelectronic correlation between on-wafer GaAs photoconductor pulse generator and sampling gate. The raw data were fit with the model due to D. H. Auston to derive the sampling aperture function and electrical pulse shape. These are also shown

Fig. 2.
Measured optoelectronic correlations taken with photoconductor sampling gates at distances of (a) 33, (b) 400, and (c) 1200 microns from the photoconductor pulse generator

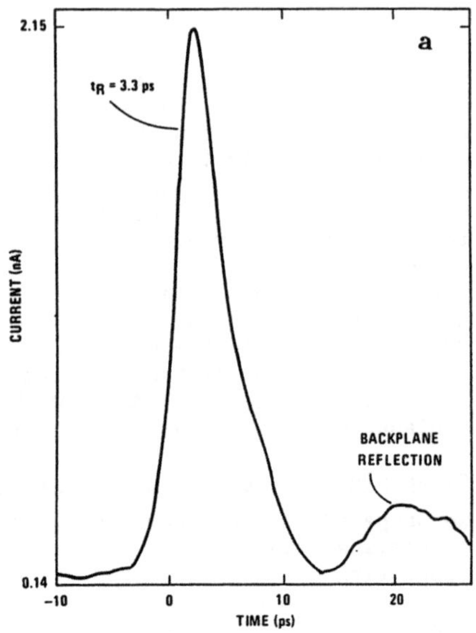

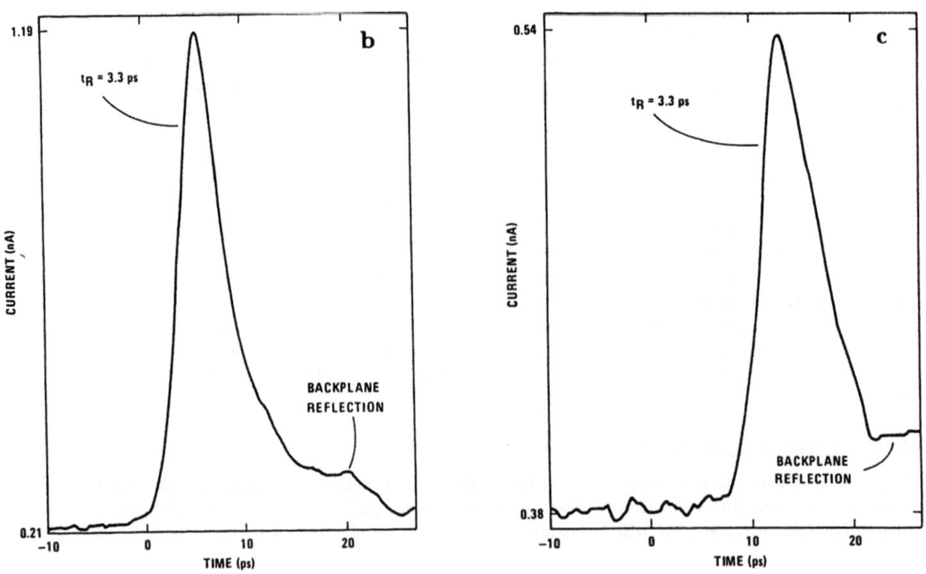

to separate the correlation into the sampling aperture function and the actual electrical pulse shape. In the fit we allowed the carrier relaxation in the pulser and sampler to vary independently. The best fit to the experimental correlation is shown in Fig. 1, along with the derived sampler aperture function and the electrical pulse shape.

Data from our test structure permitted us to investigate pulse propagation in coplanar-waveguide transmission lines. Figure 2 shows measured correlations taken at distances of 33, 400, and 1200 micron distances from the photoconductor pulse generator. The data indicate that coplanar waveguide has very low dispersion. The data show no degradation in the 3.3 ps risetime of the correlation over the full 1.2 mm propagation length.

4. Summary and Conclusion

In summary, we have developed on-wafer GaAs photoconductors that should permit measurements of transient response in GaAs devices and circuits with overall measurement bandwidths greater than 50 GHz and time resolution of 2.2 ps. Because of the absence of jitter in these electrical measurements it should be possible to characterize integrated GaAs devices and circuits at frequencies up to hundreds of GHz. Also, we have shown that coplanar waveguide is a very low dispersion, scalable, all-planar transmission line that is very attractive for these time-domain measurements.

REFERENCE

[1] D. H. Auston, "Impulse response of photoconductors in transmission lines," IEEE J. Quantum Electron., Vol. GE-19, pp. 639-648, 1983.

Characterization of On-Chip Polycrystalline Silicon Photoconductors

D.K. Fitzpatrick [1], *D.R. Bowman* [2], *and R.B. Hammond* [3]

[1] Harris Semiconductor, Melbourne, FL 32901, USA
[2] U.S. Military Academy, West Point, NY 10996, USA
[3] Los Alamos National Laboratory, Los Alamos, NM 87545, USA

1. Introduction

Several optoelectronic approaches have been proposed for measuring the high-speed transient response of electronic devices and circuits [1]-[4]. Most of these approaches, however, cannot be used to perform on-chip measurements of silicon-integrated circuits because of process incompatibilities. This paper describes measurements and analysis of integrated polycrystalline-silicon photoconductors (PCEs) that have been developed to perform high-speed measurements on silicon substrates.

The development of integrated PCEs for ultrafast electrical pulse generation and sampling on silicon substrates has been previously reported [5]. These PCEs, when stimulated with femtosecond dye laser pulses, can be used for high-frequency characterization of silicon-integrated devices, interconnects, and circuits at frequencies up to 100 GHz, and with time resolution as short as one picosecond[5],[6].

Here we report experimental and modeling studies of on-chip photoconductor pulse generation and sampling. Specifically, we discuss a) the effect of ion-beam damage on carrier lifetime and sampling-aperture time, b) the circuit limits to pulse risetime, and c) the room-temperature stability of ion-beam-damaged sampling gates.

2. PCE Fabrication and Measurements

PCEs were fabricated as pulsing and sampling elements in an integrated circuit microstrip transmission-line test structure (see Fig. 1). The transmission line structures consist of 0.8-micron-thick aluminum conductors in 50 and 100 micron widths on both p and n-type silicon substrates. Substrate resistivities vary between 6 and 14 ohm-cm. The PCEs themselves are fabricated in a 0.5 micron undoped polysilicon layer deposited on a field oxide by low-pressure-chemical-vapor-deposition.

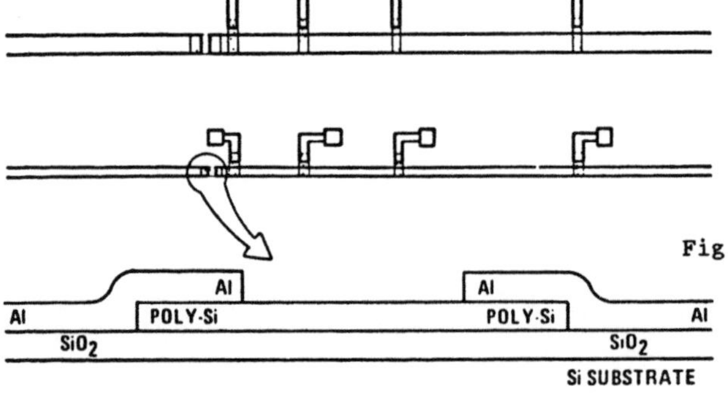

Fig. 1. Schematic diagram of on-chip-photoconductor interconnect test structure

A PCE is a rectangular area of polysilicon with metal lines contacting opposite sides of the rectangle. High dark resistance is achieved by using undoped polysilicon. After the polysilicon layer is patterned, the wafer is subjected to a high temperature anneal (1150°C for one hour in an Argon ambient) to increase the average grain size and improve carrier mobilities within the photoconducting region [7],[8]. Aluminum transmission lines are deposited and patterned following this anneal.

To study the effects of ion-beam damage on carrier lifetime and sampling aperture time of the PCEs, the PCE photoconductive regions were irradiated with 250 KeV silicon to doses of $1.5X10^{14}$, $3.0X10^{14}$, and $5.0X10^{14}$ cm^{-2}. Implant doses of these magnitudes are not sufficient to amorphize silicon, but produce a point defect density of approximately 5 X 10^{22} cm^{-3}. Silicon is used as an implant species because it is electrically inert within silicon and has a sufficiently low critical energy (162 KeV), E_c, to be within the capability of common ion implanters. E_c describes the implant energy for a particular species above which the induced damage is primarily the result of nuclear collisions, which cause the most lattice disorder, and hence the fastest carrier relaxation times. Because the amount of damage is only a weak function of the implant energy above E_c, the number of trapping sites (and, thus, the photoconductor recombination rates) can be tailored within a limited range by controlling only the dose for a particular implant species and energy [8].

3. Experimental Results and Data Analysts

The measurement technique utilized in this work is the same as reported in [6],[8]. Here we use the characteristics of the correlation signal to characterize the PCE response. The risetime of the correlation signal is used to determine the effect of the ion damage and circuit environment on the performance of the PCE. All correlation measurements are taken at the second sampler (signal propagation distance = 500 microns, see Fig. 1) in order to avoid local modes near the pulse generator as well as dispersion [9].

The microstrip test structures were fabricated in both 50 and 100 micron widths. These line widths, along with the physical characteristics of the substrate, produced transverse electromagnetic (TEM) impedances of approximately 92 and 76 ohms, respectively. Both structures use a 10-micron gap in the transmission line for the active region of the pulsing photoconductor.

Figure 2 shows the measured risetime of the correlation signal plotted versus transmission-line width. The measured data is presented with sampler implant dose as a parameter and contains experimental errors estimated to be +/-0.5 picoseconds. Prior experimental work [6],[8] has shown an inverse relationship between the dose and recombination rates in the PCE. This is also confirmed by these measurements. In going from a sample dose of 1.5 x 10^{14} cm^{-2} Si to a dose of 3 x 10^{14} cm^{-2} the risetimes of the correlation signals decrease for both the 50 and 100 micron-wide test structures. If the dosage is increased further to 5 x 10^{14} cm^{-2} circuit limits in the 100 micron-wide line mask any decrease in risetime due to shorter carrier lifetime. In contrast, the 50 micron-wide line shows a decrease in risetime for this higher dose. From this observation we conclude that for the 100 micron-wide test structure the optimum dose of 250 keV silicon is between 3 x 10^{14} and 5 x 10^{14} cm^{-2}. The fastest response possible for the 50 micron-wide structures remains to be determined.

The results of the measurements on the 50 micron-wide line indicate that the structure is not circuit limited,while the 100-micron-wide line

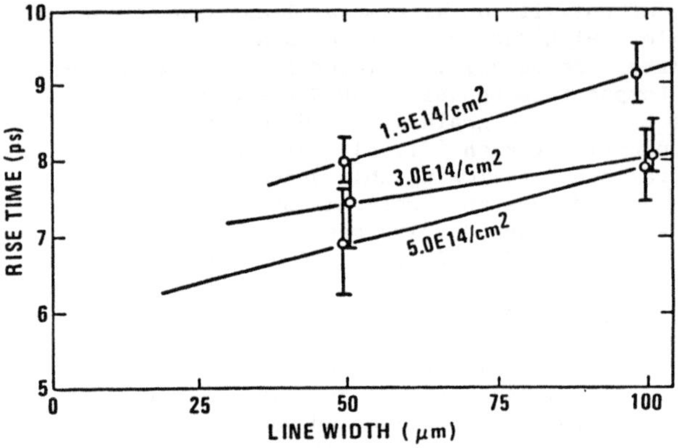

Fig. 2. Plot of correlation risetimes vs line width for 3 implant doses

structure becomes circuit limited at 5 X 10^{14} cm^{-2}. In our previous work, carrier relaxation times within the PCEs limited observed response. Then we approximated PCE response as an initial rising step that decayed exponentially. The rising edge of the correlation signal was due only to carrier relaxation within the sampler. In the circuit-limited case, this assumption is invalid. We must use more rigorous techniques to analyze the data. We accomplish this by implementing available PCE [10] and microstrip models in an analysis program, MIPCEM (Microstrip and Photoconductive Element Modelling program [8]).

We performed computer simulations of our experiments with MIPCEM to assess the validity of the models and to determine the relaxation times in the sampling gates. In the fits only the recombination rates within the sampler and pulser PCEs were varied. The results of these simulations are summarized in Table I. The simulations agree very well with the data. In all cases, it was possible to extract the recombination parameters of the PCEs from the 50 micron-wide structures and use them to predict the measured response on the 100 micron-wide structures within experimental and model error.

Figure 3 shows a series of measurements taken six months after the data of Fig. 2, plotted in comparison with the initial measurements. The increase in the response time of the PCEs is due to the room-temperature annealing of the implant damage over a six-month period. The data show a wide variation in response.

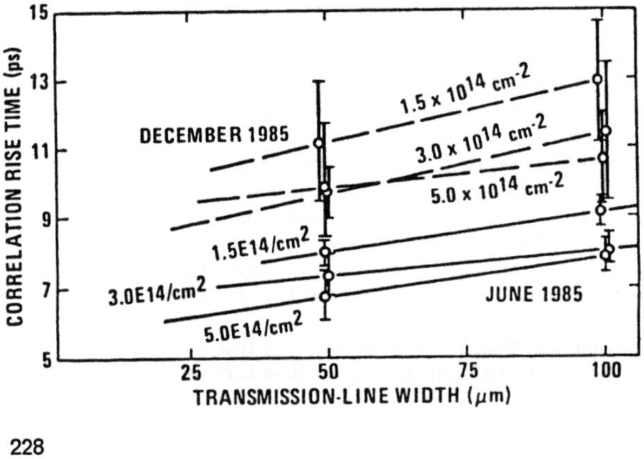

Fig. 3. Same as Fig. 2 except results taken six months after damage implant are also plotted

TABLE I - MEASURED VERSUS SIMULATED CORRELATION RISETIME

Implant Dosage	Risetime 50-μm	Risetime 100-μm
$1.5X10^{14}cm^{-2}$		
Measured	8.0 ps	9.3 ps
Simulated (τ=3.7ps)	7.8 ps	9.0 ps
Relative Error	3.0 %	3.2 %
$3.0X10^{14}cm^{-2}$		
Measured	7.3 ps	8.1 ps
Simulated (τ=3.0ps)	6.9 ps	8.1 ps
Relative Error	6.0 %	0.0 %
$5.0X10^{14}cm^{-2}$		
Measured	6.7 ps	7.9 ps
Simulated (τ=2.8ps)	6.7 ps	7.9 ps
Relative Error	0.0 %	0.0 %

REFERENCES

[1] N. S. Nahman, "Picosecond-domain waveform measurements," IEEE Proc., Vol. 66, pp. 441-454, 1978.

[2] N. S. Nahman "Picosecond-domain waveform measurements," IEEE Trans. Instru. Meas., Vol. IM-32, pp. 117-124, 1983.

[3] J. K. Everad and J. F. Carrol, "Practical comparison of opto-electronic sampling systems and devices," IEE Proc., Vol. 130, pt. I, pp. 5-16, 1983.

[4] P. R. Smith, D. H. Auston, A. M. Johnson, and W. M. Augustynaik, "Picosecond photoconductivity in radiation-damaged silicon-on-sapphire films," Appl. Phys. Lett., Vol. 38, pp. 47-50, 1981.

[5] D. R. Bowman, R. B. Hammond, and R. W. Dutton, "Polycrystalline-silicon integrated photoconductors for picosecond pulsing and gating," IEEE Electron Device Lett., Vol. EDL-6, pp 502-504, 1985.

[6] D. R. Bowman, R. B. Hammond, and R. W. Dutton, "New integrated polysilicon photoconductor for ultrafast measurements on silicon," Tech. Dig. Int. Electron Devices Mtg., Washington, D.C., December, 1985, pp. 117-120.

[7] R. B. Hammond and N. M. Johnson, "Impulse photoconductance of thin-film polycrystalline silicon," J. Applied Physics, Vol. 59, pp. 3155-3159, 1986.

[8] D. R. Bowman, PhD dissertation, Stanford University, Stanford, CA, 1986.

[9] K. W. Goosen and R. B. Hammond, "Time-domain analyses of picosecond-pulse propagation in microstrip interconnections on si integrated circuits," Manuscript in preparation.

[10] D. H. Auston, "Impulse response of photoconductors in transmission lines," IEEE J. Quantum Electron., Vol. GE-19, pp. 639-648, 1983.

Index of Contributors

Alexandre, F. 172
Andre, J.-P. 148
Ankri, D. 151
Antonetti, A. 101,204
Asano, T. 177
Auston, D.H. 184

Bacot, C. 151
Baratoff, A. 24
Baratte, H. 140
Bareikis, V. 28
Bauser, E. 97
Beha, H. 200
Bending, S. 24
Beneking, H. 123,144
Besombes, C. 151
Bimberg, D. 218
Blacha, A. 200
Böttcher, E.H. 218
Bowman, D.R. 226
Bresse, J.F. 156
Briere, A. 148

Cabon-Till, B. 108
Capasso, F. 50
Castagné, R. 2
Castagné, R. 35
Chambery, P. 148
Chang, C.Y. 132
Chen, Y.K. 168
Chomette, A. 101,104
Clauberg, R. 200
Cloos, J.M. 123

Dangla, J. 151
Daoud-Ketata, K. 156
Desrousseaux, P. 151
Deveaud, B. 101,104
Dolfus, P. 35
Dubon-Chevallier, C. 151,156
Duchenois, A.M. 151

Eastman, L.F. 168
English, J.H. 19
Enquist, P.M. 168
Etrillard, J. 156

Fernholz, G. 144
Ferry, D.K. 136,210
Fitzgerald, E.A. 168
Fitzpatrick, D.K. 226
Forchel, A. 97
Frank, D.J. 140
Freire, V.N. 32
Fujimoto, M. 79
Furukawa, S. 177

Ghibaudo, G. 108
Gibbs, A.J. 223
Gibbs, H.M. 204
Gloanec, M. 172
Goldstein, L. 164
Goodnick, S.M. 116
Gossard, A.C. 19
Gruhle, A. 123
Guéret, P. 24

Hamaguchi, C. 112
Hammond, R.B. 223,226
Hayes, J.R. 19
Heiblum, M. 11
Henry, M. 172
Hesto, P. 164
Hillmer, H. 97
Hiyamizu, S. 72
Höpfel, R.A. 93
Huebener, R.P. 160
Hulin, D. 101,204

Ikegami, T. 79
Ishibashi, T. 79
Ishiwara, H. 177

Kamdem, J. 172
Kash, J.A. 200
Kastalsky, A. 62
Kavanagh, K.L. 168
Ketterer, K. 218
Kibickas, K. 28
Kim, J.H. 214
Klein, J.P. 172
Knoedler, C.M. 140
Kobayashi, T. 88
Kuivalainen, P. 40

La Tulipe, Jr., D.C. 140
Lambert, B. 104
Lehovec, K. 127
Levi, A.F.J. 19
Li, S.S. 214
Liberis, J. 28
Lindberg, K. 40
Lippens, D. 44
Liu, W.C. 132
Löchner, K.S. 97
Lugli, P. 116
Luzzi, R. 32

Mannhart, J. 160
Marclay, E. 24
Marso, M. 123
Matulionis, A. 28
Mayer, G. 97
Meier, H. 24
Migus, A. 204
Miliušytė, R. 28
Miyatsuji, K. 112
Mizutani, T. 79
Mougel, B. 35
Mouis, M. 35
Mourou, G.A. 191
Muto, S. 72
Mysyrowicz, A. 204

Najjar, F.E. 168
Nieruchalski, J.L. 44
Nuss, M.C. 184

Okisu, N. 88
Osman, M.A. 210
Oudar, J.L. 101

Paršeliūnas, J. 28
Parisi, J. 160
Paulter, N.G. 223
Pelouard, J.L. 164
Peyghambarian, N. 204
Pône, J.-F. 35
Požela, J. 28
Praseuth, J.P. 164
Py, M. 24

Ramberg, L.P. 168
Ravaioli, U. 136,210
Regreny, A. 101,104
Roentgen, P. 144
Romestain, R. 104

Sakalas, P. 28
Sambe, Y. 88
Seitz, H.K. 200
Shah, J. 93
Shannon, J.M. 122
Smith, P.R. 184
Solomon, P.M. 140

Tanimoto, H. 112
Tsutsui, K. 177

Vasconcellos, A.R. 32
Vescan, L. 123

Wang, Y.H. 132
Wolny, M. 148
Wright, S.L. 140

Yamasaki, K. 79
Yokoyama, N. 72

Springer Series in Solid-State Sciences

Editors: **M. Cardona, P. Fulde, K. von Klitzing, H.-J. Queisser**

Volume 63

Electronic Properties of Polymers and Related Compounds

Proceedings of an International Winter School, Kirchberg, Tirol, February 23–March 1, 1985
Editors: **H. Kuzmany, M. Mehring, S. Roth**
1985. 267 figures. XI, 354 pages. ISBN 3-540-15722-0

Volume 62

Theory of Heavy Fermions and Valence Fluctuations

Proceedings of the Eight Taniguchi Symposium, Shima Konko, Japan, April 10–13, 1985
Editors: **T. Kasuya, T. Saso**
1985. 106 figures. XII, 287 pages. ISBN 3-540-15922-3

Volume 61

Localization, Interaction, and Transport Phenomena

Proceedings of the International Conference, August 23–28, 1984, Braunschweig, Federal Republic of Germany
Editors: **B. Kramer, G. Bergmann, Y. Bruynseraede**
1985. 125 figures. IX, 264 pages. ISBN 3-540-15451-5

Volume 60

Excitonic Processes in Solids

By **M. Ueta, H. Kanzaki, K. Kobayashi, Y. Toyozawa, E. Hanamura**
1986. 307 figures. XII, 530 pages. ISBN 3-540-15889-8

Volume 59

Dynamical Processes and Ordering on Solid Surfaces

Proceedings of the Seventh Taniguchi Symposium, Kashikojima, Japan, September 10–14, 1984
Editors: **A. Yoshimori, M. Tsukada**
1985. 89 figures. XII, 195 pages. ISBN 3-540-15108-7

Volume 58

The Recursion Method and Its Applications

Proceedings of the Conference, Imperial College, London, England, September 13–14, 1984
Editors: **D. G. Pettifor, D. L. Weaire**
1985. 42 figures. VIII, 179 pages. ISBN 3-540-15173-7

Volume 57

Polycrystalline Semiconductors

Physical Properties and Applications
Proceedings of the International School of Materials Science and Technology at the Ettore Májorana Centre, Erice, Italy, July 1–15, 1984
Editor: **G. Harbeke**
1985. 159 figures. VIII, 245 pages. ISBN 3-540-15143-5

Volume 56: **T. Moriya**

Spin Fluctuations in Itinerant Electron Magnetism

1985. 98 figures. X, 239 pages. ISBN 3-540-15422-1

Volume 55: **D. C. Mattis**

The Theory of Magnetism II

Thermodynamics and Statistical Mechanics
1985. 40 figures. XII, 177 pages. ISBN 3-540-15025-0

Volume 54

Magnetic Excitations and Fluctuations

Proceedings of an International Workshop, San Miniato, Italy, May 28–Juni 1, 1984
Editors: **S. W. Lovesey, U. Balucani, F. Borsa, V. Tognetti**
1984. 114 figures. IX, 227 pages. ISBN 3-540-13789-0

Volume 53

Two-Dimensional Systems, Heterostructures, and Superlattices

Proceedings of the International Winter School Mauterndorf, Austria, February 26–March 2, 1984
Editors: **G. Bauer, F. Kuchar, H. Heinrich**
1984. 231 figures. IX, 293 pages. ISBN 3-540-13584-7

Springer-Verlag
Berlin Heidelberg New York
London Paris Tokyo

Springer

MIX
Papier aus verantwortungsvollen Quellen
Paper from responsible sources
FSC® C105338

If you have any concerns about our products,
you can contact us on
ProductSafety@springernature.com

In case Publisher is established outside the EU,
the EU authorized representative is:
Springer Nature Customer Service Center GmbH
Europaplatz 3, 69115 Heidelberg, Germany

Printed by Libri Plureos GmbH
in Hamburg, Germany